Beyond Bakelite

Lemelson Center Studies in Invention and Innovation

Joyce Bedi, Arthur Daemmrich, and Arthur Molella, general editors

Arthur P. Molella and Joyce Bedi, editors, *Inventing for the Environment*

Paul E. Ceruzzi, *Internet Alley: High Technology in Tysons Corner, 1945–2005*

Robert H. Kargon and Arthur P. Molella, *Invented Edens: Techno-cities of the Twentieth Century*

Kurt Beyer, *Grace Hopper and the Invention of the Information Age*

Michael Brian Schiffer, *Power Struggles: Scientific Authority and the Creation of Practical Electricity before Edison*

Regina Lee Blaszczyk, *The Color Revolution*

Sarah Kate Gillespie, *The Early American Daguerreotype: Cross-Currents in Art and Technology*

Matthew Wisnioski, Eric S. Hintz, and Marie Stettler Kleine, editors, *Does America Need More Innovators?*

Kathryn D. Sullivan, *Handprints on Hubble: An Astronaut's Story of Invention*

Joris Mercelis, *Beyond Bakelite: Leo Baekeland and the Business of Science and Invention*

Beyond Bakelite

Leo Baekeland and the Business of Science and Invention

Joris Mercelis

The MIT Press
Cambridge, Massachusetts
London, England

This book was set in Stone by Westchester Publishing Services. Printed and bound in the United States of America.

Library of Congress Cataloging-in-Publication Data

Names: Mercelis, Joris, author.
Title: Beyond Bakelite : Leo Baekeland and the business of science and invention / Joris Mercelis.
Description: Cambridge, Massachusetts : The MIT Press, 2020. | Series: Lemelson center studies in invention and innovation | Includes bibliographical references and index.
Identifiers: LCCN 2019027006 | ISBN 9780262538695 (paperback)
Subjects: LCSH: Baekeland, L. H. (Leo Hendrik), 1863-1944. | Industrial chemists—United States—Biography. | Science and industry. | Inventions. | Plastics—History.
Classification: LCC TP140.B3 M47 2020 | DDC 660.092 [B]—dc23
LC record available at https://lccn.loc.gov/2019027006

10 9 8 7 6 5 4 3 2 1

Contents

Series Foreword

"I just want to say one word to you. Just one word… Plastics." That line from the film *The Graduate* has come to symbolize the hubris, promise, and disappointment embodied in one of the world's most ubiquitous materials. As this book goes to print, plastics are cheap, widely used, and durable. But that durability means that plastics persist in the environment for decades. Images of swaths of the ocean or beaches awash in plastic trash are regularly in the news. To address the ever-growing mass of plastic waste, entire countries are banning single-use plastic items, and solutions for recycling plastic effectively and for creating eco-friendly plastics are active fields of research.

A little over a century ago, however, Leo Baekeland's invention of the first synthetic plastic—Bakelite—earned him the then laudatory moniker "Father of Plastics" in the press. Bakelite proved to be an extremely useful product, with applications ranging from radio cabinets and electrical insulators to colorful Art Deco jewelry. Yet there is much more to Leo Baekeland's story than Bakelite. In his exploration of Baekeland's long career, Joris Mercelis traces Baekeland's life from his youth in Belgium through his education and immigration to the United States, and highlights Baekeland's other best-known invention, Velox photographic paper. Throughout, Baekeland maintained his American and European professional connections and continued to cross boundaries between academia, business, and industrial research. Mercelis frames Baekeland as a scientific entrepreneur and focuses on themes of intellectual property and hybrid scientific-industrial networks, casting new light on the relationships and interdependencies between science, innovation, and industry in the late nineteenth and early twentieth centuries.

Invention and innovation have long been recognized as transformational forces in American history, not only in technological realms but also in politics, society, and culture; they are arguably more important than previously thought in other societies as well. Since 1995, the Smithsonian's Lemelson Center has been investigating the history of invention and innovation from an interdisciplinary perspective. Books in the Lemelson Center Studies in Invention and Innovation extend this work to enhance public understanding of humanity's inventive impulse. Authors in the series raise new questions about the work of inventors and the technologies they create, while stimulating cross-disciplinary dialogue. By opening channels of communication between the various disciplines and sectors of society concerned with technological innovation, the Lemelson Center Studies advance scholarship in the history of technology, engineering, science, architecture, the arts, and related fields and disseminate it to a general interest audience.

Joyce Bedi, Arthur Daemmrich, and Arthur Molella
Series editors, Lemelson Center Studies in Invention and Innovation

Acknowledgments

The research project leading up to this book originated at the Department of History and the Museum for the History of Sciences at Ghent University in Belgium. I remain particularly grateful to Christophe Verbruggen, who, in collaboration with Jan Art and Danny Segers, gave me plenty of stimulating advice and encouragement during my years as a PhD candidate. I also benefited greatly from the feedback of the members of my doctoral advisory and dissertation reading committees, Ernst Homburg, Lewis Pyenson, Ann Johnson, Raf Vanderstraeten, and Eric Vanhaute. I am, moreover, still thankful to the members of Ghent University's research units Social History since 1750 and Communities Comparisons Connections, as well as to my other colleagues and friends in Belgium, the Netherlands, and Germany who helped and motivated me in the early stages of this project.

Most of the primary research behind this book was, however, done in the United States. This was made possible thanks to fellowships or grants from the Smithsonian Institution, Science History Institute, and Hagley Museum and Library. I gratefully acknowledge the financial support of these institutions, as well as from the Fund for Scientific Research–Flanders. For the same reason, I am indebted not just to the archival and library personnel of Ghent University and the Belgian State Archives but also to their counterparts at Johns Hopkins University, the US National Archives, the Library of Congress, the Science History Institute, the Hagley Museum and Library, Columbia University, and the George Eastman House. Above all, the staff of the Archives Center at the National Museum of American History, Smithsonian Institution, was more than generous in making available the Leo H. Baekeland Papers and in drawing my attention to related collections. During my stays at the Smithsonian, I also received valuable comments and suggestions from, among others, Art Molella, Paul Forman, Debbie Warner, Eric Hintz, Pete Daniel, and Jia-sun Tsang.

As intellectual property gradually became a more central theme to my research project, it was highly stimulating to become part of a community of historians who shared an interest in this theme. I owe a lot to Alessandro Nuvolari, Astrid Kander, and Ben Gales for organizing an ESTER seminar on The Economic History of Patents and Innovation. On this as well as later occasions, I learned much from many of the seminar participants, including Stathis Arapostathis, Berris Charnley, Christine MacLeod, Patricio Sáiz, and Jochen Streb. At a related event, I also had the fortune of meeting with Christopher Beauchamp, who generously helped me identify relevant legal archives in the United States and shared some of his extensive knowledge on the history of intellectual property.

With respect to another central topic of this book, scientific entrepreneurship, I am especially indebted to Anna Guagnini. Shortly after I had finished my dissertation in Ghent, Anna involved me in what would develop into a comprehensive, collaborative research project on the histories of academic entrepreneurship and the commercialization of academic science. At the earliest stage, this entailed many stimulating discussions with Matteo Serafini, Gabriel Galvez-Behar, and Kenneth Bertrams, among others. Later on, I also gained valuable insights from the other participants in conferences in Ghent and Lille and the contributors to special journal issues of *History and Technology* and *Management and Organizational History*, particularly Dan Wadhwani.

It has been a great honor to complete this book project as a member of the faculty of the Department of History of Science and Technology at Johns Hopkins University. The department has a long tradition of supporting research on science-industry relationships, including several seminal studies on the history of industrial research. Collaborations with Bob Kargon and Bill Leslie, and conversations with Susan Morris and many other colleagues as well as graduate and undergraduate students, have provided me with new perspectives on these themes.

My manuscript has improved significantly thanks to the suggestions and encouragement of Katie Helke and the editors of the Lemelson Center Studies in Invention and Innovation series at the MIT Press, especially Joyce Bedi. I would also like to thank Jeffrey Meikle and the anonymous reviewers of the book proposal and manuscript for their valuable comments and criticisms, and Erin Davis and Hal Henglein of Westchester Publishing Services for their careful production and copyediting work. Earlier versions of chapters 3, 4, and 7 appeared in, respectively, the *Journal of Belgian History*, the edited volume *Knowledge Management and Intellectual Property* (edited by S. Arapostathis and G. Dutfield), and *Technology and Culture*. Edward Elgar and the Johns Hopkins University Press most helpfully permitted me to reuse parts of these earlier publications.

At different stages in my research, my work on this book has also been supported by numerous other colleagues, including Brigitte van Tiggelen, Corinna Schlombs, Kathryn Steen, Paul Israel, Mila Davids, Pierre-Yves Saunier, and Erik van der Vleuten. Kristel Wautier, Hugh Karraker, Karel Julien Cole, Carlo Martino, Hank Stroobants, Reindert Groot, and the Chemie-Freunde Erkner e.V. have been generous in guiding me to relevant sources and in sharing their perspectives on Baekeland and Bakelite. My parents, siblings, in-laws, and friends have been especially appreciative of what it takes to finish a project of such large scope. More than anyone else, though, I would like to thank Greet for her patience, love, and tireless support. Her willingness to serve as a sounding board on matters ranging from title choices to more fundamental "so what?" questions has improved the manuscript in many respects. She has also enabled the completion of this study in various other ways, not least by moving with me to the United States. It is to her and to our book-loving daughter and book-eating baby son, Nora and Oscar, that I dedicate this work.

1 Introduction

The most popular displays at the New York world's fair of 1939–1940 evoked a "world of tomorrow," but one of the exhibition's frequent visitors was repeatedly reminded of his own past. Leo Baekeland (1863–1944), a Belgian-born American chemist, inventor, and entrepreneur, would return to the fair more than 25 times after his initial visit on 7 July 1939 (figure 1.1). Baekeland's interest in the event in part reflected the technological enthusiasm with which he had been imbued from a young age: the diary notes he took in 1939–1940 reveal the same passion about new technology that he had previously expressed in relation to major innovations such as the telephone, automobile, radio, and airplane. The world exhibition also provided Baekeland, as well as many other visitors, with a temporary nostalgic escape from concerns about day-to-day affairs and, especially in the fair's second season, about the consequences of the new war that had broken out in Europe.[1]

However, these factors alone do not explain why Baekeland, who had attended several world's fairs in Europe and the United States, starting with the Paris exposition of 1889, regarded this one as "by far the best conceived, the largest and most impressive of any of its predecessors here and in foreign countries."[2] To understand this assessment, it is at least as important to keep in mind that the New York fair coincided with the ending of Baekeland's half-century career as a scientific entrepreneur. Three months after the opening of the exposition on 30 April 1939, Baekeland and his fellow board members of the Bakelite Corporation, the last of several companies Baekeland had founded in the course of his life, decided to sell the firm to the Union Carbide and Carbon Corporation. The shareholders of Union Carbide and those of the Bakelite Corporation ratified the transaction in August and September,

Figure 1.1
Baekeland (right) and electrical engineer Ernst Alexanderson at the 1939–1940 New York world's fair.
Source: Manuscripts and Archives Division, New York Public Library.

respectively, marking the retirement from business of the latter firm's first and only president.

Given these circumstances, it was logical that Baekeland would be interested in finding out how the New York fair presented the legacy of his career. On the first of several inspections, Baekeland had entered his own company's display at the world exhibition (figure 1.2) "like any stranger,"[3] but the responsible curator in the Hall of Industrial Science naturally identified him as the man whose well-known contributions to plastics technology he was expected to highlight. In fact, consistent with contemporary media accounts of the importance of Baekeland's work, the Bakelite Corporation's exposition did not just praise Baekeland for the invention of Bakelite, the first fully synthetic plastic, in 1907. It also drew a direct line from this achievement to the development of the entire synthetic plastics industry up to the late 1930s, illustrating "the profound effect of one man's discovery on the lives

Figure 1.2
Bakelite exhibition in the Hall of Industrial Science at the 1939–1940 New York world's fair.
Source: Manuscripts and Archives Division, New York Public Library.

of millions" and staging Baekeland as the "father of plastics"—a title not yet tainted by concerns about profligate consumption, health hazards, and environmental waste.[4]

Elsewhere at the fair, the broad adoption of Bakelite, and many other synthetic plastics, was visible, too. For instance, in another section of the Hall of Industrial Science, Rohm and Haas displayed a "new, beautiful plastic material more transparent than glass," known chemically as polymethylmethacrylate but trademarked as "Plexiglas."[5] At the Eastman Kodak exhibition in the Production and Distribution Zone, Baekeland, who had been active in the photographic industry for over a decade, was struck by the "marvelous enlarged color photos" on display in the Great Hall of Color, but it is likely that he also observed cameras constructed of molded Bakelite and newly created applications of cellulose acetate plastics.[6] In the Japanese pavilion, which Baekeland visited on 25 July 1939, he was certainly happy to notice Bakelite gears in silk-spinning equipment as well as Bakelite tea trays. Previously, Baekeland had also watched his son demonstrate boat and airplane parts made from phenolic plastic resin on the occasion of the fair's "Bakelite Day,"[7] and he must have been pleased that a Bakelite pin depicting the fair's theme center—consisting of the obelisk-like "Trylon" and the "Perisphere," a colossal hollow globe—was available in the souvenir shops. In short, even if Baekeland had not frequented the stand of his own corporation, he would have had plenty of opportunity to assure himself that millions of fair visitors could get a sense of the widespread adoption of the material associated with his name.

Moreover, those who could not make it to the New York exposition were reminded by the press of who had given his name to Bakelite. During the second year of the fair, in particular, *Time* featured a short biographical profile of Baekeland that was simply entitled "father of plastics."[8] *Time* published the piece because the Franklin Institute had just presented Baekeland with its prestigious annual medal. Five years earlier, when Baekeland was elected to the US National Academy of Sciences, British journal *Chemical Age* had not even considered it necessary to explicate whom it considered to be the "father of plastics."[9] Since the invention of Bakelite in 1907, Baekeland had also received multiple other honors and awards, including the Willard Gibbs Medal of the Chicago Section of the American Chemical Society (ACS), the Perkin and Messel medals of the Society of Chemical Industry, several honorary doctorates, and election as president of the ACS in 1924. In some cases, the awarding bodies cited achievements outside the plastics industry among the reasons justifying their choice. Nevertheless, it is beyond doubt that Baekeland was primarily honored for his seminal work on synthetic plastics.

How was it possible that a single individual received so much credit for the establishment of a sector to which numerous researchers, inventors, and firms inside and outside the United States had made significant contributions? As this book demonstrates, to fully answer this question, it is necessary to examine how the relationships between science, technology, and industry had changed during the course of Baekeland's life and how Baekeland and his collaborators had both contributed to and responded to these transformations. More specifically, I argue that it is essential to closely consider the themes of intellectual property (IP) and scientific entrepreneurship through a transatlantic lens—a research approach that sets this book apart from previous studies of science-industry connections and interdependencies in the late nineteenth and early twentieth centuries. Before outlining the book's argument and methodology in more detail in the following sections of this chapter, however, it is worth noting some initial clues detectable at the New York world's fair. By 1939, for example, it had become common to present the phenolic resin industry—the segment of the synthetic plastics sector initiated by the commercialization of Bakelite—either as an "entirely American" or "great American" sector, as the Bakelite Corporation did in its display.[10]

As historian Kathryn Steen has shown, this promotional strategy was also evident in other segments of the US chemical industry. Particularly since the outbreak of World War I, America's chemists and chemical firms had sought to demonstrate that they were able to compete with their European, and especially German, counterparts, who had been dominating the production of synthetic organic chemicals and pharmaceuticals.[11] In this respect, that Baekeland, rather than plastics innovators from Germany or other European countries, was celebrated as the "father" of a rapidly expanding, science-related industry benefited the entire American chemical community.

To acquire the "father of plastics" title, it had also helped Baekeland that he had social and financial resources, as well as previous experience in business, to fall back on to commercialize Bakelite effectively and claim credit for its invention. While earlier plastics inventors, such as John Wesley Hyatt and Alexander Parkes, had a very limited formal education in science, Baekeland held a PhD degree from a respected university. Moreover, he had already become a leading figure in the American chemical community before he and his patent agents prepared the first patent applications for Bakelite. By 1907, Baekeland had also become independently wealthy, mainly through the sale of the Nepera Chemical Company, a photochemical firm Baekeland had cofounded, to George Eastman of Eastman Kodak in 1899. Perhaps Baekeland was reminded of this episode when observing Nepera's most successful product, a class of photographic papers branded as "Velox," at the 1939–1940 world's fair. The display of Velox pictures in the Japanese pavilion certainly made him conclude, "All this with my introduction of Hooker Electrochemical cells for soda, hypochlorites and chlorine [as a representative of the Hooker Electrochemical Company on a trip to Japan in 1914] puts me on record of having introduced four industries in Japan."[12] While this claim needs to be qualified, it is true that Baekeland had made significant contributions to electrochemical technology in addition to his accomplishments in the fields of synthetic plastics and photography. Moreover, Baekeland's background in photographic chemistry and especially electrochemistry not only helps explain why he was able to invent Bakelite, as historian of technology Wiebe Bijker has argued.[13] Baekeland's business experiences in these areas also informed his approach to commercializing the invention, which contributed in no small measure to the recognition he would receive.

What is more, that a US-based industrial chemist had led the introduction of a new synthetic plastic also resonated well with a broader movement in America to highlight the promise of industrial research. As several historians have noted, the New York world's fair was exceptional in that it emphasized the ingenuity of corporate laboratories to the point where "science" was almost reduced to industry-based research.[14] Harold Urey and other scientists at Columbia University, whose Chemical Engineering Department had engaged Baekeland as an honorary professor in 1917, strongly objected to this bias. Baekeland, however, had always remained the head of a research and development (R&D) intensive company in addition to being an academic. Accordingly, he saw no problem in identifying scientific progress with industrial R&D.

The Science-Industry Nexus

The fact that Baekeland had not even commented on the world exhibition's emphasis on the R&D work of large industrial corporations such as AT&T, Du Pont, and General Electric reflects fundamental changes in the relationships between science and industry that had taken place since Baekeland's student days in Belgium. To begin with, the integration of laboratory-based research by firms like those celebrated in 1939 for creating a "world of tomorrow" had not really started until the late 1870s. Initially, moreover, this process had largely been limited to a few countries and industries, including the German synthetic organic chemicals sector and the electrical industry in both Germany and the United States.[15] In fact, not just in industry but also in academia and government, the spread of collaborative laboratory research was, in large part, a twentieth-century phenomenon.[16]

The rise of industrial research was only one of various developments that contributed to the fundamental interdependence between science and industry that is still with us today. In the field of chemistry and a limited number of other branches of knowledge, the close and systematic intertwining of scientific, technological, and economic pursuits predated the second half of the nineteenth century.[17] Even in such areas, however, the science-industry nexus was fundamentally transformed in the late nineteenth and early twentieth centuries as the result of a combination of quantitative and qualitative changes in the development of the economy and the sciences. One such change was that, as part of a process known as the "second industrial revolution," the system of industrial capitalism that had initially emerged in Britain in the decades around 1800 came to rely more heavily on scientific knowledge and labor. The invention of the industrial research laboratory was one indicator of this development. At least as important, however, was the employment of scientists and engineers in industrial positions other than in R&D, including management, manufacturing, and sales. As a case in point, only a small minority of early twentieth-century industrial chemists were working in central research laboratories of the type that has received a disproportionate amount of attention in the historical literature on industrial R&D.[18]

These changes in the economy coincided with alterations in the organization of the sciences that were at least as "revolutionary." As an example, in the course of Baekeland's lifetime, there was a dramatic expansion of the higher education sector, especially its scientific and technological segments. This was perhaps most clearly reflected in the rise in student enrollment in the United States from approximately 33,000 in 1860–1861 to well above one million by the beginning of the 1930s, a 36-fold increase! But, in Europe, too, academic institutions experienced above-average rates of growth, partly driven by the strong belief of governments and private donors that effective industrialization and modernization efforts required investments in scientific and technological instruction.[19] The gradual spread of laboratories and other research facilities within academic institutions on both sides of the Atlantic was another indicator of this expansion. There was also a rapid increase in the number of academic teachers and researchers, who often came to dominate the professional scientific societies that were established from the mid-nineteenth century onward.[20] In addition, the creation of new research and testing establishments, such as the Imperial Institute of Physics and Technology (Physikalisch-Technische Reichsanstalt) in Germany, the National Physical Laboratory in Britain, and the Bureau of Standards in the United States, contributed to a diversification of the sites of scientific and technological knowledge production.[21]

It has been argued that these developments facilitated the pursuit of science of a relatively "pure" nature; that is to say, of scientific activities aimed at creating and disseminating knowledge for its own sake. Therefore, and considering that the university became a (if not the) primary site where this type of science was practiced, it may seem paradoxical that the late nineteenth and early twentieth centuries also witnessed an intensification of academia-industry relationships. In reality, however, the expanding higher education systems of this period could simultaneously accommodate various regimes of knowledge production and diffusion, as scholars Terry Shinn and Dominique Pestre have emphasized.[22] The practically oriented, "utilitarian" mode was most relevant to the development of the science-industry nexus. Among other things, it was reflected in the initiation of teaching and research programs in fields of strong economic relevance, such as agricultural science and a range of new engineering specialties, including aeronautical, chemical, electrical, and fire protection engineering.[23] However, even the knowledge production regime with the strongest disciplinary orientation at least indirectly contributed to science-related industrial development. Indeed, irrespective of a focus on disciplinary or practical questions, academic establishments played a significant role in the emergence of economies and societies that were more knowledge intensive by training scientists and engineers in ever-larger numbers.

Another notable feature of academia-industry relationships in this time period was that collaborations between academic scientists and industrial firms were being reconfigured through a range of institutional changes inside and outside academia. The first set of innovations, including the industrial research fellowship and the cooperative engineering course, formalized relationships between individual higher education institutions and firms. Other changes, also aimed at fostering academia-industry relationships that would be more enduring, occurred in the broader institutional scientific infrastructure.[24] Several professional scientific societies, for example, created separate sections or committees concerned with the interests of their industrial members. Some national science academies and funding agencies established similar setups, including the Science-Industry Bureau of the Belgian Fund for Scientific Research (founded in 1928) and the industrial section of the Parisian Académie des Sciences. In addition, new science institutes that were physically separate from higher education institutions but of neither industrial nor fully academic character likewise emerged in this period. The best-known examples of such "third-sector" establishments were the Kaiser Wilhelm Institutes and the Imperial Institute of Physics and Technology in Germany, but they could also be found elsewhere.[25] Finally, there were innovations in the management of intellectual property generated in academia. The available literature primarily documents their occurrence in the United States and, to a much smaller degree, in Germany, Britain, and France. These studies show that, by the end of Baekeland's career, most American research universities had introduced formal intellectual property policies concerning inventions by their faculty and staff. In addition, some schools had already established patenting and patent-licensing facilities somewhat similar to today's technology transfer offices.[26]

Individually or collectively, these different types of institutional changes were among the factors that set the science-industry nexus of the late nineteenth and early twentieth centuries apart from science-industry relationships in previous decades. There were, however, also important elements of continuity that need to be taken into account. Perhaps most relevant, since long before the nineteenth century, "hybrid" experts had pursued knowledge about nature while concurrently addressing technological problems in the spheres of manufacturing and commerce. Moreover, since the 1700s, if not earlier, these individuals had included members of the teaching staffs of higher education institutions.[27] In the late nineteenth and early twentieth centuries, despite the formalization of academia-industry collaborations mentioned earlier, individual academic scientists continued to get involved in industrial work on an informal, case-by-case basis. As I discuss in more detail, this included work not only as consultants and patentees but also as founders, partners, or directors of business enterprises.

It is also important to keep in mind the heterogeneity of the time period and the nation-states under study and to acknowledge that the relationships between science, technology, and industry remained more intense in Baekeland's discipline of chemistry than in most other fields. For example, in most European countries other than Germany, it is debatable whether a comprehensive science-industry nexus was already in place on the eve of World War I. More generally, irrespective of the "Great War" (whose impact on science-industry-government relationships is addressed in chapter 6), the connections and interdependencies between science and industry had distinctive features in specific national or regional settings. Still, the range of countries that experienced an intensification of science-industry linkages, in combination with the adoption of institutional innovations of the type mentioned earlier, was broad and highly diverse. By the late 1930s, indeed practically all European and North American countries were counting on science and technology to stimulate economic growth, regardless of vast ideological differences.

A comparable pattern of similarity in heterogeneity applies to the spectrum of sciences involved. In coal- or oil-rich places such as England and the United States, for example, geology played a more prominent role in forging science-industry linkages than in continental European countries without domestic supplies of these resources. Likewise, some countries were quick to embrace electrical engineering yet relatively slow to develop new chemical industries.[28] Still, a similarity that has remained underappreciated concerns the extent to which, on both sides of the Atlantic, chemistry was at the forefront of the strengthening of the science-industry nexus. I am referring not just to organic chemistry, whose contributions to the synthesis and production of new chemicals, especially dyes, have been extensively documented.[29] What is much less well known is that, even in countries where organic chemistry was relatively underdeveloped (such as in the pre–World War I United States), chemists typically represented the largest group of industrially employed scientists and industrial consultants in the late nineteenth and early twentieth centuries.[30] Despite the low status of some of these chemists, their job duties were hardly limited to the performance of routine tasks. Before World War I, for example, chemists and chemical engineers had already conducted seminal R&D and advisory work in a diverse range of areas outside the synthetic organic chemicals industry.

As one of the most prominent chemists of his generation (table 1.1), Baekeland was not only affected by but also contributed to several of the developments described here. For instance, in the decades following his university graduation in 1884, Baekeland was involved in early R&D work in the photographic, electrochemical, and plastics industries as a research chemist, chemical consultant, and corporate executive. In the latter function, moreover, he and his associates at the General Bakelite Company, the predecessor of the Bakelite Corporation, contributed to the formalization of academia-industry collaboration by initiating an industrial research fellowship at Columbia University (where Baekeland also lectured to future chemical engineers). At the same time, a notable feature of the Bakelite companies' corporate culture was that scientists and engineers not only conducted research work but were also employed in manufacturing, sales, and top executive positions. In short, the expansion of the science-industry nexus was a development central to Baekeland's career that needs to be taken into account when interpreting his accomplishments and setbacks.

Table 1.1
Selected career highlights of Leo Baekeland.

Year	Honor or achievement
1884	Earned PhD in natural sciences, University of Ghent
1887	Laureate of concours universitaire, Royal Academy of Belgium
1895	Commercial introduction of "Velox" photographic paper
1899	Sale of Nepera Chemical Company to George Eastman
1904	President of Chemists' Club, New York
1907	Invention of the synthetic plastic "Bakelite"
1909	Paper presentation on "The Synthesis, Constitution, and Uses of 'Bakelite'"; President of American Electrochemical Society
1910	Formation of Bakelite GmbH in Germany; General Bakelite Company in the United States
1912	President of American Institute of Chemical Engineers
1913	Awarded Willard Gibbs Medal, Chicago Section, American Chemical Society (ACS)
1914	Chandler Lecturer, Columbia University
1915	Elected to Naval Consulting Board
1916	Awarded Perkin Medal, US Section, Society of Chemical Industry; Received honorary doctorate, University of Pittsburgh
1917	Named honorary professor, Columbia University
1918	Acting chairman, Committee on Patent Reform, National Research Council
1924	President of ACS
1925	Appointed to Chemical Advisory Committee to Department of Commerce
1929	Received honorary doctorate, Columbia University
1935	Elected to National Academy of Sciences
1937	Received honorary doctorate, University of Edinburgh
1938	Awarded Messel Medal, Society of Chemical Industry

Scientific Entrepreneurship and Intellectual Property

Two themes at the heart of the science-industry nexus that are especially relevant for understanding Baekeland's status as "father" of the synthetic plastics industry are intellectual property and scientific entrepreneurship. Previous studies of Baekeland, which have tended to focus on his actions and accomplishments as an inventor rather than as a scientist and entrepreneur, have paid comparatively limited attention to these topics.[31] More notably, they are also largely absent from the extensive body of literature on the relationships between science, technology, and industry at the time of the second industrial revolution, on which this chapter's previous section has drawn.[32] From reading nineteenth- and early twentieth-century biographical accounts of academic scientists, one might get the sense that this absence is justified. If the scientists in question obtained patents or were involved in the founding and running of business enterprises, these activities are likely to be mentioned only in passing, if at all. In contrast, these biographies go into considerable detail with respect to these scientists' contributions to "pure science," which were held in much higher cultural esteem.[33] As a result, some scholars have seriously mischaracterized the priorities of scientists from this period. British physicist Oliver Lodge, for instance, has been mentioned as an example of a "pure" scientist, allegedly supporting the claim that "the traditional ethos of science did not permit... [eroding] the boundary between science and private, profit-seeking-business."[34] In reality, however, as historian Anna Guagnini has shown, Lodge was profoundly involved in commercial work as a patentee and a business entrepreneur in the course of his academic career.[35]

Like these nineteenth- and early twentieth-century biographers, historians have long held reservations about highlighting, or even acknowledging, scientists' engagement in entrepreneurial activity and other "distractions" that were perceived to be immaterial to the development of scientific thought.[36] Over the past half century, however, there has been a gradual move away from studying science as a supposedly impersonal, detached body of knowledge—an assumption that rendered the personalities and nonscholarly activities of scientists largely irrelevant.[37] Earlier doubts about the legitimacy of scientific entrepreneurship as a topic of historical analysis seem also to have disappeared. Consequently, recent studies no longer treat the industrial occupations of prominent scientists such as William Thomson (better known as Lord Kelvin), Louis Pasteur, Henry Rowland, Arthur Noyes, and Marie Curie as impertinent diversions. Instead, they explore the sometimes fundamental connections between these activities and the scientific accomplishments for which these academics are best known.[38]

Because of a lack of systematic surveys, it is still too early to tell how common it was for late nineteenth- and early twentieth-century scientists to obtain and commercialize patents and to participate in the creation of new firms. As historian Paul Lucier has suggested, these activities may have especially appealed to individuals who did not hold salaried teaching or research positions but nevertheless held ambitions to make a living doing science.[39] At the same time, however, it is becoming increasingly clear that, even when it came to academic scientists in both Europe and the United States, scholars have been wrong to describe the participation in commercialization activities as exceptional or fundamentally incompatible with the ethics of science. For instance, with respect to patenting, an often-repeated claim by historian Charles Weiner has become untenable. In a seminal 1987 article, Weiner wrote that academics rarely patented before the 1920s, because "there was widespread fear that the effects of patenting would commercialize the academic environment, corrupt the traditional nature of the university, and impinge on the public's trust and support of university researchers." Recent studies, however, suggest that, from the late nineteenth century until the present, the share of scientists who took out patents during their academic careers may have remained remarkably stable.[40] For scientists participating in the creation of new firms in the course of their academic employment, it has likewise become clear that the few cases that are often mentioned in the scholarly literature were hardly exceptional. For example, besides William Thomson's (Lord Kelvin's) relatively well-known involvement in a consulting firm, Anna Guagnini has identified two other British physicists, as well as seven engineers, who founded or led at least one business enterprise while holding a professorship at a higher education institution.[41] Similarly, in addition to Baekeland, whose university start-up firm is discussed in chapter 3, there were at least eight other academic employees who participated in the creation of photographic businesses in the 1880s or 1890s, with or without the consent of their employing institutions. Small as they may seem, these numbers are significant if considered in proportion to the size of the professional groups to which these individuals belonged.

Crossing Boundaries

My analysis of Baekeland's entrepreneurial and intellectual property activities is not, in any case, restricted to his involvement in start-up firms and his strategies for managing legal intellectual property rights such as patents. This book also examines issues of entrepreneurship and IP connected with the major historical transformation discussed in the previous section of this introduction: the creation of a comprehensive science-industry nexus.[42] In particular, I consider these themes in relation to the crossing, reconfiguring, and blurring of boundaries between the spheres of science, technology, and industry in the late nineteenth and early twentieth centuries. With regard to entrepreneurship, this broadened focus of analysis rests on the assumption that scientific entrepreneurs such as Baekeland faced specific risks, as well as opportunities, associated with their dual roles as industrialists and members of disciplinary associations and communities. Some of the challenges concerned the criteria by which their business activities were evaluated: besides the standards of quality and profitability all entrepreneurs had to take into account, they also had to come to terms with specific expectations of their scientific peers. In 1923, for instance, Baekeland expressed his ambivalence about a promotional publication on the history of Bakelite by claiming that "we chemists do not want advertising as film stars or actors or writers or authors" and that "any such thing does us much harm just among our professional friends."[43] Similarly, his strong 1926 refutation of a fellow chemist's assertion that the Bakelite Corporation failed to appreciate the value of scientific research points to the importance that both Baekeland and his disciplinary colleagues attached to this point.[44] At least as numerous, however, were the opportunities for taking advantage of a position at the intersection of the spheres of industry and science. Scientific entrepreneurs, for example, could try to exploit their academic and professional credentials and affiliations to facilitate the hiring of scientifically trained personnel and the accessing of academic knowledge and research facilities. Alternatively, they could attempt to translate sociocultural authority into financial profits by associating themselves with science in fundraising and public relations efforts. When pursuing such indirect, "ornamental" uses of science, they had to keep in mind, however, the potential reputational harm brought about by scientific "window-dressing."[45] The opportunities and risks faced by scientific entrepreneurs were indeed often related.

With regard to intellectual property, this book's focus on the relationships between science, technology, and industry necessitates a differentiation between two distinct but related kinds of IP, each of which shaped Leo Baekeland's career and legacy. The first type comprises the legal tools on which authors, inventors, and firms could rely to control the use and dissemination of creative works, including copyright, patent, trademark, and trade secret law. Of these different instruments, patents and trade secrets were by far the most central to Baekeland's career, as well as to the development of a comprehensive science-industry nexus. They are, accordingly, the formal intellectual property rights that receive the most attention in this book.

The second form of IP concerns another category of private intangible assets, the intellectual credit attributed for scientific and technological firsts. This type of IP can take the form of prizes and honorary professorships, besides other accolades. It has often been characterized by priority claims and disputes and can, but need not, go together with a reliance on formal IP law. As Robert Merton observed in his mid-twentieth-century writings on the sociology of science, scientists sometimes refer to their intellectual accomplishments in terms of "property."[46] Chemist Antoine Lavoisier, for instance, wrote in 1793 that the aim of the maker of a scientific discovery "has been achieved when he has secured his property, when it has become established that it [the discovery] is really his."[47] Merton, as well as later scholars such as economists Partha Dasgupta and Paul David, contrasted this form of intellectual property with the profit-oriented use of formal IP rights in commercial environments. In particular, they opposed the "definition of technology as 'private property' in a capitalistic economy" with the scientific practice of freely releasing findings into the public domain. In their view, the latter behavior did not primarily depend on altruism. Rather, the rapid publication of scientific results was necessary in order to be recognized for one's distinctive contribution to the development of one's science. It was, in other words, a prerequisite for obtaining intellectual property of the second kind.[48]

How did these two types of IP relate to each other in the late nineteenth and early twentieth centuries? Merton has rightly been criticized for contrasting scientific openness with corporate secrecy without having carefully examined the knowledge-sharing practices of academic, and especially industrial, scientists. He also did not pay close attention to how the term "intellectual property"—as well as related concepts such as "scientific property"—had been used in the years and decades prior to his famous 1942 essay on the ethics of science.[49] These are pertinent issues, as the IP phrase was not popularized until the late twentieth century. In the US context, Baekeland had been one of the first authors to adopt it in journals ranging from *Science* to *Industrial and Engineering Chemistry*. As part of patent reform efforts analyzed in chapter 7, he repeatedly praised intellectual property as "the most logical, the most legitimate personal property," which should be treated with at least as much respect as property rights relating to real estate or chattels.[50] Baekeland's use of a rhetoric of intellectual ownership may have partly reflected his upbringing in Belgium, where advocates of "industrial property" (propriété industrielle) had, at least since the mid-nineteenth century, advocated the existence of a "true right of property in the intellectual realm, just like in the material realm."[51] Both in Belgium and in the United States, however, other commentators on these countries' patent systems argued that "owning" an idea for a limited amount of time was fundamentally different from owning a house. In a 1931 talk promoting industrial R&D, Bakelite research director Lawrence Redman even contrasted patents with "intellectual property," which he defined as the "unused new knowledge" held by an organization. When it came to trade secrets, moreover, the relevance of the IP notion was (and remains) at least as controversial.[52]

This book, however, does not so much want to uncover alternative traditions of thinking about patents and secrecy as draw attention to how academic and industrial scientists' reliance on these tools affected their chances for getting intellectual recognition for their creative accomplishments. In addition, I explore the extent to which the patenting and withholding of technological knowledge and artifacts complicated university-industry collaboration. I demonstrate that, while the intensification of science-industry relationships in the late nineteenth and early twentieth centuries created significant tensions, the intertwining of capitalist and scientific incentive systems also benefited individual scientific entrepreneurs and their companies. This book indeed relates Baekeland's status as "father of plastics" to his familiarity with distinct reward systems and his ability to exploit them to his own advantage. Unlike with Velox, the unpatented class of photographic papers that, in Baekeland's own words, "made my reputation," Baekeland and his legal advisers had comprehensively patented Bakelite inside and outside the United States (see chapters 4 and 7). In both cases, Baekeland had delayed the publication of relevant scientific and technological findings so as not to undermine his formal IP position. He did not, however, abstain from seeking credit as a scientific author. In fact, this book suggests that, throughout his career, Baekeland remained sensitive to the customs and expectations of his almost exclusively male professional peers. As a result, claims attributing him intellectual "fatherhood" of the synthetic plastics industry were more readily accepted.

A Transatlantic Life

Besides the focus on intellectual property and scientific entrepreneurship, as defined previously, an additional feature that sets this book apart from previous studies of science-technology-industry relationships in the late nineteenth and early twentieth centuries is the transatlantic perspective. This approach capitalizes on the fact that Baekeland not only crossed boundaries between the spheres of science, technology, and industry—to the extent that they were separated. As he migrated from Belgium to the United States at the age of 25 but maintained contacts with family, friends, and colleagues in Europe, his life also provides what one historian has described as "a key to unlock those histories usually compartmentalized into discrete nationally-defined studies."[53] Above all, by examining Baekeland's career, it is possible to throw new light on the significance of transatlantic connections and relations to the making of a comprehensive science-industry nexus in the United States.

The existence of relevant linkages is undeniable. It is, for example, well documented that, in the final decades of the nineteenth century, hundreds of American students traveled to Germany to obtain doctoral degrees or postdoctoral training in the sciences, especially in chemistry. It is also well known that the foreign experiences and impressions of these students helped bring about significant changes in the domestic development of the sciences in the United States. In particular, early research universities such as Johns Hopkins University (founded in 1876), Clark University (1887), and the University of Chicago (1890) were modeled, in part, after German counterparts, whose admiring (but often inaccurate) representation in the United States also lent more legitimacy to the ideal of pursuing science for its own sake.[54] American students' stints in Germany, moreover, contributed to the introduction of several new disciplines on the other side of the Atlantic, ranging from physical chemistry to experimental psychology, as well as to the formation of scientific societies emulating European examples.[55]

What is less known, and seemingly contradicts the contemporaneous advocacy of German style "pure science," is that it was also not unusual for American visitors to single out the industrial exploitation of science and science-related technology as a particular German strength. Previous historians have primarily documented this in relation to the creation of early corporate research laboratories in the United States. They have suggested that German linkages, as well as concerns about competitive threats posed by German and other European firms, contributed to the decisions of some—but by no means all—large American R&D pioneers, including Du Pont and Eastman Kodak, to establish relatively autonomous research laboratories.[56] However, unlike with the institutionalization of research in American academia, the role of European connections in the emergence of such industrial laboratories is generally seen as relatively unimportant. Even where transatlantic linkages are recognized, historians have, by and large, attached more significance to institutional developments within the United States. Especially consequential was the stricter enforcement of the Sherman Antitrust Act of 1890 after the turn of the twentieth century, which limited firms' possibilities for purchasing externally developed technologies and for taking over competitors. In addition, the creation of in-house research laboratories was part of a more general tendency toward incorporating economic activities into ever-larger industrial enterprises instead of relying on market transactions.[57]

These well-documented changes might seem to support one scholar's claim that, although industrial research activities in Germany and the United States were concentrated in the same industries, "U.S. enterprises did not regard German firms, their business strategies, their research and development structures, or their methods and goals of innovation as exemplary. Accordingly, they did not attempt to transplant strategic and organizational experiences from Germany to the U.S."[58] This book, however, challenges this interpretation by focusing on a Belgian-born immigrant entrepreneur who *did* present German science-related firms as exemplary and who collaborated with German Americans to commercialize his inventions. I start from the position that, while transatlantic connections may have been relatively unimportant to the emergence of the industrial research laboratory, their significance to other aspects of the science-industry nexus has remained underexplored and, accordingly, potentially underestimated. Chapter 10, for example, analyzes whether the American Bakelite companies' strategies of technical selling reflected the German backgrounds of several of Baekeland's main associates. More fundamentally, I argue that a transatlantic perspective is necessary to appreciate the role of scientific entrepreneurship and intellectual property in the development of the science-industry nexus. For instance, many of the early US chemical firms that developed new strategies for exploiting science and technology for business purposes were run by immigrant entrepreneurs from Europe, especially from Germany. The extent to which their leadership drew on its European experiences and connections when determining approaches to, among other things, the protection of knowledge and intellectual property generated by their employees has not yet been examined comprehensively.[59] The case of Baekeland and the immigrant entrepreneurs with whom he collaborated, however, points to an intense circulation of relevant knowledge, know-how, and intellectual property across the Atlantic. In fact, several chapters of this book demonstrate how Baekeland's inventions, like various other science-related technologies of the second industrial revolution, were developed and commercialized in a transatlantic space that was fundamentally interconnected.[60]

Sources

This interconnectedness is documented in detail in the sources on which this book is based, particularly in Baekeland's extensive personal diaries and correspondence. A private comment by American legend Thomas A. Edison can serve to introduce the challenges faced by the historian studying Baekeland's actions and accomplishments on the basis of these materials. When confronted with contradictory claims by Baekeland in the context of a several-year patent dispute, Edison noted in 1916 that "Dr. B. [Baekeland] inherits from his mother the inalienable right to change her mind, which right every woman works to the limit. Pre-natal influence. Anyone that tries to remember such an ungodly name as Hexamethylenetetramine is sure to suffer mental fatigue and absentmindedness."[61] Throughout this study, it will become clear that Baekeland's writings and public pronouncements were, in effect, riddled with apparent paradoxes and inconsistencies. Hence, in this sense, Edison (who, needless to say, had never met Baekeland's mother) was not totally unjustified in making this rather misogynistic remark. In order to explain these discrepancies, though, it is surely more helpful to consider Baekeland's efforts to proactively shape the ways in which he was perceived by others. Baekeland's accomplishments were, to be sure, impressive in their own right. Born into a family of relatively modest means, he had earned a PhD in science at the age of 21. By the end of the nineteenth century, moreover, he had gained a fortune of about $215,000 through the sale of the Nepera Chemical Company, the second photographic enterprise that he had cofounded, to George Eastman of Eastman Kodak. In a 1929 letter to his friend Edouard Remouchamps, Baekeland indicated that he had been "prepared to die" since his thirty-sixth birthday; that is, from the year of the Eastman takeover onward.[62] From this point—not uncoincidentally—he certainly qualified as a "self-made man," a winner in "the race of life" in a society committed to the liberal dream that any white male could make it in business if only he tried hard enough.[63] But Baekeland also took considerable pains to ensure that others considered him to be a "deserving" rich person, as opposed to the "idle consumers" who had inherited their wealth or had earned it through stock market manipulation and other forms of "robbery."[64] For example, when informing a correspondent that the winter residence that he had purchased in 1929 was located in a part of Florida that "proportionately to its population…counts more men whose name is in 'Who's Who in America' than any other town large or small in Florida," Baekeland immediately added that "no possessors of money can get their name in Who's Who, unless they have other achievements than money."[65]

According to a letter by Baekeland to the *Miami Herald*, this claim concerning the number of *Who's Who* listings per capita was supported by calculations of "one of my assistants."[66] In several instances that are far more important to this study, however, Baekeland did not mention any collaborator or associate. Historian Susan Morris, for example, has highlighted how Baekeland consistently—and misleadingly—used the "I" form when discussing milestones in the history of Nepera. Morris quotes the following passage from an address that Baekeland delivered at Johns Hopkins University in 1932: "I had just built a little factory in Yonkers for the manufacture of Velox photographic paper.…I was about ready to produce this paper on a commercial scale, when the intense [economic] panic of 1893 broke out.…I had also unforeseen manufacturing difficulties, but I gradually managed to overcome them.…After Velox papers had become a well-paying enterprise, I sold out my interests to the Eastman Kodak Company."[67]

Similar statements abound in Baekeland's published and private writings, with regard to Velox as well as to Bakelite, and have doubtlessly contributed to the spread of inaccurate information concerning his career. Baekeland's claim that "I sold out my interests" to Eastman Kodak, for example, seems to correspond with previous historians' assumption that he received most, if not all, of the $750,000 Kodak agreed to pay and that he had been Nepera's main negotiator. In reality, however, Baekeland earned less than a third of this amount, as he had to divide the sum with his partners, Leonard Jacobi and Albert Hahn, and, as he put it in his diary, "got my share after paying my debts and interests to Jacobi," who had been Nepera's chief financier and the primary negotiator of the transaction.[68]

How, then, can Baekeland's actual accomplishments be differentiated from the ones for which he falsely claimed credit, and how can accurate descriptions be separated from bragging? The question is especially pertinent because Baekeland started keeping a diary on 3 April 1907, roughly two months before he would enthusiastically write down his first notes about the invention of a new material that he initially called "Bakalite." As I discuss in chapters 7 and 8, Baekeland became rapidly aware of the broad range of possible applications for this fully synthetic plastic, and the transformational impact they might have. Accordingly, documenting and highlighting this historic development while it was unfolding—from the initial laboratory breakthroughs in 1907 up to the sale of the Bakelite Corporation in 1939—may have been one of his aims for penning approximately 11,000 more diary pages in the following years and decades. Moreover, the fact that other inventors, chemists, and entrepreneurs did not leave behind similarly comprehensive personal papers further complicates the assessment of their contributions to the emergence of the phenolic resin industry. It would have been helpful, for example, to know more about the seminal work of electrochemist and inventor Nathaniel Thurlow, whom Baekeland hired as a research collaborator around the end of 1904. Thurlow examined the reaction between phenol and formaldehyde in the following years. By early 1907, this work had led to the invention of an impregnating substance that seemed commercially valuable in itself and, as later experiments demonstrated, could be used to manufacture Bakelite. Thurlow, however, agreed to sell his patent rights to Baekeland, thus rendering his contribution to the creation of a new industry largely invisible.[69]

Fortunately, though, it is generally possible to test facts, observations, and interpretations set forth in Baekeland's diaries, as well as in his almost equally comprehensive personal correspondence, against a whole range of other sources in archival collections on both sides of the Atlantic. Moreover, the personal nature of Baekeland's records, including the more than 15 diary volumes to which previous historians had no access,[70] makes it possible to cast light on aspects of the science-industry nexus that would be hard to discern otherwise. Baekeland's diaries, for instance, document the chemist's informal meetings with patent agents and examiners,

Figure 1.3
Baekeland on the cover of *Time* magazine, 1924. *Time* later celebrated him as the "father of plastics."

complementing the picture that emerges from his official correspondence with the US Patent Office. My primary source material also allows me to identify the motives behind Baekeland's entrepreneurial activity as well as the strategies he pursued and the real or potential conflicts of interest he faced when crossing the boundaries between the spheres of science and industry. Finally, besides documenting why Baekeland came to be regarded as the "father of plastics" (figure 1.3), my sources also reveal his indebtedness to a whole range of organizations and individuals, many of which did not belong to the "relevant social groups" identified in previous research.[71]

Chapter Overview

All this book's chapters make distinct contributions to my thesis that the expansion of the science-industry nexus in late nineteenth- and early twentieth-century Europe and the United States shaped the careers and the real and perceived legacies of Baekeland and many of his contemporaries. Chapters 2 and 3 critically examine Baekeland's statements regarding the limited economic value of his education at the State University of Ghent and explore how unusual it was for an employee of this higher education institution to establish a start-up firm. Then, in chapter 4, I relate Baekeland's legal IP strategy for Velox to the strengthening of firms' knowledge ownership rights vis-à-vis their technical personnel and discuss how this development affected Baekeland's ability to claim credit for his inventive and scientific accomplishments. Chapters 5 and 6 then turn to several little known but significant episodes in Baekeland's career, including his participation in the creation of two additional high-technology firms, his performance as an expert witness in a nuisance suit regarding the emission of chlorine gas, and his evaluation of chemical warfare during World War I and as the 1924 president of the American Chemical Society. These chapters also analyze the moral and professional contexts in which Baekeland, a practically minded innovator, became a leading figure in scientific circles.

The remainder of this book consists of four thematically organized chapters examining the interplay between Baekeland and Bakelite. I first investigate Baekeland's international and domestic IP strategies for this synthetic plastic and connect these to his efforts to reform the US patent system (chapter 7). Chapter 8 further explores tensions and opportunities associated with Baekeland's dual roles as an industrialist and as a member of the US chemical community. It shows that his publicly stated preference for industrial over decorative applications of Bakelite contrasted with his strategic decisions as a business executive. Chapter 9 then explores the importance of theoretical knowledge to the science-industry nexus. It does this by analyzing Baekeland's participation in early debates concerning the molecular structures of synthetic plastics like Bakelite, which had not yet been identified as "macromolecules" or polymers. Chapter 10 discusses the American Bakelite companies' commitment to research and development, their technically oriented sales policies, and their remuneration schemes for their scientifically trained personnel from a transatlantic point of view. Finally, in the afterword, I address the continued relevance of this book's thesis by briefly reflecting on how the relationships between science, technology, and industry have changed since Baekeland's retirement in 1939.

2 Social Mobility through Education

In honor of his accomplishments in applied chemistry, in 1916 Baekeland was awarded the prestigious Perkin Medal by the American Section of the Society of Chemical Industry. While he stated in his diary that it "is not very pleasant to have to talk about yourself," his friends insisted that he "should give my experience" in his acceptance speech at the award dinner.[1] In accordance with this request, Baekeland set out to illuminate the American chemical industry's state of development by means of a series of autobiographically inspired lessons and recommendations. The first provides the starting point to this chapter. Baekeland captured it with the heading "Practical Life as a Complement to University Education," but he seemed to suggest that this slogan might well understate the problem at stake: "My real intense education," he claimed, "only began after I had left the university, as soon as I became confronted with the big problems and responsibilities of practical life." He also expressed the desire to "remain until I die, a postgraduate student at that greater school of practical life, which has no fixed curriculum and where no academic degrees are conferred, but where wrong petty theories are best cured by hard knocks." According to Baekeland's address, he had experienced such a knock around 1887. At that time, his attempts to prepare a key photographic chemical, commercial silver bromide, had (allegedly) made him lose his "cocksure" confidence "that everything was as simple as it appears in the textbooks." Then, he added, "By the time I got fully engaged in the manufacture of silver-salt products for photographic purposes, I had come to the conclusion that instead of [the] three or four varieties of bromide or silver [discussed by Belgian chemist Jean Servais Stas in an 1874 publication], there existed perhaps a hundred varieties, but that of those hundred varieties, there was only one which might probably keep me from the poorhouse and assure me success as a manufacturer."[2]

As early as 1942, a close observer of the American chemical industry noted that Baekeland's comments on the "greater school of practical life" could have been drawn from the concluding chapter of a nineteenth-century Horatio Alger novel. His remarks also fitted into a long-standing tradition among businesspeople in general, and chemical industrialists in particular, of questioning the value of a higher education. Still, later historians mostly seem to have taken Baekeland at his word when quoting the statements just mentioned in their accounts of the difficulties encountered by Dr Baekelandt et Compagnie, the photographic firm Baekeland cofounded at the end of 1887.[3] For instance, it remains to be seen whether Baekeland really had little previous knowledge to fall back on when he ventured into business as a young academic. Moreover, Baekeland's apparent disillusionment with his university education may have been caused by things other than his use-oriented photochemical experiments, if he was ever really devoted to theory-oriented academic learning in the first place.

This chapter shows that the picture of Baekeland's educational background provided by his Perkin Medal address of 1916 is misleading in several ways. Baekeland's secondary schooling, for a start, was unusually utilitarian for a later entrant of a Belgian university. Moreover, even in what was arguably his most comprehensive scientific study—a thesis on so-called dissociation processes, which he completed prior to the establishment of Dr Baekelandt et Cie—Baekeland revealed a distaste for theoretical "speculations." In addition, even though his higher education included courses that would indeed prove irrelevant to his later career as a scientific entrepreneur, it also provided him with a knowledge base on which he could (and did) build with respect to several of his inventions, including Bakelite plastic. Last but not least, Baekeland's educational accomplishments helped him accumulate cultural and symbolic capital and establish a professional network.

At the same time, however, the juxtaposition of academic "theory" and industrial "practice" can be a helpful analytical tool for characterizing Baekeland in comparison with other scientists and inventors, and for understanding strategic choices he made at various stages in his career. As will be discussed in more detail, a case in point was his reaction to the opportunities for growth created by the emergence of new photographic processes in the 1870s and 1880s. In this context, higher education teachers and researchers on both sides of the Atlantic became involved in the photographic industry as consultants, patentees, and entrepreneurs. Some of them used the experiences, skills, and/or money thus gained to support their academic agendas, including the advancement of photochemical theory.[4] Baekeland's first photographic venture, by contrast, primarily served as a stepping-stone to a career as an independent innovator.

Baekeland's retrospective reflections on the gap between "theory" and "practice" also provide a useful perspective from which to explore how late nineteenth- and early twentieth-century educational institutions sought to contribute to industrial development, a theme that is further examined in chapter 3. Several of Baekeland's recollections on his "self-made" childhood suggest that he had internalized the ideology of the American Dream. One hypothesis could therefore be that his verbal assaults on "our self-satisfying but often superficial generalizations of the text-books" were inspired by the dislike for science for its own sake that French intellectual Alexis de Tocqueville had famously observed in the United States.[5] A comprehensive 1881 report on the state of chemistry and physics education in that country reveals that such an attitude had been widespread at the time Baekeland started his university studies on the other side of the Atlantic. The author of this report, for example, stated that "the intellectual side of education is sacrificed to the material" in American scientific schools and that a "wholly commercial spirit" pervaded many of these and other types of higher education establishments.[6] Academics such as physicist Henry Rowland challenged this attitude in the following decades as part of a transatlantic movement for "pure science." Still, fields such as organic chemistry and mathematical physics, which lent themselves to "brilliant speculation" but seemed to have a limited industrial relevance in the pre–World War I United States, continued to be neglected in comparison with their sometimes dominant position in European curricula.[7] Moreover, despite the avoidance of such subject areas, many American chemists and chemical engineers felt that their country's system of education generally "does not produce men who are prepared in the best possible way for practical work in the manufacturing establishments of the United States."[8] In short, in the US context, Baekeland was not exceptional in highlighting lessons learned through the "University of Hard Knocks," a phrase that was being popularized in the years around 1916 to stress the importance of real-life experience.[9]

In Baekeland's home country, however, the influence of several intellectual traditions that insisted on practical utility had also been tangible. In fact, before 1870, the perceived utilitarianism of Belgium's universities had been frequently complained about, and the Belgian scientific community had sought to gain international recognition by producing "reliable facts" as opposed to "daring theories."[10] This approach did not go unchallenged, but it nevertheless continued to be advocated until well into the twentieth century, including by chemists of Baekeland's generation. Meanwhile, from the final decades of the nineteenth century onward, laboratory and other experimental exercises became more common in higher-level as well as lower-level schools, and educational reformers such as Omer Buyse, the first director of the University of Labor (Université du Travail) in Charleroi, attempted to increase the status of manual labor. Therefore, to interpret Baekeland's educational background and its significance for his later career, it will be necessary to look at both American and European ideas and developments.

This is especially true because the available sources date, to an important degree, from after Baekeland's migration to the United States in 1889. To be sure, Baekeland's student notes document the different sorts of scientific and technological knowledge he was familiarized with in Ghent in the late 1870s and early 1880s, but contemporary information on aspects of Baekeland's Belgian years other than his schooling is not nearly as extensive and will therefore be supplemented with retrospective accounts. In particular, Baekeland's correspondence with his wife and with Belgian friends, as well as the diary entries that he wrote down during visits to his native country after 1907, allow the contours of his youth and early adulthood to be sketched. In this manner, new light can be thrown on the making of Baekeland as a scientist, inventor, and entrepreneur.

When I Was a Poor Boy

In his article "American Entrepreneurs and the Horatio Alger Myth," economic historian Bernard Sarachek revealed that 56 percent of the 187 entrepreneurs in his research sample of mostly nineteenth-century individuals had lacked a supportive father. Sarachek distinguished between fathers who had died prematurely, those who had separated themselves for more than a year from their sons, "inadequate" fathers, and "rejecting" fathers "who tended to thwart their sons' efforts and impress upon their sons a sense of inadequacy and ineptness." Leo Baekeland would have fallen into the third category, which comprised "cases in which the father appears as a predominantly negative authority figure."[11] The person in question was Carolus

Baekeland (1830–1902), a shoemaker and innkeeper (figure 2.1). From Leo's correspondence with his wife and his friend Edouard Remouchamps, the best available source material on this topic, it seems as if Carolus indeed served as a negative role model. For instance, on 18 July 1899, Leo wrote to Remouchamps that, if he had had more money to spend, Carolus would have risked harming his own health, suggesting that he was a drunk.[12] Moreover, two weeks later, after receiving an alarming letter from his sister Rachel (b. 1876), Leo informed Remouchamps, a doctor by profession, that:

> It is the old story with my father, who since my childhood has already caused a lot of grief and difficulties, and whom I, nevertheless, try to excuse for and forgive; first because I think that he never had the benefit of a good education, and second, because I think that something has to be wrong in his brain. [He] appears once more to make life difficult for mother and Rachel, and I truly do not know what to do in order to prevent this. When I was short of money and he did not receive any pocket money, he seems to have behaved better than during any other period. Due to weakness and natural imitation, he has always had the manners of the majority of Belgian husbands and played the little tyrant to his wife and children.... Fortunately, I have been able to exert psychological authority over him from an early age, thus saving my dear mother many unpleasant events. If he knows that I will soon come to Ghent, he will quickly change his manners, and he will be afraid that I will know about it, but in the meantime, it frustrates me that just when my loving mother and sister need rest he would be making their lives difficult.[13]

The influence that Leo exerted over his father probably helps explain why, as his wife, Céline Swarts, reminded him in October 1890, Leo had been idolized by his mother and his sisters, Rachel and Leonie (b. 1869).[14] As far as his siblings were concerned, this admiration was not reciprocal. In fact, Leo repeatedly seemed annoyed by what he regarded as their stupidity; it was only later in his life that he came to refer to them more affectionately. Yet the rumor that Leo did not get along well with his parents, causing a brief relationship crisis with Céline Swarts about half a year before the couple's marriage, was unfair in view of Leo's devotion to Rosalie Merchie (1833–1914), his "outstanding, beloved mother," whom he respected because of her honesty, thrift, and work ethic (figure 2.1).[15] Unlike her husband, Merchie, who had been a domestic servant with a more well-to-do family, had learned how to write, even though she used ordinary language that abounded with spelling errors. Significantly, shortly before her seventy-seventh birthday, she indicated to her son that "laboring is our entertainment. Believe me, it hurts me when I see someone else do work that I may or cannot do

Figure 2.1
Portraits of Leo Baekeland's parents, Rosalie Merchie and Carolus Baekeland, by Edmond Sacré, a prominent local photographer.
Courtesy of Ghent University Museum.

anymore."[16] The same letter also confirms the religious attitudes expressed elsewhere in the correspondence. Merchie had expected the newspapers that Leo had sent her to include more information about Bakelite, but, to her surprise, noticed that their only purpose had been to wrap up a paternoster inside. Furthermore, to Merchie, the fact that Leo and his family had not suffered any major accidents on their journeys signaled that God had been with them.

In this regard, Merchie certainly did not function as a positive role model. Indeed, her son came to regard the grip of Roman Catholicism on the Belgian people's lives as an impediment to societal progress, a view shared by the liberal, freethinking friends with whom he liked to spend his spare time in Ghent, his city of birth. Moreover, later, in the United States, Baekeland and Swarts sought to teach their own son "by word and example a strong and strict code of ethics, far superior to any so called revealed religion, so that he does not learn to fear evil for fear of the punishments of hell, but we teach him that heaven and hell are of one's own making right here during this life."[17] In this sense, Baekeland was clearly not a Weberian entrepreneur; that is, a Protestant (especially Calvinist) striving to accumulate earthly wealth and property as a reflection of a predestined afterlife in heaven. Still, through the figure of Benjamin Franklin, a notable link exists between Baekeland and sociologist Max Weber's much-discussed essay on *The Protestant Ethic and the Spirit of Capitalism*. To Weber, Franklin epitomized an apparently irrational, transcendental "ethic" that regarded the acquisition of ever more money for the sake of itself as a central purpose in life. Baekeland, for his part, is said to have been inspired by Franklin's memoirs. As the often-repeated anecdote goes, "Franklin's life not only had shown him [Baekeland] that a boy of humble beginnings could make his own way in the world; it also had fixed in his mind the dream of becoming an inventor and perhaps a famous and wealthy man, and implanted the further notion that America might be a place of special opportunity for men of low estate but high learning."[18]

This interpretation can probably be traced back to a speech that Baekeland delivered at the Franklin Institute in Philadelphia in 1940 upon receipt of an honorary medal. However, on this occasion, Baekeland stated solely that his schoolteacher had "fascinated" his fellow pupils and himself in 1871 "by reciting the glorious life of Benjamin Franklin."[19] Reduced to this basic form, the story is plausible, because Franklin's example surely impressed children around the world. Serbian-American inventor Michael Pupin (1858–1935), for instance, had heard about Franklin's electrical experiments in Pančevo, a city close to his native village of Idvor, located in a peripheral area of the Austro-Hungarian Empire. Moreover, according to Pupin's 1923 autobiography—a copy of which Baekeland donated to the library of the State University of Ghent (see figure 2.2)—he developed "a keen interest" in the United States as a result of this encounter.[20] As for Baekeland, however, it is not at all clear that stories about Franklin had a more inspiring effect than news about later US inventors such as Thomas A. Edison and Alexander G. Bell. According to the chemist's diary, for example, the announcement of the transmission of speech by Bell's telephone had made an "enormous impression" on him as a 13-year-old boy. Accordingly, he recollected, "Later on when I read in the publications how the telephone was constructed I set to work to make one and succeeded to make it operate faintly between the second story and the cellar of my home in Ghent." As a teenager, Baekeland was also taught that Edison had gained "world fame" thanks to his invention of a

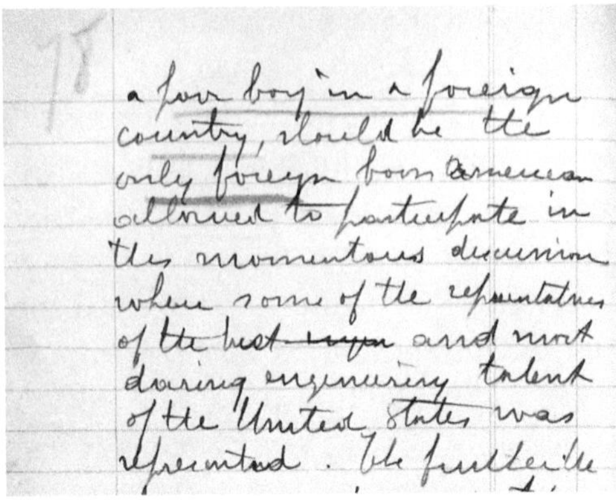
a poor boy in a foreign country, should be the only foreign born American allowed to participate in this momentous decision where some of the representatives of the best ~~foreign~~ and most daring engineering talent of the United States was represented. [illegible]

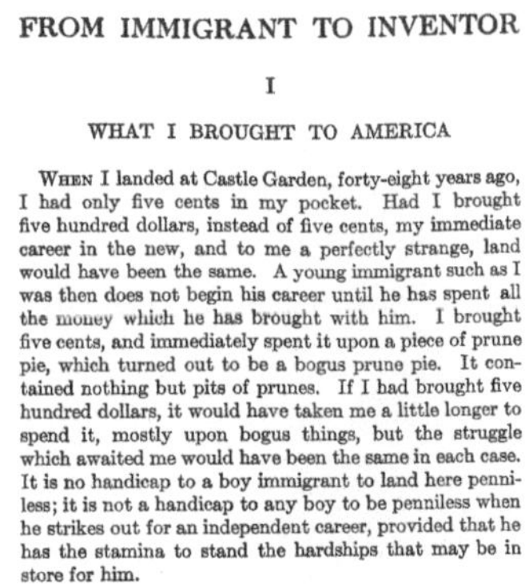
FROM IMMIGRANT TO INVENTOR

I

WHAT I BROUGHT TO AMERICA

When I landed at Castle Garden, forty-eight years ago, I had only five cents in my pocket. Had I brought five hundred dollars, instead of five cents, my immediate career in the new, and to me a perfectly strange, land would have been the same. A young immigrant such as I was then does not begin his career until he has spent all the money which he has brought with him. I brought five cents, and immediately spent it upon a piece of prune pie, which turned out to be a bogus prune pie. It contained nothing but pits of prunes. If I had brought five hundred dollars, it would have taken me a little longer to spend it, mostly upon bogus things, but the struggle which awaited me would have been the same in each case. It is no handicap to a boy immigrant to land here penniless; it is not a handicap to any boy to be penniless when he strikes out for an independent career, provided that he has the stamina to stand the hardships that may be in store for him.

Figure 2.2
February 1918 fragment of Baekeland's diary (left) and excerpt from Michael Pupin's bestselling autobiography *From Immigrant to Inventor* (Scribner's, 1923) (right), in which Pupin describes how he arrived in the United States with "only five cents in my pocket." Baekeland mailed a copy of the book to the University of Ghent, as well as to a recent immigrant to Canada who had asked him for $140 to complete an accounting course. Besides Pupin's memoirs, which revealed many parallels with his own career, Baekeland also gave this man the advice that "character, work and initiative," rather than money gifts, "lead to the right path of success." See Baekeland to Arthur P. O'Toole, 1 July 1931, in LBP, series 3, box 11, folder 4.
Courtesy of Leo Baekeland Papers, Archives Center, National Museum of American History, Smithsonian Institution, AC-0005-D24-042.

phonograph. Much later, in the context of patent disputes concerning Bakelite, he moreover noted that he had found consolation in reading an Edison biography that documented similar difficulties experienced by the "wizard of Menlo Park." In short, even though previous studies may have overestimated the specific influence of Franklin, there are many indications that the well-publicized accomplishments of "heroic" American independent inventors had appealed to Baekeland's imagination.[21]

In any case, like Franklin, the "patron-saint of self-made men," Baekeland also sometimes tended to exaggerate the "raggedness" of his initial situation as a counterpoint to his later "riches."[22] In retrospective reflections on his youth in Ghent, Baekeland put great weight on the modest conditions into which he was born and, as a corollary, on the need to take his life into his own hands from an early age. He was especially prone to evoking his relatively humble origins in three sets of circumstances. We will return to one of them—his justification for a prolonged separation from his wife, who had grown up in a bourgeois family—in chapter 3, as part of an analysis of Baekeland's first years in the United States. In that case, Baekeland brought up his modest background as a factor that complicated their living together. By contrast, under other circumstances, he, with the benefit of hindsight, seemed to regard his background as an asset. In the first place, it made him enjoy his later successes even more. In the context of his election as president of the American Electrochemical Society in 1907, for instance, Baekeland could "hardly abstain from thinking how slowly and steadily I had found my way onwards and there was quite some distance traveled which the poor unknown ... shoemaker's son would hardly have dreamt of thirty years ago." During a confidential meeting regarding the development of an aerial bomber in February 1918, Baekeland "felt elated that I[,] born as a poor boy in a foreign country[,] should be the only foreign-born American allowed to participate in this momentous discussion where some of the representatives of the best and most daring engineering talent of the United States was represented." When receiving honorary doctorates from Columbia University and the University of Edinburgh, Baekeland again realized that "the poor Flemish boy had become somebody." Likewise, on the occasion of a 1936 visit to Bakelite Limited, the British subsidiary of Baekeland's Bakelite Corporation, he commented that, "If I had been born 'with a silver spoon' in my mouth; if we [his wife Céline and he] had not met our difficult years, we might also have missed the thrill of progress and achievement."[23]

In line with the American Dream ideology set forth in a multitude of success manuals, Baekeland maintained that, in meritocratic societies like the United States or Canada, similar rewards could be achieved by anyone who tried hard enough and, in the process, developed a character as strong as his own.[24] Hence, it is only natural that Baekeland was also secondarily reminded of his youth and early adulthood when watching his own children and grandchildren grow up in environments more privileged. He especially liked to compare himself with his only son, George Washington, who was born in 1895, the year in which Jenny, the Baekelands' first daughter, had passed away. After George's fifteenth birthday, for example, Leo remarked that, "At his age I was I believe in the same mental condition as he is with this difference that he has had the benefit of better intellectual environment. I had to do everything by myself and find my own way. The only help I had was the encouragement of my beloved mother."[25]

To avoid having them develop into spoiled rich brats, Leo took pains to pass on the value of self-reliance to George and George's sister Nina (b. 1896). In 1908, he elaborated on his educational program for his two children in a letter to Remouchamps. "First of all," he began,

> Although I am financially pretty well off, I have made my will in such a way that my children shall not receive a cent when I die. So from now on they can already prepare themselves to the idea that their father will leave them no money to lead an idle life. This, of course, involves the idea that they shall be self-supporting, and thus cultivate self-respect, instead of lowering themselves by being dependent on others. In the meantime I shall give them every opportunity to have as thorough an education as possible which will fit them to become useful and self-respecting citizens, instead of snobs who think it is nice to do nothing and live on the work of others.[26]

Independence and self-respect were major values of at least some nineteenth-century European artisan communities, so Baekeland's attitude on this point may partly reflect his upbringing as a shoemaker's son. But the chemist's educational ideals were probably even more common in his country of arrival. After all, their ubiquity helps explain why, according to economic historian Naomi Lamoreaux, "If ever there was a time or place when entrepreneurs were the most admired figures in society, it was the United States during the late nineteenth century."[27] Moreover, like Baekeland, several early twentieth-century Belgian and French commentators on the American educational system regarded the encouragement of self-help, vigor, and entrepreneurialism as one of its hallmark features.[28]

In Baekeland's opinion, the fact that he and his wife had been able to send their children to mixed-gender schools reflected another merit of the US approach to education. As a result of this policy, he noted, pupils learned "how to conduct themselves towards each other and [this] takes away the stupid sex-consciousness which is the base of European education." In addition, he emphasized to Remouchamps that, contrary to common practice in Belgium and other countries, no end-of-the-year prizes were awarded to the brightest pupils so as not to discourage their less gifted classmates. Finally, Baekeland clearly looked favorably on the accomplishments of the progressive education and manual training movements on the American side of the Atlantic, as he attached "very little importance to *exclusive* book learning" and told his son to regard his manual classes as the most important.[29] Taken together, this strongly suggests that several years before his Perkin Medal address of 1916, Baekeland's views on education had in many ways already been Americanized.

Technical Training

In contrast to Baekeland's later instructional preferences, as well as those of progressive reformers in both America and Europe, the chemist's own primary and secondary education had consisted almost exclusively of formal lectures. This, however, does not mean that his schools had prioritized scientific theory over industrial practice. In fact, to some extent the opposite was true. At the municipal institute where Baekeland attended evening classes between 1877 and 1880, in particular, theoretical courses were meant to supplement students' existing practical skills and experiences.

Having opened its doors in 1833, the bilingual Industrial School (École industrielle, or Nijverheidsschool) of Ghent was one of the two oldest educational institutions of its kind in Belgium, the other one being located in Liège. By 1877, the number of Belgian industrial schools had grown to 31, but the institute in Ghent without any doubt remained one of "the most flourishing," as the author of a comparative international report had put it a few years earlier. Unlike primary and general secondary schools, or higher technical institutions such as the country's state engineering colleges, these industrial schools had a great deal of autonomy in designing their curricula and evaluation procedures. In doing so, as historian Marc D'hoker has shown, almost all of them mainly aimed to reach industrial workers, artisans, and/or miners. But they also tended to accept younger students who combined the establishments' evening or Sunday classes with daytime instruction at the primary or general secondary level.[30] In 1878–1879, for example, Baekeland pursued his studies at his École industrielle—known more colloquially as the "Indus"—in parallel with classes at the vocational department of Ghent's royal secondary boys' school. He then appears to have dropped out of the latter institute and, according to a later diary entry, instead combined his evening education with a job at a local pharmacy.[31]

Belgium's industrial schools used their relative organizational freedom to attune their offerings to local and regional job markets. Accordingly, what mining and steel were to the heavy industry cities of Liège and Charleroi, textiles were to Ghent, the "Manchester of the Continent." For instance, one of the courses that Baekeland successfully completed dealt with "Chemistry of fibers and textile materials, fabrics, etc.—Retting of flax and hemp; various processes; creaming and bleaching of the threads.—Handling of greasy wools.—Dyeing of cotton or woolen ribbons.—Bleaching and finishing of various fabrics.—Dyeing and printing of cotton, linen, woolen and silk cloths, etc."[32]

By the same token, the evolution of these schools' teaching activities can serve as a (rough) indicator of economic transformations affecting their surroundings. In Ghent, for example, it was resolved in 1873 that English and German classes would be introduced because "knowledge of these languages becomes, to anyone employed in industry or commerce, an ever more pressing necessity." In February 1880, when Ghent's photographic scene was in the midst of a process of expansion, the Nijverheidsschool's administrative bureau decided to equip a photographic studio. And in 1881 it was agreed that, in view of the recent progress in the discipline in general, and the science-related accomplishments of German and Swiss synthetic organic chemical firms in particular, the school's chemistry classes should be updated and enlarged.[33]

The young Baekeland, however, had already decided to make chemistry the main focus of his studies before the field was singled out for the promising job market prospects associated with an expanding science-industry nexus. Moreover, in the late 1870s, most employers within and outside the chemical industry were still skeptical about the value of the industrial schools' diplomas. Some of the largest Belgian chemical firms, including soda producer Solvay & Cie, in any case were hiring engineers rather than chemists.[34]

A 1913 visit to his native country confirms that Baekeland's option was likely to have been an unconventional one to follow in the context of the 1870s. During his stay in Ghent, Baekeland noticed "the fellow who was manager of [the] Lebegue-Van Paemel [a manufacturer of dyestuffs] office some 37 years ago [i.e., around 1876] and who bullied me then and sneered at me trying to study chemistry!! What a difference since then!!" Yet two days later Baekeland encountered someone who may well have inspired his choice: Dr. Jules Morel, his first teacher of chemistry. Baekeland favorably compared Morel with most of the other people whom he had met or observed during his trip in Western Europe. According to his diary, Morel

> seemed happy to meet me as much as I was to see him again. After lunch he came to the hotel and we spent all afternoon chatting together in english.... Seems to be well posted on matters american altho' he never was in America. Strange to say

> this erudite, intelligent and active man is little or not known in his country and still less in his own city. Those who refer to him treat him as an outsider of no importance. Others call him "de zotte morel" [the crazy Morel] because he happens to have ideas more advanced than the vulgar common crowd, inclusive the self satisfied, smug University Professors.[35]

Morel's classes had not only informed Baekeland and his fellow pupils about the textile-related uses of chemistry that had been the main cause behind Ghent's development into a center of the nineteenth-century Belgian chemical industry. His lectures had also taken into consideration the chemical needs of bakeries, breweries and distilleries, sugar producers and refiners, and manufacturers of acids, lacquers, oils, paper, and starch, among other local industries.[36] For example, in the section in Baekeland's student notes dedicated to the chemistry of starch, it was emphasized that the product represented "a very, very important industry"—a claim supported by the city directory, which shows that several starch manufacturers were located in Ghent in this period.[37]

The available course descriptions and/or lecture notes suggest that the contents of the other classes that Baekeland completed at the Indus were likewise shaped by a practical orientation. These included a course in industrial economics, to which Baekeland would refer during an educational reform discussion in the United States of the early 1910s as an example supporting his belief that "political economy is almost indispensable in a curriculum of chemical engineering." He was also trained in bookkeeping and technical drawing, two topics that would remain valuable throughout his later career, and he attended an elementary physics class that explained the phenomena of heat, sound, light, electricity, and magnetism in relation to the workings of steam engines, "electrifying machines," telegraphs, and phonographs, among other technologies.[38]

Baekeland's student notes demonstrate his school's approach of punctuating theoretical lectures "with examples that illustrate their useful aspects to the pupil, sustain his attention in class, and stimulate his work enthusiasm."[39] On Baekeland, the promise of practical payoffs is likely to have exerted the motivating influence aimed at; as is documented in further chapters of this book, throughout his career, his scientific work would largely be driven by use-inspired considerations. Seen in this light, it is tempting to connect Baekeland's excellence in chemistry and physics with the passion for photography that he developed as a teenager. Upon his graduation in 1880, he received medals for these two sciences, both of which contained photography-related subjects. For instance, in Morel's lectures on organic chemistry, Baekeland learned about collodion, a nitrocellulose solution that, despite its flammability, was broadly adopted in the "wet" process that dominated photographic technology between approximately 1855 and 1880. (It was also an essential component of celluloid and other early plastics.) Likewise, according to Baekeland's certificate of competence, "the theory of light and its applications. Eyeglasses, heliography, etc." had been expounded in his physics classes.[40] At least some of these topics were also treated in the school's photography lessons, which Baekeland is known to have attended, although they are not mentioned on his diploma. Since their creation in 1862, these classes had been taught by Gustave De Vylder (1824–1895), the first president of l'Association belge de photographie and an internationally renowned expert. More so than with contemporary photography courses at various other European institutions of technical education, their focus was on industrial applications.[41]

In short, despite the doubts of pre-1880 industrial employers, it is clear that Baekeland benefited in several ways from the theoretical training that he received at the Indus. It was not just that he learned about technological areas and industries that he would enter in the following years and decades, in Europe as well as in the United States. As discussed in the next section, together with the support of several mentors, Baekeland's mastery of his technical studies also paved the way for him to join the local university, where he would be exposed to a different balance between theory and practice.

State University of Ghent

Baekeland's certificate and medals from the Indus did not, however, automatically provide him with an "entrance ticket" to academic studies.[42] On the contrary, he was fortunate that admission criteria had temporarily been relaxed, for he did not possess a degree in general secondary education, and his knowledge of classical languages was at most rudimentary. Nonetheless, while he was not formally ineligible, Baekeland surely did not belong to the typical elitist "clientele" of Belgian universities in this period. Furthermore, and perhaps most important, the prospect of having to raise hundreds of francs to support further studies may have been daunting to him and his family.[43]

Earlier historians have credited Morel, Baekeland's chemistry teacher at the Indus, with having encouraged Baekeland and his parents to nevertheless attempt to take this unusual step.[44] This interpretation is certainly plausible, but another, more powerful mentor may have played at least as important a role. Auguste Wagener (1829–1886) was a man of multiple occupations and intellectual interests. Although he was a professor in classical literature and history at the State University of Ghent, he sought to increase the share of science and mathematics in secondary education at the expense of Greek as part of a prolonged controversy between social-liberal "modernizers" and conservative forces associated with the Catholic Party and Church.[45] Wagener also directed the engineering schools attached to the University of Ghent and, from 1869 until 1881, served on the Indus's administrative committee. Moreover, as the so-called administrator-inspector of the University of Ghent—a government commissioner who intermediated between the university and the minister of the interior and of public education—he had a strong impact on academic life in general and personnel appointments in particular. It is well documented that Wagener showed goodwill toward Baekeland in this function after the latter's graduation (see chapter 3), but a previously unknown letter from Baekeland seems to suggest that Wagener had already been supporting him several years earlier. When, in 1930, Baekeland's old friend Camiel De Bruyne was elected to the position formerly held by Wagener, Baekeland wrote him that, "When I think of you as the present administrator inspector of the University, my thoughts turn half a century ago [i.e., around 1880–1881] to dear old Wagner [Wagener], who was so kind and generous to me when I was so young and poor, and who, even in my student days, honored me with his kindness and good will."[46]

To be sure, the letter contains no hard evidence that Wagener encouraged Baekeland to pursue academic studies. However, given Wagener's affiliation with both the Indus and the University of Ghent, his ideological commitment to challenging class divisions through education, and the appreciation that he later showed for Baekeland's personal situation and interests, it seems fair to suggest that this influential "gatekeeper" may well have helped and advised the teenager even before his graduation from the Indus.

In any case, supported by a government fellowship, Baekeland matriculated in the university in 1880, the year in which he had completed his technical education. Compared with his earlier schooling experiences, he would be exposed to a broader range of instructional methods during his academic undergraduate and doctoral studies in the natural sciences. Starting in the early 1860s, when organic chemist August Kekulé had introduced German-style chemistry labs, the university's Faculty of Sciences had gradually begun offering practical exercises in order to develop a closer connection between teaching and research. As a result, Baekeland did not just take a "chemical manipulations" course taught by Théodore Swarts, Kekulé's former student and successor, but also had to attend experimental physics classes and an outdoor botany course.

Regardless of the class format, Baekeland consistently obtained scores of no less than 85–95 percent in his second so-called candidature year and in his two-year doctoral studies. This percentage, which was a sharp improvement over the mediocre marks that he had received in his first candidature, was all the more remarkable in light of Baekeland's unusual social and educational background.[47] He thus had plenty of reason to be proud of his doctor's degree—roughly the equivalent of a present-day master's. In fact, he cherished the title ever after receiving it on 23 July 1884. For instance, he organized dinner parties on the occasion of the thirty-fifth and fiftieth anniversaries of its granting. In addition, around the time of the golden jubilee in 1934, De Bruyne recollected that, the day after Baekeland's graduation, "somewhat recovered from your hangover, standing in front of your mirror you shouted: 'look there stands a doctor'"![48]

Notwithstanding his participation in laboratory exercises, Baekeland's abilities as a university student were first and foremost tested by means of oral examinations. As a matter of fact, new educational laws did not introduce the requirement to conduct original scientific research in order to qualify for a PhD degree until 1890–1891. These legislative reforms constituted a seminal event in the emergence of the research university and the institutional development of the sciences in nineteenth-century Belgium. In particular, besides further strengthening the link between teaching and research, they also formally established various scientific specializations. To give an example, Baekeland had simply obtained a doctorate in the "natural sciences" (sciences naturelles), which were defined in opposition to the mathematical and physical sciences (sciences mathématiques et physiques). On the degrees of later natural science graduates, by comparison, it was indicated whether their specialty had been botany, chemistry, geology, or zoology.[49] In reality, though, as is evident from table 2.1, Baekeland and his fellow students already had to focus on one of these disciplines in their

Table 2.1
Examination requirements for the doctorate in the natural sciences of the State University of Ghent.

First test: standard examinations in three of the following four subject categories	
A	Zoology proper
	Animal geography and paleontology
	Anatomy of textures
	Comparative anatomy and physiology
B	General and particular botany, including botanical geography and paleontology
C	Mineralogy
	Stratigraphic geology and paleontology
D	General chemistry
	Analytical chemistry
Second test: in-depth examination in remaining subject category. Practical experiments in this category.	

Note: Baekeland chose chemistry as his major.
Source: Ghent University, "Programme des cours année académique 1879–80," Ghent University Library.

doctoral studies. More generally, the educational reforms of 1890–1891 had been foreshadowed by the introduction of not just practical exercises but also various other initiatives, including the granting of PhDs in chemistry at the Catholic University of Leuven.[50]

After his graduation, Baekeland was able to remain at the University of Ghent as a so-called preparator in the course of general organic and inorganic chemistry, a job he had occupied since the end of 1880. Then, in 1885, he was promoted to the position of assistant in the same course, taught by Prof. Théodore Swarts, and was appointed a professor in chemistry and physics at the Middelbare Normaalschool in Bruges, a teacher training college. Not without pride, Baekeland later informed American chemist Charles F. Chandler that he owed the latter occupation to a "special intervention" of Brussels-based chemist Jean Servais Stas (1813–1891), whose advice had "quite some weight before the Belgian Government in the appointment of Professors of Chemistry." Like Morel and Wagener, Stas apparently helped "den self-made wonderman"—as De Bruyne put it in a mixture of English and Dutch—climb the social ladder.[51]

Another name that should be added to this list of gatekeepers and mentors is that of Théodore Swarts, whose daughter Baekeland began to court in the winter of 1884–1885. Given that marrying one's professor's daughter was a well-tested strategy of academic strivers, it might seem as if Baekeland's relationship with Céline Swarts was motivated by his ambition to be upwardly mobile.[52] The primary source material, however, does not support this possibility. In fact, there is evidence that Baekeland claimed too much credit for himself in connection with his "discovery" of his future wife. For example, according to his diary, he provoked "prolonged laughter and vigorous applause!" during his Perkin Medal address of 1916 by stating "that I felt particularly grateful for that old [chemical] laboratory [at the University of Ghent] because it was there that I made what I still consider privately my greatest discovery, when I found out that my Senior Professor of Chemistry had an unusually attractive daughter." Yet from the couple's correspondence it becomes clear that Céline Swarts had taken the first step: in December 1890, she reminded her husband that they had come a long way since their first meeting in the university laboratory six years earlier, and that on that occasion Baekeland had "run away when I arrived. But what a fortunate decision of mine to have come to ask you to teach me how to ice-skate"![53]

Postdoctoral Competition

As a student, Théodore Swarts had been part of an international group of up-and-coming scientists conducting original laboratory investigations under the supervision of August Kekulé, Ghent's general chemistry professor from 1858 until 1867. After Kekulé's departure to Bonn, Swarts had started directing this type of advanced study himself and, according to biologist Julius MacLeod, "not a single professor had contributed more to the emergence of a scientific spirit" among the students of the University of Ghent in the following decades. In this respect, as in others, Swarts may have held particularly high expectations for his aspiring son-in-law, for whom he and his wife appear to have envisioned an academic career (although, as discussed in chapter 3, Swarts also tried to secure an industrial position for Baekeland in 1890).[54] His professional relationship with Baekeland was, however, complicated by several factors, of which the latter's ambivalent attitude toward scientific specialization is one of the most relevant here. In a later autobiographical comment, for example, Baekeland indicated that he had always preferred to jump "from one subject to another in extreme directions. My Professor [Théodore Swarts] could not understand it, but I can, because I concentrated my attention on one subject so that I got heartily sick of it, and jumped to another subject which had no connection with it."[55]

Like Swarts's views on the value of research and research awards to academic lives and careers, Baekeland's eagerness to explore new territory may have contributed to his decision to participate in a postdoctoral competition organized by Belgium's Royal Academy, the so-called concours universitaire. Particularly before the government-mandated introduction of the research-based doctoral dissertation in 1890, this contest provided junior scientists with a rare opportunity to demonstrate their capability to conduct original scientific investigations and to present their findings in a comprehensive treatise. Nonetheless, the competition had for decades been suffering from a lack of interest, in part because it had little to offer to students with ambitions for professional careers outside academia. Another common criticism was that participants in the contest were not allowed to come up with research topics of their own. Instead, subjects were selected by the Académie royale, thus enabling the centralized determination of the direction of scientific inquiry.[56] In 1885, the academy opted for a chemistry question that could not be solved by relying on methods and insights from organic chemistry, the subdiscipline that had dominated Baekeland's training and for which his university had obtained an international reputation. The assignment was about phenomena of "dissociation," a term coined by French experimental chemist Henri Sainte-Claire Deville (1818–1881) in the context of his investigations of chemical equilibria and his measurements of vapor densities. In particular, candidates were asked first to summarize the state-of-the-art knowledge regarding these phenomena and then contribute original observations of their own.[57]

As it turned out, Baekeland was the only person to respond to this call, and the challenging nature of his task may have contributed to its appeal. To begin with, this field of research was situated at the interface of chemistry and physics, so Baekeland had to bridge a gap that had been institutionalized at his own university and many others at the time. More specifically, earlier scientists had expanded on Deville's results by applying insights from two math-intensive areas in physics—kinetics and thermodynamics—to chemistry. Their reliance on relatively advanced quantitative techniques contrasted with organic chemists' strong adherence to qualitative methods. In addition, not a single mathematics course had been included in Baekeland's natural science curriculum. Finally, the experimental advances concerning dissociation processes realized by Deville and others had rested on the availability of ingenious equipment that could simultaneously withstand temperatures of up to 1,000–1,200°C (1,800–2,200°F) and instantly cool down decomposed gases to prevent their complete recombination.[58] Baekeland, however, did not have the means to order a ready-to-use installation and had to be creative in personally designing one that would do the job.

As he clarified in his introduction, Baekeland's use of the term "dissociation" comprised "each phenomenon of decomposition in which the molecules of a compound split up, under the influence of heat, into new, generally more simple molecules, which can recombine to reconstitute the original compound, if temperatures are appropriately [i.e., not instantly] lowered." He situated Deville's research on this kind of phenomenon against the background of the prolonged discussion of chemist Amedeo Avogadro's 1811 hypothesis that, temperature and pressure being equal, an identical number of particles will be contained in gases of the same volume. In particular, as Baekeland pointed out, the work of Deville and other students of dissociation had made it possible to use Avogadro's law as a basis for determining molecular weights in cases where this had previously seemed to create anomalies.[59]

Not least because of this broader theoretical significance, Deville's results had been intensely debated, and Baekeland devoted the lion's share of his thesis to the first part of the academy's assignment, the review of prior research. His literature survey was, however, highly selective. Whereas he discussed a series of empirical case studies at length, either he showed very little interest in theoretical issues or deemed himself badly qualified to handle them. Whatever the case, in his introduction, he indicated that he would only briefly summarize the discussion on the accuracy of Avogadro's principle, as well as the related gas laws of Boyle-Mariotte and Gay-Lussac, because these intellectual exchanges would in any case remain "famous in the annals of science." More important, in a brief theoretical chapter, Baekeland claimed that he could reasonably limit himself to "giving a general idea of the theory that is currently adopted by the largest number of scholars"—the kinetic theory of matter and heat developed by Austrian scientist Leopold Pfaundler—because "In this study it is not our goal to expound and analyze the different scientific theories that have been proposed to explain phenomena of dissociation. Such an undertaking would greatly exceed the limits to be drawn. This subject, by the way, constitutes one of the most delicate questions in statistics and chemical dynamics."[60]

In his own contribution to this body of research, a case study on the dissociation of lead nitrate, Baekeland also showed a critical attitude toward theoretical suppositions. In the final part of his thesis, as well as in the two later publications in which he presented his results, he observed that the case of lead nitrate was of special interest because it concerned the reversible decomposition of a solid (PbN_2O_5) into another solid (lead monoxide, or PbO) and, contrary to all previous examples, two different gases (oxygen and nitrogen dioxide, or NO_2).[61] On the downside, though, various practical obstacles had to be overcome before the goal that Baekeland had set out to achieve—determining the maximum pressures exerted by the gases at gradually increasing temperatures—could be accomplished. For instance, as the lead nitrate available on the market was insufficiently pure for his purposes, he had to purify it himself. Likewise, because of the need to "maintain during entire months temperatures of 250°[C, or 482°F] in rooms in which temperature changes occurred very frequently," among other practical challenges, Baekeland repeatedly reconstructed and refined his measurement installation (see figure 2.3). In this manner, he was able to obtain a satisfactory degree of exactitude. His measurements were indeed taken over as late as 1964, therefore suggesting that they had stood the test of time.[62] On the other hand, Baekeland had been far less thorough with regard to the theoretical interpretation of his findings, laying himself open to criticism in this area. In fact, in 1904, a chemist at Columbia University applied two related principles, the law of mass action and van't Hoff's equation, to Baekeland's results, in a short article in the *Journal of Physical Chemistry*, because "for some reason [Baekeland] has made no application of the possible theoretical relations to his results. As the application of some of these shows much concerning the [chemical] equilibrium, which is lost by his purely qualitative treatment, and as Mr. Baekeland has evidently no intention of considering the question further, the results of such application are presented in this paper."[63]

Baekeland's sidestepping of theoretical issues did not, however, prevent him from gaining what can be seen as a specific type of intellectual property: a jury led by Stas, a practically minded scientist himself, awarded the contest's only participant marks sufficiently high to ensure him of an academic "palm of victory."[64]

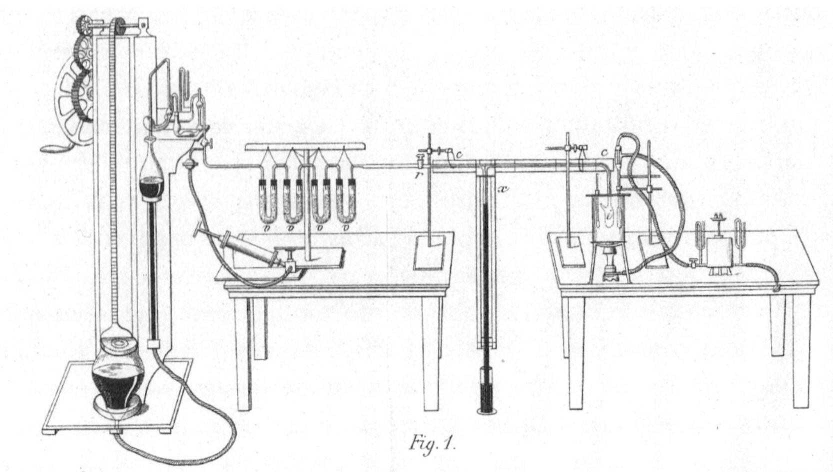

Figure 2.3
Baekeland's first installation for studying the dissociation of lead nitrate, as described in his thesis and his two later publications. This arrangement stopped being useful at temperatures of 357°C or higher, when it became impossible to prevent the nitrous gases produced from the lead nitrate in the glass container *i* from reacting with the mercury in the manometer *x*.
Courtesy of Leo Baekeland Papers, Archives Center, National Museum of American History, Smithsonian Institution, AC-0005-0000123.

Useless Knowledge?

Baekeland would never again author a scientific study that was as comprehensive as his prize-winning submission to the Royal Academy. With its focus on careful experimentation and measurement, this work represented the culmination of academic training that, in terms of course format, had included a substantial practical component. In terms of content, moreover, this research had pushed Baekeland beyond the emphasis on organic chemistry that set his education at the University of Ghent apart from the schooling received by most chemistry students in the pre–World War I United States.[65]

The criticisms that Baekeland later expressed about the limited practical use of his "entirely theoretical" academic education did not just concern his own career. In 1911, for example, he more generally claimed that the natural science program in Ghent "turned out people ... who came out with a lot of knowledge but tremendously little judgment. They made poor engineers and still poorer chemists. In fact, I cannot think of a single man who afterwards really amounted to something. They were repudiated by engineers and still more so by chemists, still they had the diplomas. Some of these fellows had to make their living by teaching in the technologic schools and universities, but in practical lines they did not succeed."[66]

Baekeland made these retrospective comments as part of a discussion on educational reform within the American Institute of Chemical Engineers, the industry-oriented professional society that he presided over in that year. His remarks were tailored to the priorities and concerns of this association, which was trying to agree on the chemical engineering curriculum that would best meet the needs of the US chemical industry.[67] But it is certainly true that, back in the late 1870s and 1880s, it had been more common for graduates from Baekeland's program to become teachers than to transition into industrial work. In fact, not just in Ghent but also at the science faculties of Belgium's other universities, the increase in student enrollments experienced in those years was partly driven by the legally enshrined status of the natural science doctorate as an entry qualification for professional teachers.[68] Moreover, as mentioned previously, Baekeland himself had been able to obtain a position as a chemistry and physics professor at the Teacher Training College of Bruges in the year following his graduation.

If Baekeland had further pursued the career of a science instructor, the relatively broad scope of Ghent's natural science program would have made it possible for him to offer classes in several disciplines in addition to chemistry and physics. But how, if at all, did this academic training prepare him for the more risky work as a photographic inventor-entrepreneur in which he instead became engaged from late 1887 until 1899? The University of Ghent did not belong to the relatively small group of higher education institutions in Europe and the United States that by the 1880s had introduced courses specifically devoted to photography. Multiple members of its teaching staff, however, shared with Baekeland an interest in photographic technology and/or related sciences such as photochemistry and optics. Théodore Swarts, for example, had encouraged Baekeland to carry out a photochemical study of the oxidation of hydrochloric acid and had advised him on this subject in late 1884 and 1885. Moreover, at least seven other professors supported the local chapter of l'Association belge de photographie, including by presenting research papers to their fellow photographers and by providing them with access to laboratory facilities. As a final example, engineer

Théophile d'Hauw translated into French several papers by Austrian photographic scientist Josef Eder, who would later claim to have anticipated Baekeland's invention of Velox photographic paper.[69]

In any case, to illustrate the alleged gap between academic theory and industrial practice, Baekeland referred to a research paper by chemist Jean Servais Stas rather than to the work of his former teachers or colleagues in Ghent. In an 1874 article in the *Annales de chimie et de physique*, Stas had distinguished between four physical modifications of silver bromide and stated that one of those, the granular, was the most light-sensitive substance of which he was aware. Academics such as Hermann Wilhelm Vogel, the director of the photochemical laboratory of the Technische Hochschule Berlin-Charlottenburg, presented Stas's findings as a scientific foundation for the development of photographic dry plate technology in the late 1870s.[70] Baekeland, in contrast, used the same findings to exemplify precisely the opposite, namely the disconnect between textbook abstractions and industrial practice. As mentioned, in his Perkin Medal address of 1916, he stated that his first commercial endeavors had made him realize that there existed many more varieties of silver bromide than those distinguished by Stas but that only one of them might probably guarantee him success. Moreover, unlike Vogel, Eder, and others trying to promote photographic science and technology as an academic subject area, Baekeland did not seem to regard the relative neglect of photographic research in both industry and academia as a matter of concern.[71] Instead, if we are to believe his comments on the theme of "Practical Life as Complement to University Education," the complexity of silver bromide emulsions encouraged him to treat all textbook generalizations with caution. In his speech, he claimed to have reasoned as follows on an unspecified day:

> Here is one of the very simplest of all chemical reactions; so simple that any chemist hardly wastes any time over it, and yet it involves phenomena so intricate, so varied in scope and effect, that we have not yet found the proper means to fathom them. Here is bromide of silver, freshly precipitated in the dark; it has the property of not darkening if brought in contact with a reducing agent—a developer—and yet the mere action of light, even for only an infinitesimal fraction of a second, brings about such a profound alteration in its constitution that henceforth it will blacken and set silver free if treated with a reducing agent, and thus produce a photographic image. Furthermore, this bromide of silver, before and after it has seen the light, is exactly the same, as far as we can decide by chemical analysis or any of our other clumsy means of observation. Suppose now

> that the human race were not provided with the sense of eyesight—just as we are probably missing several other senses—we would forever remain unaware of that peculiar effect of light on silver bromide or on chemical reactions in general.[72]

One of the first scientists to draw attention to Stas's silver bromide variations had been Désiré Van Monckhoven (1834–1882), a former student at the University of Ghent, who had become a successful entrepreneur and a leading author on the practice of photography, photographic optics, and photographic chemistry.[73] Indeed, in an 1880 publication on the gelatin–silver bromide process, Van Monckhoven had expressed reservations about the practical value of Stas's research not unlike those made by Baekeland in 1916. This is one of several indications suggesting that Baekeland had been significantly influenced by Van Monckhoven in his approach to photography.[74] In fact, given that, first, Baekeland had become interested in photographic chemistry by the time of Van Monckhoven's 1880 publication; second, that Van Monckhoven's photographic enterprise in Ghent is generally regarded as the first manufacturer of silver-gelatin emulsions on the European continent; and third, that there is some evidence that Van Monckhoven personally advised Baekeland on these matters,[75] it seems plausible that around the early 1880s Baekeland came to appreciate the complexities involved in this kind of work. This is another reason why the interpretation that Baekeland's work for Dr Baekelandt et Cie in the final years of the same decade made him question the practical use of scientific abstractions seems unsustainable.

Finally, it should be emphasized that Baekeland's case against the economic usefulness of chemical theory was of a remarkably general nature. Starting from his dissatisfaction with the chemical equation representing the formation of silver bromide, he denounced the inadequacy of any textbook generalization rather than the limited chemical and physical understanding of photographic fundamentals. To what degree, however, was it legitimate to generalize from one chemical industry to all others? Ironically, in 1915, Baekeland himself highlighted the (alleged) singularity of the photochemical industry, apparently undermining his later argumentation. In a laudatory article on Edward Weston, a British-American electrical inventor, he remarked,

> It seems rather fortunate for him [Weston] that one of the first employments he got in New York was with a chemical concern [William H. Murdock & Company] which made photographic chemicals. This was the time of the wet-plate, when photographers made their own collodion, their own silver bath, their own paper.

> Whoever went through those delicate operations knew the difficulties, the uncertainties which were caused by small variations in the composition of chemicals or in the way of using them. Photochemistry is excellent experience for any young chemist who is disposed to generalize too much all chemical reactions by mere chemical equations. Whoever has to deal with those delicate chemical phenomena, which occur in the photographic image, knows that many unforeseen facts can not easily be accounted for by our self-satisfying but often superficial generalizations of the text-books.[76]

Other photochemists would have concurred with these reflections. For instance, in his memoirs, Fritz Wentzel (1877–1963) of Germany highlighted the "extraordinary cleanliness" required in the preparation of sensitized photographic materials and claimed that "even chemists of reputation in other branches of chemistry, for example, the production of baking powder, when transferred to management positions in the photographic industry, could hardly understand the difficulties with which their subordinates had to cope." There can also be no doubt that, when preparing photographic emulsions on a commercial scale, "an unremitting attention to detail is absolutely necessary, and the slightest variation of detail upsets the whole cycle."[77]

Still, corporate photochemists were surely not alone in their dependence on practical know-how and skill, in combination with a painstaking carefulness. In a recent study on late nineteenth-century chemical technology textbooks in Sweden, for example, it was concluded that the books' industrial utility seems to have been inversely correlated with the amount of scientific theory included in them. In early twentieth-century America, moreover, the dissonance between chemical science and large-scale commercial technology was one of the factors that would encourage Baekeland and his professional colleagues to develop and promote the new discipline of chemical engineering.[78]

A European Success Story

Baekeland's life and career up to 1887 had been a remarkable success story in many ways. Thanks to the combination of his own efforts and skills and a series of favorable external factors and circumstances, such as the support he received from various mentors and the availability of study fellowships, the shoemaker's son had become a doctor in science and had been able to strengthen his academic curriculum vitae afterward. Moreover, Baekeland's relationship with Céline Swarts, an indirect result of his university studies, had boosted his integration into the local sociocultural elite and given him access to further symbolic capital.

While he later cultivated the image of himself as a "self-made man," Baekeland also occasionally acknowledged that he owed a great deal to his education, particularly to his technical training at the Indus. For example, as he wrote an old friend familiar with this municipal institute in 1899, one of the first things that he considered doing after becoming independently wealthy was to help establish "a kind of industrial school" in the New York suburb of Yonkers. Moreover, toward the end of his life, he was still convinced that, thanks to his schooling at the Indus and on-the-job training at an unspecified local pharmacy, he had been "better prepared to follow the lectures [at the State University of Ghent] than my fellow students, who had learned bad Latin and worse Greek."[79]

As for the industrial usefulness of Baekeland's university education, this chapter has offered a more mixed image. On the one hand, his academic training had provided him with knowledge, laboratory skills, and professional contacts relevant to most of the fields in which he would focus his inventive and entrepreneurial efforts in the following decades. Théodore Swarts, as an example, had shared with Baekeland his in-depth knowledge about the chemistry of camphor, a substance that contributed to Baekeland's later shift in research from photography and electrochemistry to plastics.[80] As a university student, Baekeland also became acquainted with some of the people with whom he would later launch business ventures (see chapters 3 and 5).

On the other hand, Baekeland was right to highlight several limitations of his academic schooling. First of all, to keep abreast of the latest developments in the field of chemistry, which was rapidly growing and diversifying, it would be necessary to regularly refresh, update, and expand the scope of his knowledge. As later chapters in this book document, Baekeland did this not just through self-study but also by participating in various professional societies. In addition, in 1900, he would temporarily return to academia as a student to deepen his knowledge about electrochemistry, a subject marginal to his training in Ghent.

Second, compared with European and US higher education institutions at the forefront of the second industrial revolution, such as the Technische Hochschule Berlin-Charlottenburg and the Armour Institute of Technology in Chicago, the University of Ghent was relatively slow to adjust its science and engineering curriculum in response to a changing economic and technological environment.[81] This does not mean that Baekeland's university was an "ivory tower." As I will emphasize in chapter 3, a substantial number of its employees participated in industrial activities, while others were heavily involved in the sociopolitical realm.[82] However, when it comes to evaluating the university's contributions to the economy, the case could be made that it only belatedly began to prioritize the training of industrial scientists and engineers, in addition to science teachers and civil servants.

Finally, the successes and failures of science-related firms were often determined by factors that had little to do with the education of their scientific employees, such as financing difficulties or the ability to obtain comprehensive intellectual property protection. As the following chapters demonstrate, this is something that Baekeland experienced firsthand in the years and decades after 1887, when he first became involved in the creation of a firm.

3 Crossing Boundaries: From European Academia to American Industry

In the nineteenth and early twentieth centuries, as today, higher education institutions contributed to science-related industrial development in a range of ways, including through the business ventures of their employees. Because of a lack of systematic studies, it remains hard to tell how common it was for academic scientists to combine their teaching and research work with entrepreneurial engagements. But there are strong indications that, at least in the physical, engineering, and medical sciences, academics' involvement in the creation of new firms was a significant but often neglected dimension of the strengthening of the science-industry nexus.[1]

Much of the available literature deals with scientists who held prominent positions within their own disciplines and academic institutions. For example, historians have analyzed in substantial detail how and why chemists Justus Liebig and Louis Pasteur, physicists William Thomson and Henry Rowland, and engineers Franz Reuleaux and Alois Riedler participated in business initiatives without abandoning the pathbreaking research and teaching activities for which they are primarily remembered.[2] However, as this chapter illustrates by examining Baekeland's role in a university start-up in the late 1880s, such commercial work could also appeal to early career academic scientists who did not (yet) have a long-standing reputation in their fields of study and held lower-level jobs of a nonpermanent nature.

This difference has to be kept in mind when evaluating the benefits and risks associated with academic entrepreneurs' ventures into business. In terms of personal finances, as an example, junior academics who did not have the good fortune of being independently wealthy may have been completely dependent on external sources of revenue to cover some of their living expenses. After all, even a significant number of senior professors with relatively high salaries are known to have relied on commercial work to match a culturally prestigious occupation with corresponding levels of material comfort, among other purposes.[3] At the same time, in situations where academics' business initiatives were disapproved of by their colleagues or supervisory authorities, junior scientists were in a relatively weak position to defend themselves. In general, they could be dismissed more easily than professors with permanent appointments—or they could just not be considered for a promotion.[4]

One might think that these risks were particularly high in Baekeland's case because he and his collaborators established their firm within an academic environment that, according to several studies, was characterized by an "ivory tower" mentality. In particular, late nineteenth- and early twentieth-century university scientists in Belgium, like many of their peers abroad, tended to present themselves as being primarily dedicated to "pure science."[5] However, the fact that these scientists mentioned their own or their colleagues' participation in commercial activities only in passing, if at all, should not be taken to mean that such work had been marginal to their careers. Moreover, as this chapter demonstrates, it would also be wrong to infer that an involvement in business initiatives was considered inappropriate for Belgian academics.

Zooming in on the risks that Baekeland took by becoming an entrepreneur not only casts new light on the short but turbulent history of Dr Baekelandt et Cie, a photochemical enterprise. It also results in a revised interpretation of his 1889 migration from Ghent to New York. According to most biographical accounts, Baekeland traveled to the United States mainly for academic purposes, taking advantage of the fellowship that the Royal Academy had offered him. As an indirect result of his visits to scientific institutes and laboratories, so this interpretation goes, he made the acquaintance of the vice president of E. & H. T. Anthony & Company, a leading photographic firm, who offered him a job that seemed too attractive to refuse.[6] For a variety of reasons, the authors of these studies did not make use of the archival source that provides the most detailed insights into Baekeland's actions and thoughts in the years around 1889—his correspondence with his wife. By integrating these letters into the analysis, together with various supplementary sources, this chapter provides a very different interpretation of Baekeland's life-changing transition from European academia to American industry.

Dr Baekelandt et Cie

When Baekeland coestablished Dr Baekelandt et Cie in late December 1887, his appointment as an assistant for the course in general chemistry at the State University of Ghent had just been renewed. In addition, it was still unclear how long it would take to find a suitable candidate to replace him at the Middelbare Normaalschool of Bruges, a teacher training college, where he had been lecturing on chemistry and physics since the 1886–1887 academic year.[7] On the face of it, the history of the photochemical firm, a limited partnership (société en commandite simple), seems to support popular notions as to the incompatibility of traditional universities and commercial risk taking. Baekeland, for example, achieved business success only after departing from academia, as well as from his home country. Moreover, several of the reasons that have been cited to explain the failure of the partnership, which accumulated debts and was dissolved after two years, concern the difficulty of combining academic and industrial work.

In particular, besides the previously discussed disconnect between academic and industrial chemistry, earlier studies have pointed to the (alleged) distaste for business of Théodore Swarts, Baekeland's academic boss and future father-in-law, as well as to Baekeland's lack of time because of his obligations at the university. Other reasons that have been mentioned to account for the firm's failure include Baekeland's premature abandonment of the enterprise; intense competition from other home- and foreign-based photographic enterprises; financing problems; and Baekeland's unfortunate decision to try to innovate in the area of glass plates at a time when, internationally, more flexible celluloid films were about to be introduced.[8] The following paragraphs reexamine this episode on the basis of Baekeland's personal correspondence and a series of legal documents concerning the firm's history that the chemist received from his lawyer after his move to the United States. In brief, these sources suggest that financing difficulties and a lack of business experience, as well as commitment, were more important than technical setbacks as determinants of the firm's failure. That conclusion confirms the observation of historian Susan Morris that the partnership's "struggle with technical problems should not be exaggerated."[9] Nonetheless, as Dr Baekelandt et Cie was primarily established for bringing Baekeland's self-developing plates to market, it seems appropriate to start the analysis with a brief elucidation of the invention and the technological field to which it belonged.

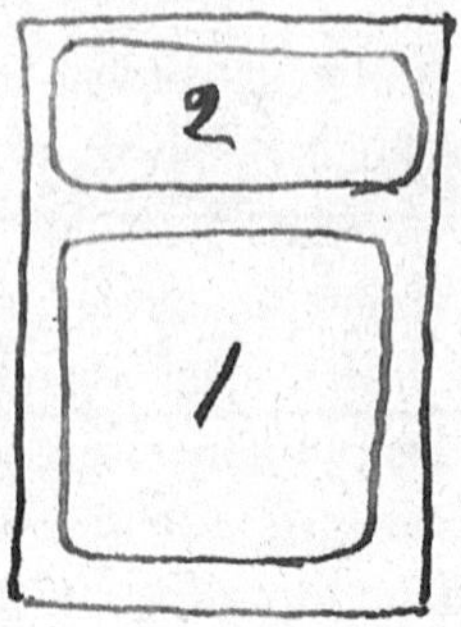

Figure 3.1
Sketch from Baekeland's laboratory notebooks. Courtesy of Leo Baekeland Papers, Archives Center, National Museum of American History, Smithsonian Institution, AC-0005-0000124.

Baekeland, a passionate amateur photographer, had aimed at making the predominant photographic process of the 1880s easier to use. That method, the gelatin silver dry plate technique, typically consisted of four steps: (1) the taking of the photograph, a process in which an exposure to light resulted in the creation of a latent image on the carrier of the gelatin silver halide photographic emulsion; (2) the chemical development of this invisible image into a visible negative; (3) the "fixing" of this photographic negative, also by chemical means, to ensure its stability; and (4) the making of prints. Between fall 1887 and spring 1889, Baekeland was granted patent protection, at least in Belgium, France, the United Kingdom, Germany, and the Austro-Hungarian Empire, for a water-developable dry plate that largely automated the second step.[10] According to an eyewitness, and as Baekeland himself later illustrated in a laboratory notebook, the academic inventor "mixes his developer in two separate portions with gum arabic, and puts it on the back of the plate in different parts of the plate [i.e., parts 1 and 2 in figure 3.1] so that the different chemicals in the developer are not mixed together until the solutions dissolve and therefore no decomposition takes place, and as the gum arabic makes a pasty mixture, there is no danger of the developer running around on the face of the plate."[11]

Baekeland assigned his intellectual property rights to this invention to a limited partnership named after him, which was registered on 3 January 1888 at the Commercial Court of Ghent. As was legally required with this type of organization, the firm's articles of association distinguished between a managing partner fully liable for possible debts (Baekeland) and a silent partner/investor whose liability was, in principle, limited to her or his initial contribution—in this case, making available a recently erected property on 120 Palinghuizen Street (rue des Aiguilles) in Ghent as well as a sum of 20,000 Belgian francs. This statutory distinction can be traced back to Napoleon's Code de Commerce and, compared to ordinary partnerships, was meant to better protect investors who were not interested in actively running the firms they cofounded against opportunistic behavior on the part of their associates. Yet Baekeland's partner/investor, Valérie Gleesener-Guequier—the daughter of Stéphanie-Constance Gleesener-Duhayon (1815–1891), a widow of considerable wealth—and her husband, Jules Guequier, were not simply uninvolved investors. Together with the managing responsibilities that they agreed to undertake came the obligation to stand surety for potential debts. In particular, the founding agreement specified that Baekeland and Gleesener were to share equally in profits and take mutual responsibility for losses at the end of each year.

A further agreement of exploitation (convention d'exploitation) established a clear division of labor between Baekeland and Guequier, who also held a junior position in the Faculty of Sciences at the University of Ghent. While Baekeland's academic colleague would chiefly take care of the commercial side of the business, technical questions were the chemist's own responsibility, at least initially, for it was understood that "as soon as it is acknowledged that Mr Jules Guequier-Gleesener possesses the know-how and competence required for running this industry, Mr Leon Baekelandt will have good reason to make those scientific journeys abroad that he may deem expedient, and to occupy himself with any scientific problem or study that pleases him."[12]

The reference to international journeys made sense in light of the travel fellowship that Baekeland had earned through his research on dissociation processes. In early 1889, he used this grant to cover trips to Germany and the United Kingdom, therefore suggesting that his knowledge about the invention was no longer needed to effectively commercialize it. In reality, however, Dr Baekelandt et Cie surely did not stand on a firm technical footing around the time of his absence.[13] Not long before his departure, as an example, he had to pass a few sleepless nights in the Palinghuizen factory because of a "foolish" malfunctioning machine—a setback that negatively affected his mood. As his fiancée lucidly reported, he had also "enraged" her family members because the gelatin on the photographic plates with which he had provided them had become detached, and it was "ridiculous" to recommend these plates when they clearly were not working properly.[14] This kind of problem continued to appear after Baekeland had left the country. As a case in point, in May 1889, the editor of the Viennese *Photographische Rundschau* declared Baekeland's patented invention unfit because of its apparent unreliability, claiming that the chemical stability of the attached developer could not be taken for granted. In addition, in September 1889, American industrialist Richard Anthony pointed to slowness of development as a major drawback of the technique. Still, Baekeland's glass plates were generally judged to be a useful simplification, particularly for amateur photographers, in the various photographic societies in which they were presented. Moreover, in spite of the invention's perceived shortcomings, Dr Baekelandt et Cie was awarded a bronze medal at the Paris world exhibition of 1889.[15]

It can also be demonstrated that Baekeland had not yet given up on his photographic plates when he made his journeys abroad. In fact, it is likely that he attempted to further his firm's interests on these trips. As an example, in England, he visited a Crystal Palace photographic exhibition and, according to his diary, was invited to inventor-entrepreneur Joseph Swan's home in Bromley, "where he [Swan] wanted to make arrangements for exploiting my [Baekeland's] first patent, water-developing dry plate."[16] On another trip, this time to Berlin, Baekeland was welcomed in the photochemical laboratory of Hermann Vogel at the Technische Hochschule.[17]

In the United States, however, Baekeland clearly intended to pursue plans of his own. Intriguingly, Dr Baekeland et Cie's statutes included the provision that future photography-related patents that might be granted to him in America, as opposed to those granted in all other countries, would remain his personal property. This provides further evidence that Baekeland had been contemplating starting a new life as a US-based inventor long before he and his wife crossed the Atlantic in summer 1889. A letter that Baekeland sent to Guequier relatively soon after his move to New York confirms that his Belgian business associates had no place in this adventure. In response to a request for help, Baekeland informed Guequier that his "stay here [in the United States] will last much longer than I initially believed and…I will not be back before next summer." Yet, he wrote, his prolonged absence should not be a reason for concern, because "you should easily be able to do without me because you know the manufacture in its entirety, you make better emulsions than I do, and, as for the selling, you can shake up your clients a bit and try to find new outlets."[18] Likewise, Baekeland

Table 3.1
Main creditors of Dr Baekelandt et Cie (January 1890).

Name of creditor	Sum (in Belgian francs)
Banque de Flandres	22,198.36
Oscar Defer	10,103.5
Eggermont	20,514.98
S.-C. Gleesener–Duhayon	35,225
Société Maris et Cie	7,621.18
	Total sum: 95,663.02
	Percentage of total debt: 88.77%

Source: Based on data included in LBP, series 8, box 34, folder 1.

wrote Guequier that he and his wife could not afford to deprive themselves of basic needs by carrying part of the financial burden, but here again, Baekeland opined that this would not be much of an issue, for "Madame Gleesener [Valérie Gleesener's mother] is quite rich and she will be able to provide you with what you need. To repeat once more, the thing is impossible for us at present, and another solution will have to be found. Could you not speed up the formation of the English company, or couldn't you find capitalists seriously interested in joining the enterprise?"[19]

Ideally, Baekeland's association with the Guequiers was supposed to last 18 years, from the beginning of 1888 until the end of 1905. However, according to their firm's founding agreement, Dr Baekelandt et Cie could be dissolved "after two years if it is recognized by the inventory drawn up at that date that the partnership has not realized profits" or any time later if it was running a loss.[20] Baekeland made use of this clause at the earliest possible moment, in January 1890, about five months after his move to New York. Significantly, by then the firm owed 58 creditors a total sum of 107,769.63 Belgian francs—or about 36 times the salary that Baekeland had been offered as a "professor in training" (agrégé spécial), the postdoctoral position to which he had been promoted in summer 1889.[21] As table 3.1 illustrates, this amount was distributed extremely unevenly: no more than five lenders accounted for almost 90 percent of the firm's total debt. The heaviest load was carried by Valérie Gleesener's mother, who had added about 15,000 francs to the starting capital with which she had already provided her relatives. The identity of the other major creditors shows that Baekeland and his partners, perhaps benefiting from their academic affiliation, could also raise large sums of money from nonrelatives in Ghent. The Bank of Flanders (Banque de Flandre), to begin with, had its headquarters in Baekeland's home city. Second, investor Maris et Cie, a producer of golden frames, employed 20 workers in Ghent by 1885.[22] Finally, the creditor listed as "Eggermont" most likely referred to Camille Eggermont (1834–1904), a local entrepreneur and philanthropist, whose son Albert was later granted an honorary membership in l'Association belge de photographie (ABP). At the same time, Baekeland and the Guequiers also managed to attract funding from outside Ghent: their creditor Oscar Defer was a merchant from near Charleroi.[23]

While Baekeland seems to have never repaid any of these financiers, the Guequiers agreed to refund Defer and Maris within five years, a pledge for which Stéphanie-Constance Gleesener-Duhayon stood surety. However, according to Céline Swarts, the widow's wealth had diminished considerably by then. It was certainly not sufficiently large to prevent the Guequiers from being "in the soup." As Swarts informed her husband in August 1890, they had been begging "at all doors for money," presumably to fulfill their obligations toward earlier creditors as well as to inject new life into their photographic firm.[24] They would continue the enterprise in the form of the ordinary partnership (société en nom collectif) V. Guequier-Gleesener et G. Van de Poele from 1891 until 1894, and as the limited partnership Valérie Gleesener et Cie until its ultimate dissolution in 1897. (When telling Baekeland that Van de Poele was joining the firm, his spouse commented that "this is all that is necessary to trigger the collapse of the business. He is a party animal who never did any studies and who I think knows nothing about commerce and perhaps less about photography than your little wife.") Whatever the cause, it is evident that the Guequiers were not able to break out of the negative spiral, or to recover their earlier losses, for in 1931 their daughter Adrienne blamed Baekeland for having ruined her family.[25]

How could Baekeland sidestep the unlimited liability to which general partners in a société en commandite were normally exposed? Remarkably, as part of the partnership's dissolution and liquidation, Guequier and Gleesener personally acquitted him of this obligation, enabling their associate to get off relatively lightly. Their motives for not holding Baekeland accountable for his responsibilities in "this nasty affair" are not documented in the archival record.[26] They may have taken into consideration his limited means, but later they clearly came to regret the arrangement. For instance, according to a March 1891 letter of Céline Swarts, the Guequiers had purposefully caused complications by not informing her US-based husband of sums that he was supposed to have paid off, nor of the subsequent visit of a bailiff. As a result of the ill will of his former partners, Baekeland was fined, but his wife appears to have convincingly explained the situation in a beseeching letter to King Leopold II, written, as she noted to her husband, to ensure "that you won't be busted upon your return."[27] More fundamentally, Valérie Gleesener's mother denied ever having approved the exemption for Baekeland and sought to reclaim up to about 15,000 francs.[28] Gleesener's own lawyer, however, admitted that the legal basis on which his client's claims rested was precarious. Instead, he proposed "that Mr. Baekelandt commit himself to not engaging, either personally or through third parties, in the manufacture of photographic plates in Belgium during a term of twenty years"[29]—a request to which I return in chapter 4.

Baekeland's role in the Guequiers' financial downfall is said to have caused him public embarrassment upon a return to his home city in 1900.[30] Whether or not this actually happened, his and his wife's analysis of what went wrong provides an important alternative perspective on his first steps as an entrepreneur. According to Swarts, the Guequiers deserved no pity whatsoever, as they had been squandering Gleesener-Duhayon's fortune, "living it up and cheating on clients."[31] When she vented her opinion, Baekeland reproached her with scandal mongering.[32] The couple, however, clearly concurred that Guequier was incompetent and deceitful. Yet, at the same time, Swarts sought to impress on her husband that his start-up had primarily suffered from capital shortages. In a period in which Baekeland was considering leaving Anthony & Company, the photographic firm he had joined shortly after his move to New York in 1889, she explained:

> I think it would be a mistake to work on your own.... To do something, and especially to start something new, one needs a sufficiently firm financial footing to wait and to bear the losses that are inevitable at the beginning. That is what killed your photographic plates factory. G. [Guequier] was certainly a swindler and an imbecile, but chiefly you were lacking capital. There you have had a hard lesson, and frankly, consider that especially in America, where one has concerns established with enormous capital as competitors, it would be even a thousand times more difficult to succeed.

In the same letter, Swarts advised her husband to factor in his ignorance of commerce, thereby revealing the kind of antibusiness sentiments that have been ascribed to her father, Théodore. She expressed the opinion to Baekeland that, "since you are not the kind of man who can keep careful accounts and manage his finances, in a word a real businessman [commerçant], it would be wrong to want to undertake something on your own. Work with the capital of others, let them bear the losses and the unavoidable marks and pettiness of commerce. I am sure that you will earn a lot more by offering your services as a director in something."[33]

With hindsight, it would have been more accurate to state that Baekeland had *not yet* acquired the skills of "a real commerçant," for professional colleagues and acquaintances would later consult and admire him for his administrative, business, and financial acumen. Hence, it seems fair to conclude that Baekeland had been hindered by a lack of experience rather than talent—an argument consistent with the contention of entrepreneurship scholars Mark Casson and Andrew Godley that successful entrepreneurship is positively correlated with maturity "because [commercial] judgement improves with age and experience."[34]

In the Belgian photochemical industry of the late nineteenth and early twentieth centuries, however, it was not just novice entrepreneurs who were unable to make it. As Jules Hens and Laurent Roosens have documented, of the more than 20 enterprises founded in this period, only one—Gevaert N.V.—was still in business after World War II. If one agrees with Casson and other economic theorists that entrepreneurial talent is nurtured through practical experience, it was critical that risk takers who suffered setbacks like those encountered by Baekeland not be discouraged from trying anew.[35] In light of this, it is significant that Swarts's derogatory reference to "petty commerce" seems to suggest that business occupations were looked down on in Baekeland's academic surroundings in Ghent. This chapter's next section takes a closer look at this as part of an examination of the explicit and implicit rules, motives, and norms that possibly influenced the chemist's decision to settle down in the United States, a country in which entrepreneurial risk taking was more strongly encouraged than perhaps anywhere else.[36]

A Transatlantic Escape?

Any attempt to pin down Baekeland's decision to build a new life in the United States to a single moment in time would be doomed to failure. As discussed in chapter 2, his captivation with American inventors such

Figure 3.2
Céline Swarts and Leo Baekeland during early stays in the United States. The photo of Swarts was taken in Central Park, New York, in January 1890.
Courtesy of Leo Baekeland Papers, Archives Center, National Museum of American History, Smithsonian Institution, AC-0005-0000003.

as Edison, Franklin, and Bell dated back to at least his teenage years, and the exclusion of possible US patents from Dr Baekelandt et Cie's intangible assets suggests that he had set his mind on trying his luck on the other side of the Atlantic no later than the end of 1887, when the partnership was formed. The stakes were high, however, because Baekeland's transatlantic ambitions risked jeopardizing his relationship with Céline Swarts, who was not as adventurous as her fiancée at this stage of her life and was more emotionally attached to her family and friends. Eventually, Swarts agreed to join her partner, and on 10 August 1889, two days after their wedding, the couple embarked on the steamer SS *Westernland* in Antwerp (figure 3.2).[37] But homesickness and a pregnancy made Swarts prematurely return to Ghent the following summer, and she and Baekeland lived apart almost constantly until spring 1891. During this period, Swarts regularly reminded her husband that, though prepared to leave Ghent, she would prefer to stay in Europe.[38] Baekeland, for his part, was clearly not satisfied with his job at Anthony & Company, his New York employer. Around late November 1890, he resigned from the firm, and afterward he proved unable to earn a comfortable living for himself, his wife, and his mother, whom he advised to move to the countryside.[39] He actually began to accumulate debts. Why, then, did he nevertheless remain in the United States?

The remainder of this chapter provides a revised interpretation of this outcome, demonstrating that it almost certainly had more to do with Baekeland's eagerness to leave his home region far behind than with the quality of his new life in the United States. In other words, I make the case that so-called push factors, perhaps even irrespective of Baekeland's troubled relationship with the Guequiers, outweighed the elements that "pulled" the Belgian chemist to America. This, however, does not mean that the institutional and sociocultural environment in which Baekeland's science-related entrepreneurship was embedded was a binding constraint. Most relevant, the regulations governing the outside activities of state universities' employees were not enforced in the 1880s, and there are no signs that Baekeland's colleagues in Ghent considered entrepreneurial business engagements inappropriate for academics.

Push and Pull

Are standard accounts of Baekeland's first transatlantic journey correct in connecting it with the travel fellowship that the Royal Academy had awarded him in 1887? There can be no doubt that officially Baekeland was to carry out a scientific mission on behalf of Belgium's Ministry of the Interior and of Public Education, which financially supported his crossing to New York. However, it is obvious that Baekeland did not travel to the United States for purposes that were primarily academic in nature. Instead, his personal correspondence suggests that his focus was on arranging "his affairs" in New York, which he discussed with a man named Lufbery in or near Paris and clearly concerned his self-developing glass plates.[40] Baekeland may have been introduced to Lufbery by Jules Lechat, a rubber manufacturer at whose plants in Ghent and Lille the aspiring inventor seems to have regularly conducted laboratory experiments.[41] For his part, Lufbery, a person "generous to active and enterprising people," was "the one who concluded the transactions with Anthony," the photographic firm that hired Baekeland shortly after his arrival in New York.[42] This clearly suggests that Lufbery arranged a meeting for Baekeland with the Anthony company, contrary to the common assumption that this appointment was an indirect result of the chemist's visits to scientific institutes after his arrival in the United States.[43]

At any rate, when Richard Anthony, the firm's vice president and Baekeland's self-proclaimed American "discoverer,"[44] returned to his office in early September 1889, he "found Dr. Backelandt of Belgium here, waiting to show his self developing dry plates." But Anthony, as well as the technical advisers who witnessed Baekeland's demonstration with him, were less favorably impressed with their visitor's plates than with his skills as a photographic chemist. In a letter to the company's secretary, Anthony enthused about Baekeland's abilities and the ways in which he could add to his firm's bottom line. Given the scarcity of primary sources concerning this part of Baekeland's career, including the kind of work that he carried out at Anthony & Company, Richard Anthony's judgment is worth quoting at some length. In his opinion,

> He [Baekeland] proves to be a very thorough chemist and a practical emulsion maker. He says that he can make emulsions of extreme rapidity and absolutely the same every time, without the uncertainties that we have been laboring under, and I suggested that he go down to the factory and make some for us, which he agreed to do, simply to show what he can do.... He does not use the centrifugal [machine] at all in making his emulsions, and after explaining his method to Mr. Cossett [probably Franklin M. Cossitt, a photographic inventor whom Anthony had lured away from the Eastman Dry Plate and Film Company] and Dr. [Arthur H.] Elliott, they both are strongly of the opinion that his method is superior to that by the use of the centrifugal machine, and both say that it is impossible to see how there can be any variation in his emulsions if made after this method. He is extremely particular about every detail of the process, and Cossett says he has never seen a man so clean in his methods as he.... He says that he can make Aristotype paper perfectly well also, and Cossett is very anxious that we should make an arrangement with him to enter our employ regularly.... Cossett says he has already learned more from him about emulsion making than he ever learned from any one else in his life; and if he proves all that I hope, I feel strongly inclined to make an arrangement with him.[45]

This letter shows that Anthony was impressed with Baekeland's practical skills as a maker of photographic emulsions rather than with his academic knowledge or credentials. In addition, like others before him, Anthony also appreciated Baekeland's work ethic and his "push and energy," traits that resonated with the liberal "go-aheadism" prevailing in the United States at that time.[46] Hence, it is understandable that he was eager to secure the chemist's services, even though his company was experiencing unprecedented financial difficulties.[47] Cossitt expected Baekeland to sign up for approximately $1,500 a year. This would have represented an increase over his earnings as an academic in Belgium yet would still have made him "a very cheap man" from Anthony & Company's point of view. The figure probably was a slight underestimation, as Swarts recollected in 1891 that her husband had been ready to join the firm "for $2,000 or even for $1,750." However, she had evidently pushed him to bargain hard, for eventually he had been granted a salary of $3,000 a year, a royal sum for an industrial chemist in those days.[48]

The comparatively high American wages are one factor that possibly lured Baekeland to the United States and encouraged him to build a new life there; as his wife "always repeated," to the exasperation of both of them, he certainly had plenty of reasons to try to make "a lot a lot [*sic*] of money."[49] According to estimates that Swarts made in September 1890, Baekeland should have been able to cover his living expenses and support his parents with about half his $3,000 salary, allowing her to repay Lechat, the rubber manufacturer to whom she seemed to be "forever" in debt, with the remainder. However, in the next month, Swarts left her parental house in Ghent for a place of her own, and the money that she needed to rent and furnish it, as well as to run a household with a servant, had not yet been factored in. Moreover, after his resignation from Anthony around late November, Baekeland failed to secure an equivalent source of income for a prolonged period of time, forcing him to accept short-term jobs incommensurate with his qualifications.[50]

In this context, Swarts posited that "it is better to have a decent position of 8 to 10,000 fr [Belgian francs] [per year] here [in Europe] than to become a worker in America, where life is very expensive." Baekeland himself had previously pointed to the high cost of living in the United States. Yet, he nevertheless disagreed: in an embittered letter, he countered that he would "in no way return to Europe for a miserable petty position of 5 to 6,000 francs."[51] The discrepancy in wage level expectations is significant, for Baekeland did seem ready to move to about anywhere in the world if only he were duly remunerated. For example, shortly after earning his PhD, he appears to have considered participating in a colonial scientific mission to Congo.[52] In addition, in early 1891 he came very close to relocating to Paris or Valparaiso, Chile, for professional purposes. At the same time, though, he warned his wife that, if his assignment in France was confirmed, he would try to return to the United States as quickly as possible so as to avoid unpleasant reunions with European acquaintances. Thus, while financial considerations certainly were an important determinant of Baekeland's professional trajectory, they do not tell the whole story.

In fact, to his wife's chagrin, another of Baekeland's arguments was the sweeping generalization that "a few exceptions aside, all Europeans are more or less morons." Baekeland also expressed this attitude in later letters to Remouchamps, as well as in travel notes, in which he made it a game to reveal the "general mentality of nations."[53] Most of the examples that he gave to demonstrate Europe's "backwardness" did not directly concern his own career. To his relief, for instance, in the United States, the works of Emile Zola and other "decadent," "demoralizing" European writers were being censored, and women seemed to be treated decently there—a point to which his wife was receptive.[54] Still, it is pertinent to ask how Baekeland's own transition from European academia to American industry related to his later complaints about the "permanent exhibitions of stupidity, weakness of character, greed, selfishness, brutality, petty conceit and general lack of efficiency ... [hidden in] ... the cloak of art" that he observed in Western Europe in general and his hometown in particular.[55] Most relevant, had he partly been drawn to the United States, a country known for its business dynamism, by a sense that his native environment was insufficiently supportive of his entrepreneurial initiative?

In view of Belgium's educational legislation, Baekeland certainly had little reason to complain. An 1849 law obliged employees of state universities to obtain governmental permission before engaging in remunerated outside activities, but in Ghent this provision was barely enforced during the 1880s.[56] Indeed, it is doubtful whether Joseph Devolder, the minister of the interior and of public education from 1887 until 1890, was ever informed of the existence of Dr. Baekelandt et Cie. Someone who could have notified him was the university's administrator-inspector, Auguste Wagener, with whom Baekeland had a good personal relationship. Yet, Wagener limited himself to praising Baekeland as a gifted scientist who, in addition, had "the stuff to be an inventor, which is rare."[57] In November 1889, Wagener indicated to the minister that Baekeland was offering his resignation as a professor in training (agrégé spécial) because he had been given the opportunity to commercially develop his photographic invention in the United States. The administrator-inspector's accompanying advice to allow Baekeland to retain an honorary affiliation to the university provides further insight into his role as a mentor. Moreover, it is significant that Wagener explicitly referred to the chemist's "exceptional spirit of initiative in scientific matters," suggesting that he regarded this as something to encourage and admire.[58]

Although Wagener seems to have been favorably impressed with Baekeland's inventiveness and entrepreneurial attitude, there are indications that other people in Baekeland's environment disapproved of entrepreneurial risk taking. For instance, after having been informed that her husband had dissolved his contract with Anthony & Company, Céline Swarts indicated that, "I won't mention it to anybody, as you know how cautious, conservative etc. people are," even though the financial impact of her spouse's decision remained veiled and her uncommunicativeness forced her to take on debts from outsiders.[59] In 1898, Baekeland himself noted to Remouchamps that Europeans could hardly grasp that a professor could make a good industrialist, perhaps suggesting that he had personally met with prejudice on account of his involvement in Dr Baekelandt et Cie.[60]

In this connection, the claim that Théodore Swarts disapproved of his son-in-law's commercial endeavors because he would have liked him to fully concentrate on his academic pursuits is important.[61] This interpretation is supported by Swarts's lobbying efforts to have Baekeland promoted to the position of agrégé spécial prior to the wedding with his daughter in August 1889. It is also in accordance with the higher social status of Belgium-based academic chemists in comparison to their professional colleagues in industry. Moreover, it is confirmed by a letter in which Céline paraphrased her father, saying that it had "always been his goal and intention before your settling down in the US" to help Baekeland make a career at the University of Ghent.[62] However, on the other hand, when Céline finally informed her parents about her husband's departure from Anthony and his difficulties in finding an equivalent source of income, Théodore seemed eager to try to arrange for Baekeland "a decent position worthy of you and me here in Europe in whatever industry."[63] Indeed, he had previously recommended his son-in-law to the Société Générale de Maltose, a Brussels-based enterprise that had acquired the rights to a fermentation process patented by a prominent local chemist, Jean Effront.[64]

Like most of the science-related business initiatives examined in this book, the case of Maltose concerned a truly transatlantic endeavor. It was Théodore Swarts who heard in fall 1890 that the firm was trying to make arrangements for the exploitation of its US IP rights.[65] Baekeland confirmed that he would like to get involved, but was informed that Maltose would not make a decision before the end of the year. Then, on 12 January 1891, Théodore met again with representatives of the enterprise, and upon his return Céline was able to extract the following account from her tired, incommunicative father. According to Céline's letter,

> They [Maltose] were in contact with an Englishman who wanted to exploit the invention in England and the United States, but as he did not seem to be in a hurry, and only had an option until the middle of January, there was talk of dropping him. But in Paris there is an American ambassador or consul Du Thilly is his name, I believe, who got in touch with American financiers who would buy the process for 1 million, + 100,000 which would have to be paid to this gentleman. These investors ask an option for 4 months, but today it was found that 4 months was a long period of time and if one had to give $100,000 to this sir that would be a complication. In brief, a discussion on this started and here the boss [Prof. Swarts] talked about you, saying that you had gained a foothold in scientific circles there [in the United States], that you knew people and that you could be approached. To which they responded that a pure chemist was not exactly what they were looking for, that plainly someone to manage affairs and finances was needed, but that the combination of the two would be even better and they urged the boss to write you to bring you in contact with Carez [Constant Eugène Carré] (the engineer) so that you could get together with the capitalists whom I mentioned above.[66]

If Théodore Swarts indeed presented Baekeland as a "pure chemist" in his dealings with the Société Générale de Maltose, he may have overestimated the firm's interest in academic knowledge and credentials. But Swarts's recurrent get-togethers with representatives of Maltose, and his intention to involve his son-in-law in the project described here, do put into question the antibusiness attitudes that have been ascribed to him. At the same time, though, the outcome of the affair points to the risks of having to depend on Swarts's goodwill. The University of Ghent's chemistry professor continued to meet up with Maltose negotiators during the remainder of January and in early February, and his son-in-law was impatiently looking forward to hearing further details. But, to Baekeland's unpleasant surprise, Swarts appeared not to have sent him a letter—a fact that seemed to support the now 22-year-old Céline in her view that her father was "the most negligent and selfish being that I know of."[67]

From Céline's letters, Théodore, the popular professor, appears as an authoritarian and disloyal father and spouse, who contributed to his daughter's low esteem for Belgian husbands other than her own. Ironically, around the same period in which Théodore was elected to the Order of Leopold—one of the highest public honors in Belgium—Céline and her siblings had been assisting their mother in her efforts to obtain a divorce. Their preference was to arrange this by mutual consent, as detailed evidence of misconduct would have been needed in the case of a judicial divorce and, as Céline put it, "mother has never taken much fun in following or surveilling this monster."[68] But Théodore, assisted by legal advisers of his own, informed the lawyer of his family members that he would "not at any price accept a divorce by mutual consent." He even threatened that "if mom persisted in this idea [of divorcing]...he would send Bibi [Céline's then 15-year-old sister] to a boarding school and ruin the career of Frederic [Céline's brother Frédéric] by forcing him to resign [from the University of Ghent]."[69] I have seen no indications that Baekeland directly entered the dispute, but he doubtlessly advised his wife by letter and in person upon a brief return to Belgium in summer 1890. Moreover, before his departure to New York he had already indicated to Wagener that he did not get along well with his father-in-law, therefore suggesting that he was unlikely to side with "the boss." In addition, if Baekeland had relocated to his native country, Théodore Swarts could have intimidated him, just as Théodore was threatening his own son Frédéric, who had been working under his direction since September 1885.[70]

As if this were not yet enough, Baekeland's relationship with Frédéric Swarts, "who does not possess an easy character," risked becoming equally explosive as his relationship with Théodore.[71] For instance, in December 1890, a date by which Théodore's attempts to reconcile himself with his family had met with some success, Céline anticipated that Frédéric was "predestined" to succeed his father as a chemistry professor at the University of Ghent. In this respect, it was helpful that Baekeland had left academia, "because such situations are so delicate and cause so much trouble, and it is easy to say that the son always comes first and that would have been irreparable scars."[72]

Still, in a sense, to claim that Baekeland primarily desired to leave Belgium far behind because of the Swartses—or even the Guequiers—would overrate the importance of these families. In effect, Baekeland appears to have been entirely fed up with his social surroundings in the "bunch of rubbish that one calls Belgium and where everyone is an idiot," and unwilling to forgo his newly acquired sense of liberty and independence.[73] According to a letter to his wife, for example,

> The mere idea of having to return to Belgium and being once again in the company of "friends and acquaintances" is already giving me a sense of malaise. Visits and visiting cards and straitjackets and everyone meddling in your affairs no no

> no, I know too well that my life of boredom and mutual annoyances would then start all over again. My character is not social enough for that anymore. The fewer people I see the better.... [D]o not believe... that because of my present difficulties I have become tamable or could be turned into the banal, domestic animal that a society man is.[74]

Whereas Baekeland had previously apologized somewhat playfully for his pessimism "à la Schopenhauer,"[75] the tone of the letter just quoted was truly embittered, suggesting that he was suffering from depression. But his message to his spouse was certainly consistent with what he had written before; in fact, two recurring themes in the couple's correspondence had been the different social dispositions and backgrounds of each partner and their dissimilar opinions as to what constituted the basic necessities of life. For instance, Baekeland repeatedly grumbled that Swarts expected him to be working for "a good, well furnished house with servants and your parents and this delightful city of Ghent with all your girlfriends," whereas a life like that did not at all appeal to him.[76] In response, Swarts on the one hand admitted that in her social circles the view that "happiness was impossible where money was lacking" had been only natural and that she could barely stand the thought of her husband having to make up his bed himself.[77] She also agreed that their characters differed, writing him, "You are a little bear who knows how to live by himself and enjoys this solitude, but I am not like that."[78] On the other hand, though, she charged Baekeland with attempting to create an emotional gulf between them, in addition to the existing physical one. In her view, he did this by intentionally focusing on what separated them instead of what they had in common, as well as by exaggerating actual dissimilarities.

Swarts also expressed concern that her husband would permanently develop into the kind of antisocial loner that he depicted himself as in the letter just quoted. To Baekeland, however, this was certainly not an incentive for returning to Europe. By contrast, he countered that, "Living two years in a free and independent country without being bothered constantly by friends and relatives has made me appreciate the advantages of this kind of life. Here at least if I don't care about people they don't care about me and I do not have the smallest interest in involving myself in the chicaneries and annoyances of 'society' again."[79]

As previously observed by Carl Kaufmann, the scars caused by Swarts and Baekeland's quarrels and prolonged separation never fully healed.[80] Yet

Figure 3.3
Céline and Jenny Baekeland on the left (c.1893), George, Céline, Nina, and Leo Baekeland in the family photo on the right (c.1899). Tragically, Jenny died in 1895, shortly after the birth of her brother, George. The Baekelands' final child, Nina, was born in 1896.
Left photo courtesy of Leo Baekeland Papers, Archives Center, National Museum of American History, Smithsonian Institution, AC-0005-0000122. Right photo courtesy of Leo Baekeland Papers, Archives Center, National Museum of American History, Smithsonian Institution, AC-0005-0000115.

their marriage did not collapse and Swarts, together with Jenny Baekeland, the daughter to whom she had given birth on 5 November 1890, finally rejoined her husband in New York around May 1891 (see figure 3.3).

Relatively little is known about what happened between this reunion and the foundation of Baekeland's second photochemical start-up firm, the Nepera Chemical Company, in the second half of 1893. Baekeland surely attempted to hedge his bets by working on several distinct (potential) inventions, including an antiseptic compound for food preservation, for which he filed a patent application in June 1892.[81] It is also noteworthy that he promoted a process by Effront, the biochemist who had assigned his patent rights to the Société Générale de Maltose, to American alcohol manufacturers and scientists in an article in the *Journal of American Chemistry*.[82] This might suggest that Baekeland still had ambitions of becoming the US representative of the Brussels-based enterprise. In addition, he seems to have taken to heart his wife's earlier advice to actively participate in photographic clubs and societies. In 1892, for example, he presented an introductory paper on carbon photography to the Society of Amateur Photographers.[83] Finally, it is undisputed that Baekeland conducted further photochemical experiments until he was struck by appendicitis and "barely escaped the clutches of death."[84]

With hindsight, it is evident that Baekeland's encounter with Leonard Jacobi (1850–1931), a German Californian Jew, constituted a key event that helped break the downward spiral in which he found himself. Baekeland and Jacobi got to know each other in Europe, where Jacobi, a former stockbroker with an international network of professional contacts, is said to have been spending a fortune earned in finance on not less than 14 years of "travel and recreation."[85] Then, in June 1893, while Baekeland was recovering from his illness, they met again in Yonkers, the New York City suburb in which Jacobi's family had rented a mansion for the summer. Baekeland had just relocated from Manhattan to Yonkers and, apparently not deterred by the severe economic depression that would keep unemployment rates above 10 percent until 1898, he and Jacobi went into business later that year.[86]

Unlike Baekeland and Jules Guequier, Jacobi had no background in science, but his thoroughness and commonsense approach to the day-to-day running of a business proved indispensable, as Baekeland would later recollect.[87] This is not to say that Baekeland himself had not drawn any lessons from his previous problems and setbacks. Perhaps most important, he seems to have almost single-mindedly dedicated himself to company affairs. There is considerable evidence that he worked long hours, limited his vacation time—particularly during the firm's crucial first four years of existence—and did not engage in any other occupation.[88] Among other things, this meant that Baekeland now refrained from combining academic and industrial pursuits in the same way he had been doing in the late 1880s.

Negotiating Academia-Industry Boundaries

At the time he became involved in the creation of an academic start-up firm, Baekeland could not have anticipated that this type of work would be subjected to actual scrutiny in the following decade. An 1896 request by Belgium's education minister, Frans Schollaert, for example, resulted in a comprehensive survey of the remunerated outside engagements of the employees of the State University of Ghent. Schollaert had asked Gustave Wolters, Auguste Wagener's successor as administrator-inspector, whether the combining of academic duties with professional or commercial work by members of the school's administrative personnel "is not inconvenient and, in your opinion, incompatible with the position of these individuals in the university."[89] To properly address these questions, Wolters considered it appropriate to also include the paid extramural activities of the establishment's teaching staff in his investigation. His report showed that these activities ranged from educational assignments at other schools to work in politics, industry, and entertainment. (To the dismay of the administrator-inspector, two "service boys" [garçons de service] supplemented their relatively low academic incomes by performing as cabaret artists!) In addition, the survey revealed that remunerated outside engagements were particularly widespread among both senior and junior members of the faculties of law and medicine. But Wolters identified commercially active employees across the entire university, including former colleagues and acquaintances of Baekeland such as Jules Guequier and Frédéric Swarts.[90]

Did Wolters, an engineer by profession, share the critical attitude toward paid extramural work that seemed implicit in the phrasing of the minister's request? This administrator-inspector on the one hand suggested intervening in the substantial number of cases where this type of activity aroused the jealousy of nonacademic competitors or negatively impacted the university's educational work. Yet on the other hand, he urged Schollaert to proceed prudently, for "the numerous precedents, existing interests, and the scientific value of the civil servants are serious motives for modifying the current state of things only with great caution and without causing friction." His superior, however, opted for a more radical approach. Consequently, by 1898, Wolters once again had to remind the members of the university's teaching staff that they needed to obtain the minister's permission before engaging in remunerated outside activities, while emphasizing to the lower-level administrative personnel that they were generally not allowed to carry out such work.[91]

These were issues with which higher education institutions on both sides of the Atlantic had to grapple in the decades around 1900, when the interactions between academic establishments and the spheres of industry, commerce, and government intensified.[92] When it came to the State University of Ghent, as with other public universities, some of these concerns resulted from the status of its employees as civil servants, who had to maintain an aura of neutrality and were not supposed to compete for money with private citizens and firms. But, as chapter 9 illustrates by examining Baekeland's role at Columbia University, which engaged him as an honorary professor in 1917, private universities also had to defend and negotiate the boundaries separating them from industry and other realms of society. To understand Baekeland's return to academia, however, it is first necessary to consider the development of his career, as well as of the relationships between science and industry, in the first years and decades following his migration to the United States. More specifically, it is essential to explore how intellectual property issues became more central to the science-industry nexus.

4 Velox: Innovation without Patents

As late as 1915–1916, patents were rarely brought up in discussions about knowledge transfer between academia and industry in the United States. During these years, the New York Section of the American Chemical Society organized a series of meetings to promote the "necessity of cooperation between the universities and the industries." The 13 presenters of papers mainly consisted of representatives of Columbia University and the Massachusetts Institute of Technology (MIT) but also included two chemical industrialists, an industrial research chemist, and a member of the Rockefeller Institute of Medical Research. Almost all of them shared the organizers' premise that more collaboration was desirable, so the discussion mostly focused on the advantages and disadvantages of various strategies for bringing it about. Still, only two speakers, Elon H. Hooker of the Hooker Electrochemical Company and William Walker of MIT, referred to patents. Moreover, Walker, an experienced consultant and designer of a chemical engineering curriculum oriented toward basic industrial processes and problems, did this only to question their relevance.[1]

By the mid-1930s, in contrast, the policies of academic institutions regarding patents and their commercial exploitation had become a pressing issue, to which the US National Research Council (NRC) dedicated an entire conference. In European countries such as Britain, France, and Germany, moreover, the question of whether and how to restrict the autonomy of academic and/or government employees in obtaining patents also started to receive more attention in the interwar period.[2]

It would be wrong to infer from this that, as several historians have claimed, patenting violated the norms of academia in the decades before World War I and that this resistance gradually weakened from the 1920s onward.[3] To some extent, this argument holds true for medical patents, which had long met with strong moral opposition on account of their real or potential use as tools for monopolizing life-saving drugs.[4] In fields ranging from physics and chemistry to engineering, however, the patenting activities of individual academics seem to have been largely uncontroversial since at least the 1880s, if not much earlier. This was certainly the case at the State University of Ghent, where the patents of not only Baekeland but also several of his more senior colleagues in the science faculty and engineering schools do not appear to have aroused any objections. The available evidence, moreover, suggests that other higher education institutions not just in Belgium but also in other countries, ranging from Germany to Britain and the United States, had no major reservations about academic patenting at that time.[5]

The photographic industry, in contrast, was one of a number of economic sectors whose histories had been shaped by patent controversies.[6] When publicly announcing the invention of the daguerreotype process in 1839, scientist-politician François Arago had presented the technology as a French "gift to the world," whose use Louis Daguerre had not wanted to restrict via patents. The reality was more complex, but the effects of patents on the public accessibility of photographic knowledge and know-how would remain an important theme throughout the remainder of the century. From the 1850s until the late 1870s, for example, photographers on both sides of the Atlantic recurrently challenged claims that they had illegally adopted patent-protected inventions. In part because of such controversies, which sometimes led to more fundamental criticisms of the rationales and functioning of different patent systems, it remained relatively uncommon for photographic firms to apply for patents during these decades. This started to change in the 1880s, when such companies began to depend less on knowledge of external inventors and technology users and initiated their own patent-centered strategies for innovation.[7] This trend, which had its parallels in the chemical and electrical industries, did not necessarily lead to a quantitative rise in patent litigation.[8] However, it increased the stakes of the lawsuits that did take place, as their outcomes determined who could control transformative technologies ranging from photographic film and synthetic dyes to the internal combustion engine, the telephone, and the wireless telegraph, in markets of an unprecedentedly large size. This development also inspired qualitatively new uses of the US and other patent systems, including several that have been characterized as "unproductive" because they were meant to stifle, rather than stimulate, technological innovation.[9] In addition, it was reflected in the creation of patent departments within large industrial corporations and, particularly in the American context, in the redefinition of employee inventions as corporate rather than individual property. As a further step to protect and lay claim to crucial technologies, moreover, firms such as Eastman Kodak erected "walls of silence" and relied on trade secret law, raising new concerns about the accessibility of photographic knowledge and know-how.[10]

This chapter examines Baekeland's decision not to patent Velox photographic paper in relation to several of these broader developments. When publicly discussing his reasons in the 1910s, Baekeland primarily pointed to his inability to engage in prolonged litigation against large and capital-rich corporations. This explanation served his early twentieth-century efforts to reform the US patent system, as analyzed in chapter 7. To understand his entrepreneurial and IP strategies in the late nineteenth century, however, the rise of corporate IP on both sides of the Atlantic was much more significant.

A contemporary of Baekeland also gave an alternative explanation that is highly relevant to this book. Josef Eder, a photography and photochemistry professor in Vienna, charged that the technology behind Velox had not been patent eligible because of Baekeland's heavy reliance on the prior art. In particular, Eder claimed that Baekeland had simply copied instructions given in the published work of himself and his collaborator Giuseppe Pizzighelli, who should therefore be considered as the actual inventors of the silver chloride paper. Eder's allegations and the reactions that they triggered can cast light on the relationship between the two main forms of intellectual property (IP) between which this book differentiates: legal IP, such as patents, and IP in the form of scientific and technological credit. Most pertinent, the dispute provides insight into the ability or inability of industrial scientists to get intellectual recognition for their professional accomplishments without jeopardizing the interests of their firms. As I will demonstrate, Baekeland did find ways to present his company's innovations in the form of research papers. Moreover, similar to how his involvement in different chemical and engineering societies later shaped the amount of credit he received for the invention of Bakelite, his personal acquaintance with other photographic scientists indirectly helped him in countering Eder's charges.

At the same time, however, it is undeniable that the emergence of a comprehensive system of corporate knowledge ownership provisions strongly limited Baekeland's ability to interact with other scientists. To appreciate this, it is crucial to have a closer look at the chronology of Baekeland's photochemical research. He typically claimed to have invented the paper that would become known as Velox several years before the founding of the Nepera Chemical Company in 1893. Because of a lack of sources, it is almost impossible to determine which of his sometimes contradictory statements on this issue are the most plausible. It is, however, clear that his work at Nepera partly built on his earlier photochemical investigations. This raises pertinent questions about Baekeland's contractual arrangements and legal skirmishes with his partners of Dr Baekelandt et Cie and employers at E. and H. T. Anthony & Company. After all, by the 1890s, American companies had begun to have relatively effective legal means at their disposal to claim ownership of their employees' inventions. As legal historian Catherine Fisk has demonstrated, this was a new and consequential development, which was not completed until the 1920s.[11] In the European context, Germany had experienced a similar transition at an earlier date. Moreover, in Belgium, too, firms tried to restrict the knowledge ownership rights and mobility of creative workers, as Baekeland himself had experienced.[12] Baekeland's personal and business correspondence indeed reveals this type of restriction with respect to his time at both Dr Baekelandt et Cie and Anthony & Company, thus supporting this chapter's argument that the corporate control of innovation was a major factor behind his selective secrecy about Velox.

Reasons for Not Patenting

Insofar as historians have dealt with Baekeland's IP strategy for Velox, they have generally accepted Baekeland's own arguments.[13] His first recorded comments on this topic were made in congressional hearings in 1912 on the Oldfield Revision and Codification of the Patent Statutes, of which he was a critic (see chapter 7). On this occasion, as on others, Baekeland downplayed the role of his partners at Nepera, particularly of the firm's president, Leonard Jacobi.[14] More to the point, he implied that inventors had essentially two mutually exclusive options available: (1) patenting an invention, which involved the detailed disclosure of its characteristics in a publicly accessible document; or (2) relying on secrecy. According to Baekeland's testimony, the first choice was, in principle, always preferable from both the inventor's and society's points of view. In reality, however, it was not a sensible option for inventors who, like Baekeland himself in his first years in the United States, lacked the funds to defend an American patent in court. In his own words, Baekeland maintained that, "My first successful invention here in this country was a new photographic process, called 'Vellox' [*sic*] paper. At that time I was not rich enough to take out a patent and to defend it at the same time. So I knew that my competitors would have swamped me by infringing the patent, and there would have been very little chance of defending my rights. So I practiced it secretly, and after I made business a success, then I sold out to a large corporation at my terms."[15]

Four years later, Baekeland repeated this explanation in his Perkin Medal address, with the qualification that it had not simply been a question of wealth. Under the heading "uncertainties of chemical patents," he maintained that "at that time I was already aware how difficult and expensive it is to carry on patent litigation and I realized too well that we could neither afford the money nor the time connected with patent fights."[16] While he elsewhere generally denounced the length and cost of patent litigation, on the Perkin Medal evening he specifically connected it with the nature of inventions like Velox. As a further reason for not patenting the technology, he clarified that "chemical process patents which can be followed out in secrecy can be too easily infringed with impunity because it is too difficult ... to establish these infringements before our courts of law."[17] Baekeland thus presented a potential advantage—the relative ease with which new chemical processes could be kept secret—as a drawback, pointing out that one's competitors could equally effectively hide their copying them. In addition, he emphasized that chemical inventions were more complex than "purely mechanical patents" and hence harder to grasp for judges who were not technically trained.[18]

These points may well have helped Baekeland in gaining support for patent reform proposals that he considered important, particularly for the creation of a single Court of Patent Appeals (see chapter 7). They are, however, untenable as explanations for his IP strategy for Velox. For example, Baekeland's criteria cannot clarify why even the wealthiest photographic firms did not patent all their inventions. The chemist's own laboratory notebooks, moreover, reveal that he himself had engaged an attorney to apply for a US photographic patent around 1890. The resulting application covered the use of starch in photographic emulsions and was, to Baekeland's disappointment, rejected because of its alleged lack of novelty.[19] While this invention was not obviously related to Baekeland's work on the photographic papers that would be branded as "Velox," it did meet all his conditions under which it allegedly made no sense to secure patent protection.

But what about Josef Eder's alternative interpretation? In a 1930 article in the British publication *Photographic Journal*, Eder asserted that "Baekeland had created nothing new with his silver chloride paper, but merely copied, possibly unwittingly, the old.... This Velox paper can be shown to be pure silver chloride-gelatine paper, with chemical development. Since this was only the old Eder-Pizzighelli paper, this firm [the Nepera Chemical Company] could not take out any patents."[20] The timing of Eder's claim to priority of invention seems odd. To begin with, Nepera Chemical had been taken over in 1899, and Baekeland had remained outside the photographic industry ever since. The seminal publications to which Eder referred to substantiate his claim, in contrast, dated back to 1881. Furthermore, Baekeland had discussed his work on so-called gelatin silver chloride developing-out papers prior to 1930. His presentations in the photochemistry section of the International Congress of Applied Chemistry in Berlin in 1903, above all, should have made clear to Eder and other conference participants that he considered his contribution to the field much more significant than the "mere" application of a process of the prior art. In his papers, which had been published in 1904, Baekeland had not even mentioned the work of Eder and Giuseppe Pizzighelli, an official in the Austro-Hungarian army, who, at the time of his collaboration with Eder, had been in charge of the photography department of the Viennese Technical Military Committee (Technischen Militär-Comité).[21] (Pizzighelli had died in 1912 and therefore did not play an active role in the priority dispute.)

While there are no indications that Eder had taken issue with Baekeland's own publications, he did dispute author Carroll Neblette's description of Velox in his 1930 study *Photography: Its Principles and Practice*. Neblette had written that the paper had been introduced "from the formula of Dr. Leo Baekeland."[22] According to Eder, though, Neblette's clarification showed that the Velox process was identical to the one disclosed by himself and Pizzighelli in 1881. Eder emphasized that he and Pizzighelli, not being interested in patents and industrial profits, had purposefully made available their research results "freely... to the world." But Eder cared strongly about IP in the form of intellectual credit and therefore claimed reputational rewards for himself and the late Pizzighelli, who had "died in modest circumstances."[23]

A commitment to the rapid and broad dissemination of new photographic knowledge and a concern about the attribution of credit had both been characteristic of Eder's distinguished career as a teacher, researcher, administrator, and author of not just scientific and technological works but also historical ones. As with Baekeland's thesis on dissociation processes, Eder's professional prospects had initially been boosted by prize-winning research on the photochemical properties of chromium compounds.[24] However, whereas Baekeland had later left academia and tried to establish himself as a chemical innovator in the private sector, Eder always kept working in public institutions. In fact, he played a leading role in the creation and direction of a new graphic arts establishment, the K.k. Graphische Lehr- und Versuchsanstalt, whose generously funded photography program was widely regarded as the best one available anywhere in the world in the decades before World War I.[25]

Part of the mission of the Lehr- und Versuchsanstalt was to develop commercially applicable photographic techniques and to conduct experimental studies and tests commissioned by firms. However, consistent with Eder's quoted emphasis on knowledge sharing, it was made clear in the institute's statutes that all research results obtained at its laboratory would be published, once approved by its director (i.e., Eder).[26] Moreover, starting in 1889, the year after its opening, the Lehr- und Versuchsanstalt had at regular intervals put in print the main technological processes that it had been applying. In the context of Eder's dispute with Baekeland, US photographer R. W. Harrison questioned the effectiveness of such an approach, stating that "this type of generosity…, while very beautiful in the Shakespearian and fairy tale sense, is not what the world needs or will use."[27] Like Baekeland, Harrison seems to have been convinced that the prospect of obtaining exclusive property rights was a precondition for attracting private investments in R&D work. There are, however, no signs that Eder's insistence on the free dissemination of research findings decreased the demand for the commercial services offered by his institute.

This is not to say that such open forms of knowledge sharing could not go hand in hand with the acquisition of patents. If used in specific ways, patents could indeed be turned into tools for ensuring the broad accessibility of new technologies. This held true even for the United States, the country whose legal framework provided the fewest restrictions on the rights of patent holders.[28] In the decades before World War I, as an example, multiple US academics and government scientists dedicated their patents to the general public with the aim of making their inventions available to any potential investors and users. In a similar vein, some of the earliest North American universities entering the business of patent management justified this in part by arguing that it was necessary to ensure that critical products such as vitamin D and insulin could not be monopolized by profit-oriented firms.[29]

Moreover, patents could also be used for bolstering priority claims such as those made by Eder upon the appearance of Neblette's book. In a recent study on pre-1920s physicists and electrical engineers, historians Stathis Arapostathis and Graeme Gooday relate this practice to specific features of the British patent regime. In particular, they argue that the notion of intellectual "property"—which Baekeland helped popularize among US chemists and engineers—emerged relatively slowly in the United Kingdom. In their view, it would therefore be misguided to interpret late nineteenth- and early twentieth-century patent disputes among British inventors as quarrels about legal property rather than about intellectual credit.[30]

The Austrian patent system differed in several respects from its British and US counterparts, but, at least in theory, also allowed for the pursuit of either one of the objectives just mentioned. Eder, however, clearly believed that patents were not necessary to achieve these goals. As his quoted comments on Velox illustrate, he was aware that the public disclosure of new technologies could prevent their monopolization, for inventions that had entered the public domain generally stopped being patent eligible. On at least one occasion, Eder had indeed succeeded in invalidating a patent application of someone who had overlooked Eder's own published work on exactly this ground.[31] At the time of Eder and Pizzighelli's research on silver chloride emulsions, moreover, photographers in and far beyond Vienna had mostly turned to other mechanisms to attribute inventive credit. In particular, nineteenth-century photographic societies, journals, and exhibitions served as forums for scrutinizing priority claims and honoring inventors with reputational rewards, medals, and, occasionally, financial compensation. Some photographic societies even explicitly challenged the fundamental principles of their country's patent law by defining photographic knowledge and inventions as "common property," as opposed to the exclusive possession of their originators.[32] These were all principles to which Eder paid more than lip service. To the Viennese Photographische Gesellschaft, for instance, Eder's commitment to publication had been a reason for awarding him its Voigtländer Medal in 1898.[33] It was not until after World War I, when Eder was forced to request food relief and to sell personal research samples and equipment, that he appears to have put aside his reservations about securing legal IP protection. The first two holders of Germany's most prestigious photographic science professorship, in contrast, had obtained patents throughout their decades-long careers.[34]

But what about the substance of Eder's claim that Velox would not have been patentable? Eder and Pizzighelli's 1881 disclosures on silver chloride emulsions had indeed been welcomed as a significant breakthrough.[35] There are no reasons to believe that Baekeland had not been aware of the much-discussed research. For instance, he stated in 1930 that he had started to experiment with such emulsions "shortly after Eder and Pizzighelli made their first publications on this subject." The magazines that had been covering Eder's work had, moreover, included the house journals of l'Association belge de photographie (ABP) and of Anthony & Company. As a case in point, less than a year after the release of Eder and Pizzighelli's seminal research in Austria and Germany, the Industrial School of Ghent's photography lecturer Gustave De Vylder had published a French translation in the ABP's journal.[36]

It is also important to keep in mind that, as historian Paul Israel has indicated, in late nineteenth-century America it had been "commonly accepted in patent law that the original conception of an invention was the mark of true inventive genius and that the work of reducing it to practice involved little more than ordinary mechanical skill, which could not be invention."[37] Certainly, Baekeland may not have regarded the preparation of photographic emulsions as Nepera's main accomplishment or as the critical factor that had slowed the advance of its technologies.[38] Rather, archival records document how Baekeland and his collaborators had struggled to control climatic influences and to ensure sufficiently large raw photographic paper supplies, among other challenges. In his response to Eder, Baekeland accordingly returned to the scaling-up problem discussed in chapter 3. He emphasized that "there is an enormous difference between preparing a few sheets of paper in the laboratory for one's own use, and a successful manufacturing process which can be carried on day by day in wholesale, steady commercial production; which must furnish a paper uniform in quality, and which can be kept indefinitely in storage even in our trying American summer climate, when heat and moisture conditions are similar to tropical conditions."[39]

Still, predominant notions of the nature of invention and innovation may have made plausible Eder's claim that Nepera had "merely" introduced an already known process on a large, commercial scale. If Eder could demonstrate that Baekeland's initial process had been identical to the one he had cocreated, he would have a hard-to-refute case.

A closer look, however, reveals that Eder's statement about the sameness of the processes was erroneous. One important difference was that Baekeland, contrary to Eder and Pizzighelli, had not considered it desirable to increase the light sensitivity of his gelatin silver chloride emulsions by carefully heating them. In photographic terms, this means that he had not "ripened" his emulsions. Baekeland, moreover, had also left his emulsions unwashed. Already on the 1916 Perkin Medal evening, chemistry professor Charles F. Chandler had referred to Baekeland's complete omission of the washing as a "photographic heresy."[40] The expression, perhaps suggested by Baekeland while briefing Chandler in the weeks leading up to the event, certainly seems exaggerated.[41] In fact, Chandler's own collection of photography-related clippings demonstrates that the use of unwashed emulsions had been recommended by no less than Désiré Van Monckhoven, the photographic researcher and entrepreneur Baekeland had known in his student days. In an influential 1879 lecture, Van Monckhoven had stated that "the water of our good city of Ghent is so chalky—caused by the nature of the soil—that I was obliged to find some method of doing away with washing my [silver bromide] emulsions."[42] Baekeland himself, moreover, had advised against thorough washing in a research article on the merits and drawbacks of using hydroquinone, an organic chemical compound, as a photographic developer.[43] Still, this is not to say that Baekeland's prospects for obtaining patent protection had been discouraging. For instance, two leading photographic chemists, Rodolfo Namias from Italy and Eastman Kodak's Kenneth Mees, independently corroborated his dismissal of Eder's contestation. The opinion of Namias, who had met Baekeland at several international conferences, was the best supported: before 1930, personal research experiments already had made him conclude that it was impossible to obtain images with Velox-like qualities on the basis of Eder and Pizzighelli's instructions. These failed experiments strengthened his view that Baekeland's gelatin silver chloride emulsion had been novel and nonobvious, and therefore patentable. This was not only because of the forgoing of the washing but also because of the distinct conditions under which silver chloride precipitated in Baekeland's emulsion, among other differences from Eder and Pizzighelli's methods.[44] Needless to say, these dissimilarities could have been highlighted in patent applications.

Selective Revelation

A possibility that neither Eder nor Baekeland discussed was that, irrespective of the legal enforceability of patent claims, it might simply have been more effective to keep inventions like Velox secret. When considering this option, it is important to realize that the common assumption that patenting and secrecy were, and are, mutually exclusive choices is inaccurate.[45] This was partly because the claims of strong patent applications were drafted in such a way as to increase the likelihood of being granted the broadest possible protection. By the late nineteenth century, this specialized work was mainly being performed by a separate class of patent professionals. As Alexander Bell and Thomas Edison experienced, these agents or lawyers sometimes put special emphasis on features of an invention that, from a technological point of view, were of secondary importance. Moreover, even patent applications that did accurately represent an inventor's main ideas could be highly secretive documents from which crucial details had been strategically omitted.[46] Conversely, the protection of technologies through secrecy did not preclude the revelation of carefully selected types and pieces of information. Baekeland, for example, published several articles concerning Velox in which he promoted the paper's qualities and instructed photographers how to use it. He also praised Velox in public lectures in the United States and Europe.[47] But in none of these accounts did he reveal the emulsion formulas that his firm had been using.

Baekeland's silence on this topic was characteristic of the American photographic industry of the 1890s. As late as 1883, a professional photographer from Chicago had predicted with confidence that "just as sure as there is any real improvement in dry-plate emulsion processes worthy [of] the notice of first-class workers, just so sure will it be made known to all free of charge."[48] Exactly the opposite had happened, however, by the time Baekeland moved from Ghent to New York. By the end of the century, moreover, not just Nepera (figure 4.1) but also all other US manufacturers of photographic dry plates, paper, or film were using their own "secret formulas," which they kept carefully hidden from company outsiders as well as from most of their own employees.[49]

Figure 4.1
Laboratory space and equipment of the Nepera Chemical Company. By the end of the nineteenth century, it had become common for photographic firms to tightly control access to their laboratories and the knowledge produced therein, including by using code letters and storing emulsion formulas in a locked safe.
Courtesy of Leo Baekeland Papers, Archives Center, National Museum of American History, Smithsonian Institution, AC-0005-0000116.

Besides withholding their emulsion formulas, Baekeland and his associates at Nepera also took several additional steps to monopolize their manufacturing processes. For example, initially they did not disclose the ingredients of the solutions they recommended for developing Velox photographs. They presumably did this to discourage users from preparing developers themselves instead of purchasing the ready-made ones. This policy, however, aroused protest and was rapidly abandoned: from early 1896 at least, Nepera regularly published developing formulas. At the same time, customers continued to contribute suggestions of their own, as skilled photographers had been doing for decades.[50]

During Nepera's six-year existence, Baekeland and his collaborators also appear to have never revealed the basic fact that Velox was a silver chloride

(a)

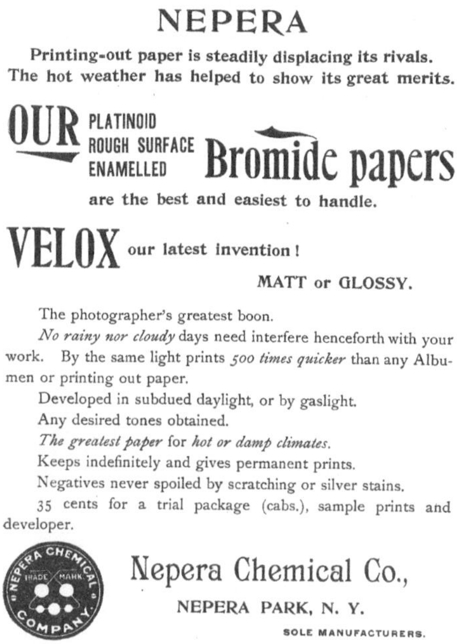

(b)

Figure 4.2

Advertisements of Nepera that did not disclose the basic fact that Velox was a silver chloride photographic paper. Reproduced from (a) Edward L. Wilson, *Wilson's Cyclopaedic Photography* (New York: 1894), and (b) *The Brochure Series of Architectural Illustration* 4 (1898).

paper. Their uncommunicativeness on this point is especially intriguing because it was in tension, if not at odds, with their brand-building efforts. For instance, when launching "Velox" in the second half of 1895, Nepera emphasized that its latest product "prints 500 times quicker than albumen paper." This claim was repeated in the *American Amateur Photographer* and in *Photographic Mosaics*, as well as in later advertisements (see figure 4.2).[51] Yet customers and competitors alike believed Velox was based on a silver *bromide* or, less frequently, chloro-bromide emulsion. One article in the photographic press even explicitly contrasted Velox with "any of the many chloride papers." This is significant because, as Mees wrote in 1920, "Velox is about a thousand times *slower* than bromide paper."[52] Thus, ironically, the innovation came to be identified with slowness instead of velocity.

Although the provisions of Nepera's sale to George Eastman included the requirement of continuing to maintain secrecy about the Velox process,

Baekeland briefly communicated more openly after the turn of the twentieth century. Having become a rich man freed from business obligations, he then seemed hungry to exchange knowledge with other academically trained chemists, as opposed to photographers "who were not sufficiently acquainted with the strict methods of laboratory investigation."[53] He was, moreover, perhaps even more eager to increase his professional reputation. At the international applied chemistry congress in Berlin in 1903, in particular, Baekeland presented no less than 8 of the 22 papers featured in the conference's section on photographic chemistry. (Almost all the other participants, who included Eder and Namias, presented only one paper each.) Typically speaking in the first person, without even mentioning his company or any of his former associates or employees, Baekeland claimed credit for a whole range of "firsts," as discussed in this chapter's final section. On this occasion, Baekeland also clarified that Velox papers "have been erroneously described in several publications as slow bromide papers or as chlorobromide papers, whereas their basis is chloride of silver prepared under particular conditions."[54]

Even though Baekeland still left his audience in the dark as to what those conditions might have been, this correction seems to have upset Nepera's president, Leonard Jacobi. Significantly, in the aftermath of the Berlin conference, Jacobi sought legal advice about whether he might be held liable if Baekeland violated his contractual requirement not to "disclose or divulge... any formulas or process of trade" to anyone other than Kodak. Baekeland, too, may have regretted his words, as he seemed anxious to avoid causing similar worries in the future.[55] For instance, it was not until the Perkin Medal evening of 1916 that Richard Anthony, Baekeland's former employer, found that the chemist "had succeeded... because he did not do as much washing as others had done."[56] As these two men had stayed on friendly terms after Baekeland's departure from Anthony & Company and had collaborated on other projects in the years before 1916, this indicates that Baekeland had taken seriously his contractual obligations to Kodak. This attitude may have saved him and Jacobi a great deal of trouble because, as the next section of this chapter illustrates, legal provisions of this type had become easier to enforce in the United States by the start of the twentieth century.

Legal Complications

As a consequence of his not patenting Velox, Baekeland was not legally required to demonstrate that he was the "first and true" inventor of the technology. Perhaps that partly explains why his accounts of the timing of his research on silver chloride papers were ambiguous, if not contradictory. Historians have typically assumed that Baekeland invented Velox around 1893, shortly before he and Jacobi started their photographic enterprise. This interpretation is in agreement with Baekeland's description of Velox as his "first successful invention here in this country [the United States]" in the 1912 Oldfield hearings. On other occasions, however, he stated that Velox was "invented in 1882" or 1883, or that he "invented it in the middle eighties of the last century; while I was still assistant-professor of Chemistry at the University of Ghent (Belgium)."[57] According to still another account, Baekeland's "conception of a chloride (gas light) paper took place" at the laboratory of Anthony & Company during Baekeland's employment there in 1889–1890.[58]

The example of Henry Reichenbach, the first research chemist of the Eastman Dry Plate and Film Company, makes clear how consequential the difference could have been. In late 1891, Reichenbach and two of his coworkers were planning to start an enterprise of their own, building on secret processes that they had codeveloped at the Eastman company.[59] Because of these plans, however, George Eastman, "a vigorous advocate of company ownership of employee innovation," fired Reichenbach and "sought and received an injunction in the New York Supreme Court to stop Reichenbach, [Carl] Passavant, and [Gustave] Milburn from using or disclosing any of the trade secrets they had taken from the company."[60] Reichenbach had weakened his position by signing a contract that required him to transfer the rights to inventions made during his employment to the Eastman company, which was reorganized as Eastman Kodak in 1892. As legal historian Catherine Fisk has demonstrated, such clauses would hardly have been enforceable prior to the late nineteenth century, when American courts had consistently defended skilled labor's IP ownership rights. Through the rise and efforts of large, knowledge-intensive companies such as Kodak during the second industrial revolution, however, the judiciary became more responsive to calls to protect corporate investments in R&D.

Baekeland's employment relationship with Anthony supports Fisk's argument that this development shaped the career options and choices of entrepreneurially minded knowledge workers. In particular, from Baekeland's correspondence with his wife, it is apparent that the prospect of having to relinquish ownership rights to his inventions—not to speak of the potential profits resulting from them—encouraged him to leave the firm around late November 1890. Tellingly, Swarts, afraid that Anthony would exploit her husband, urged Baekeland to consult a lawyer before extending his contract in early autumn 1890. Nevertheless, Baekeland apparently failed to negotiate a satisfactory outcome: the new agreement included "a clause telling that all photographic patents [would] belong to them [Anthony & Company]."[61] To his employer's surprise, the chemist resigned less than two months after signing it.

Baekeland's experience also throws some light on the legal restrictions faced by technically skilled workers in Belgium, a topic that has not yet been studied in detail. As we have seen, Baekeland dissolved his partnership with the Guequiers in January 1890; that is, at the earliest possible moment. Stéphanie-Constance Gleesener-Duhayon and her lawyer then claimed that Baekeland should be, if not legally, at least morally obliged to assume liability for part of the firm's debts. As mentioned in chapter 3, Gleesener-Duhayon's counsel indicated that his client would be willing to compromise if Baekeland retired from the Belgian photographic plates industry for at least 20 years. While it is questionable whether this lengthy a restrictive covenant would have been legally enforceable, the very fact that it was considered a legitimate option for protecting investments is significant. Nine years later, George Eastman would make Nepera's executives sign a similar 20-year non-compete clause as part of his takeover of their company.[62]

Both Anthony & Company and Baekeland's former Belgian partners thus had good reasons for envying Nepera's success. Seen in this light, Baekeland's reluctance to disclose details of his photochemical research and his ambiguous statements about the invention of Velox take on a new meaning. If he had been more open, Anthony or Guequier and Gleesener may have been encouraged to claim part of his earnings. They may even have attempted to put Nepera out of business, as Eastman had preemptively done with Reichenbach's start-up.

To assess Baekeland's indebtedness to his former associates and employer, a closer look at the chronology of his photochemical investigations is required. A reading of the Belgian photographic press reveals that Baekeland already had been working with silver chloride emulsions in the late 1880s, after the establishment of Dr Baekelandt et Cie, in connection with his self-developing glass plates.[63] As Anthony had not been interested in manufacturing dry plates, Baekeland had then modified his formulas to make them suitable for use on celluloid films.[64] While these film emulsions had not been based on silver chloride, other research that he conducted during his employment at Anthony may have been directly related to his later work on the class of papers that were popularized under the "Velox" trademark. In effect, Céline Swarts would later state that her husband had left Anthony after having conceived of a silver chloride paper, had then formed Nepera, "and worked day and night at the birth of Velox." Approximately three months before Baekeland's departure from Anthony, moreover, Swarts had informed him that his oldest brother-in-law "has already tried… your paper."[65]

Although I have found no conclusive evidence that locates the invention of Velox in this period of time, it is crystal clear that (1) Baekeland departed from Anthony to pursue projects of his own choosing and (2) that his work at Nepera partly built on knowledge that he had previously gained as a salaried employee. As an example, in the most comprehensive photographic experiments that Baekeland conducted in 1892, he started from an emulsion formula that Anthony's Franklin Cossitt had used for making gelatin-emulsion "printing-out" paper—a category of paper that did not require the development of a latent image as photographic impressions became directly visible on it. Baekeland then systematically tested possible alternatives to Cossitt's formula, with the goal of creating a printing-out paper with a fully insoluble coating. He had to abandon those experiments because of his life-threatening appendicitis but continued this line of work after his recovery and fortunate encounter with Jacobi. It is therefore no coincidence that Nepera's first two products, advertised as "Rex" and "Nepera," were also gelatin printing-out papers.[66]

In addition, Céline Swarts's letters suggest that "the chromos affair," which most likely involved her husband's experience with dichromate printing processes, was a further avenue through which he counted on earning a living when he left Anthony around late November 1890.[67] Significantly, she considered telling her parents that the prospects for this affair were "so beautiful that you should absolutely file a patent [application] as quickly as possible… and that you have had to leave Anthony in order to file that patent [application]."[68] Swarts's letters unambiguously connect the decision whether to file for patent protection to IP ownership issues rather than to the amount of resources available to her spouse or the novelty of his inventive work. Ironically, this suggests that, until the 1899 sale to Kodak, Baekeland may have been at least as concerned about the divulgence of the Velox process to former colleagues who could have contested its legal ownership than to Nepera's direct competitors.

The Photographic Paper That Made Baekeland's Reputation

In his eight paper presentations at the applied chemistry conference in Berlin, Baekeland made nine different intellectual property claims. First of all, he presented himself as the inventor of Velox, although he did not put special emphasis on this achievement. More notably, he characterized the printing-out paper on which he had been working in 1892 as "the first gelatine paper with an insoluble film [i.e., coating] which could be toned and handled like albumen paper." In addition, he maintained that he had been the first person ever to have produced raw photographic paper in the United States instead of importing it from Europe; to have used refrigeration machinery to dehumidify air, facilitating the production of photographic paper under challenging climatological conditions; and to have developed reliable methods for quantifying the amount of silver contained in photographic paper and for predicting the permanency of different types of photographic prints. To continue this list of "firsts," Baekeland also stated that no one before him had drawn attention to the suitability of a mixture of two chemicals, sodium thiosulfate and alum, for adjusting the tints of photographic images. Finally, he demanded credit for having developed a chemical explanation for the effect of metallic particles contaminating sensitized photographic paper, as well as for providing insight into why latent images sometimes disappeared—a phenomenon that he designated "photoretrogression," a neologism.[69]

What are we to make of all these claims? As with the invention of Velox, the evaluation of several of them was, and still is, complicated by Nepera's heavy reliance on secrecy. Most notably, according to Baekeland's testimony in a later antitrust suit against Eastman Kodak, he and Jacobi had made painstaking efforts to prevent Nepera's employees, let alone company outsiders, from realizing that they had started using raw photographic paper manufactured in the United States. As Baekeland explained, they had done this to avoid negative publicity, for it was widely believed that there were only two places in the world—both of them located in Europe—where high-quality paper could be produced. Accordingly, Baekeland and Jacobi had agreed to store their US paper in boxes of those two European manufacturers, with whom

Eastman Kodak was trying to conclude a restrictive cartel arrangement. As another precautionary measure, Nepera's leadership began planning to diversify into the production of celluloid film, Kodak's most profitable product.[70]

From a business point of view, there were sound reasons for maintaining strict secrecy about such initiatives. Kodak, in particular, was a formidable competitor that did not refrain from using its market power to get rid of much smaller rivals such as Nepera. As a case in point, besides attempting to limit Nepera's access to raw materials, Kodak also requested that photographic dealers stop selling Velox.[71]

It was, moreover, not far-fetched to think that Nepera's employees might leak confidential information to Kodak or other firms in the industry. In fact, as early as December 1894, press coverage of patent-infringement suits initiated by Kodak prompted a money-seeking resident of Nepera Park, Yonkers, to promise George Eastman's company that "if at any time you wish to know of the manner on which a certain Paper House conducts the coating of 'their' Bromide Paper, … I can furnish you any information which you may require." While it is impossible to verify whether the author of this letter actually knew damaging facts, his offer indicated the emergence of a market for industrial espionage services meant to protect corporate IP. By the early 1890s, for example, several private detectives could afford to specialize in patent disputes.[72]

Notwithstanding such threats and legal constraints, as this chapter has shown, Baekeland was able to earn intellectual credit for his contributions to the development of photographic technology. Those rewards overwhelmingly concerned Velox, the class of papers that, as he put it toward the end of his career, had "made my reputation." The invention and/or commercialization of Velox was indeed Baekeland's only accomplishment that would earn its place in the historiography of photography, where Velox is rightly presented as a "gaslight paper" that enabled hobbyist photographers with daytime jobs to develop their images without the use of natural light.[73] Still, at least a few of his IP claims that were not directly related to this technology were also accepted by contemporaries. As an example, his findings regarding the occasional disappearance of latent images were quoted at great length in the *Journal of Physical Chemistry* and also found their way into Austrian, French, and German publications.[74] Moreover, besides being allowed to give so many talks, Baekeland had also been invited to chair one of the sessions on photographic chemistry at the Berlin conference.[75] This suggests that his ventures in the photographic industry had increased, rather than diminished, his standing in at least this part of the chemical profession.

To some extent, the recognition that Baekeland received seems to have resulted from the misperception that Nepera had been a one-man enterprise. Baekeland's personal dislike of chemist Albert Hahn, who had joined him and Jacobi as Nepera's third partner in 1894, doubtlessly contributed to his lack of generosity in acknowledging his indebtedness to collaborators. In a long diary entry triggered by George Eastman's tragic suicide in 1932, Baekeland emphasized that he and Jacobi had been anxious to get rid of Hahn's "depressing personality," adding that Hahn "had become a nuisance by his meddling and laziness since the first days he joined us." In laboratory notebooks from the 1890s, however, Baekeland had outlined photographic experiments carried out by Hahn in a way that suggests that they had been of value to him. Hahn, who had studied at Cornell University and the University of Freiburg in Germany, also worked on different developing solutions for Velox, including Nepera's popular "MQ" developer, based on metol and hydroquinone.[76]

A brief sketch of Nepera's growth since its creation with a capital stock of not more than $25,000 can also help counter the misguided idea that the firm owed almost everything to just one "heroic" inventor-entrepreneur. After having survived the difficult first four years of its existence, Nepera had begun to expand rapidly in the United States as well as in international markets. By the time of the Eastman takeover, the firm had established branch offices in Chicago and Paris, besides its sales agencies in Brussels, Dresden, and London. In addition, while new manufacturing facilities in Nepera Park, Yonkers, had only just been erected, further extensions were being contemplated.[77] Nepera had also begun to bring out its own magazine, *Photographic Life*, under the direction of Walter E. Woodbury, who it had lured away from the *Photographic Times*.[78] In short, what used to be a small start-up firm had become a much larger and more complex organization, depending on the services of a whole range of individuals. Three of those, Jacobi, Baekeland, and Hahn, had the privilege of dividing among themselves the $750,000 that Eastman eventually offered for the enterprise. However, it was only Baekeland who gained not only financial resources but also reputational benefits that, as the following chapters demonstrate, paved the way for a remarkably successful career as a scientific entrepreneur.

5 New Opportunities and Risks

As discussed earlier, Baekeland's former business associates at Nepera panicked about the possibility that the chemist would violate the terms of the sale to Eastman Kodak by presenting photographic research papers at the 1903 applied chemistry conference in Berlin. There are, however, no indications that Baekeland himself experienced the deal's highly restrictive provisions, including the requirement to stay out of the photographic industry for not less than 20 years, as a significant constraint. The available sources indeed demonstrate that, despite some motivational problems resulting from his having achieved his American dream of becoming a "self-made man," he was eager to move in new directions. According to a later recollection, for example, Baekeland was "glad to get out of this branch [photographic chemistry] which formerly when I was an amateur photographer was a delight [but] which later on developed in a bore, when it became commercialism with all its soul-deadening phases."[1] Consistent with this comment, he had declined a job offer that George Eastman had made in the aftermath of the takeover of Nepera.[2]

One of the projects that Baekeland took up instead in the following years, the examination of the reaction between phenol and formaldehyde, is discussed later in this book. This chapter instead concentrates on various other post-Velox lines of activity that, although they did not directly concern Bakelite, are also significant for understanding how the relationships between science and industry developed in the late nineteenth and early twentieth centuries and why Baekeland would come to be seen as the "father of plastics." First, I demonstrate that, by early 1902, Baekeland had already participated in the creation of three new corporations in his hometown of Yonkers: the American Retting Company, the American

Fireproofing & Cement Company, and the Continental Flax Company. None of these firms is mentioned in earlier accounts of Baekeland's career, partly because he carefully avoided referring to them in his public autobiographical reflections. As this chapter shows, however, they provide new insights into his motives and strategies as an entrepreneur and point to further transatlantic dimensions of a growing science-industry nexus.

At least from 1904 onward, Baekeland also established himself as a chemical "expert" offering a comprehensive range of consulting services to a diverse group of industrial clients.[3] This included the provision of scientific expert testimony in lawsuits, a task that he would also perform multiple times in relation to Bakelite. In both Britain and the United States, scientists' role as courtroom witnesses had come under public scrutiny during the previous decades. This was because corporations seemed to have little trouble identifying experts who, as suggested by press accounts highlighting their apparently contradictory testimonies, seemed to subordinate the search for truth to the search for profit. As one historian has put it, "The courtroom experience of being hired to support a partisan case could exact a heavy toll on the dignity and reputation of scientific witnesses."[4] Baekeland's integrity, however, was not questioned in light of his testimony on behalf of one of his major patrons, the Development & Funding Company, an electrochemical firm.

Serial Entrepreneurship

Baekeland's participation in the formation of at least three new firms in the very first years of the twentieth century raises doubts about the characterization of him as a reluctant entrepreneur in several earlier studies. According to this interpretation, he would have preferred to spend as much time as possible in his private research laboratory in the barn of his family's newly purchased riverfront mansion in Yonkers but was forced to take up various business activities to control the commercial introduction of his inventions and to prevent others from appropriating his work.[5] Even when it came to Bakelite, as later chapters of this book will show, Baekeland's motives were more complex and multifaceted than this. But in the cases of the American Retting, Continental Flax, and American Fireproofing & Cement companies, this explanation did not apply at all, as these firms were not even meant to commercialize the chemist's own inventions.

Baekeland's involvement in these corporations is also relevant to this study because it provides additional examples of how late nineteenth- and early twentieth-century inventors, scientists, and businesspeople on both sides of the Atlantic collaborated to introduce new technologies. More precisely, like many other immigrant entrepreneurs, Baekeland drew on knowledge that he had acquired, as well as on contacts that he had made, in his country of birth.[6] In fact, there is strong evidence that his involvement in all these firms resulted from friendships and family ties that dated back to his years at the University of Ghent.

Two of the enterprises, American Retting and Continental Flax, intended to fundamentally transform the processing of linen and flax. Baekeland's native city of Ghent had long been a world center of this industry. Accordingly, it had occupied a prominent place in the curriculum of the vocational school that he had attended as a teenager. However, Baekeland's knowledge in this area appears to have mattered less than his acquaintance with one of the initiators of the firms, his wife's cousin Honoré De Swarte. In addition to being a relative of Céline Swarts, De Swarte had been the treasurer of the liberal student society 't Zal Wel Gaan, of which Baekeland and most of his closest Belgian friends had also been members. Moreover, as late as 1927, Baekeland described De Swarte as a "splendid" man "who stand[s] vividly in my memory of our student years," confirming that he had personally known the merchant and chemical manufacturer from the maritime city of Nieuwpoort.[7]

De Swarte and Georges Loppens, a Ghent-trained engineer whom Baekeland may likewise have met as a university student or assistant, had patented a "process of retting or steeping flax or hemp" in Belgium and multiple other countries around the world, including America.[8] On behalf of a business consortium that included De Swarte, Loppens then spent several years in the United States to promote and further develop the technology. By November 1900, Baekeland had become involved in these efforts, as he and Loppens established the American Retting Company in Yonkers.[9]

Baekeland's correspondence with his wife shows that he was convinced of the commercial potential of Loppens and De Swarte's invention and therefore willing to risk part of his newly acquired money to support its commercialization. By February 1902, however, he had become more cautious. As he explained to Céline Swarts, he was trying to build a solid foundation for the expansion of the enterprise while limiting his own future responsibilities and securing substantial compensation for the support that he had already provided. To accomplish these goals, Baekeland considered it desirable to incorporate a new firm in which he played a more secondary role. He was able to raise funding from some wealthy acquaintances familiar with the track record of Nepera, including several relatives of his former partner Leonard Jacobi.[10] As a result, the Continental Flax Company could be established with capital of $200,000. As Baekeland calculated, this new firm would have to

> give $75,000 worth of stock to the American Retting Co. Of these $75,000 I first request $19,000 in stock for myself in addition to the 1/2 [of the remaining amount] to which I am entitled, which means that Loppens and De Swarte each will receive $14,000 worth in stock and I $28,000 + $19,000 [i.e., a total of $47,000]. With those 19,000 I have to reimburse Seligsberg [i.e., the investment manager Franklin L. Seligsberg, who had also put money in American Retting].... I find that the $15,000 that will be returned to me in addition to the $28,000 are a small compensation for my moral responsibility and all the effort I have put into this affair. If my friends had not known me and had not believed what I had told them with respect to the undertaking, it would not have been possible at all to find subscribers.[11]

Despite the size of these numbers—the total sum that Baekeland was expecting to receive would have amounted to well above a million in today's dollars—American Retting and Continental Flax on the whole do not seem to have generated any additional wealth for him. In fact, notwithstanding his success in bringing in extra financiers, the venture seems to have come to an end by 1903. By that year, Loppens had permanently returned to Belgium, and there are no indications that Continental Flax remained active, let alone profitable, in the absence of this inventor-engineer.[12] Indeed, in the following decade, Baekeland would refer to losses resulting from his association with Loppens as one of the reasons why he had decided to make his then 18-year-old son George aware of the seemingly ever-growing financial costs of his adventurous pastimes.[13]

Considering the nature of the arrangements outlined here, the other investors are likely to have fared even worse. Moreover, Baekeland had convinced Leonard Jacobi and several other financiers to support not only Continental Flax but also an additional firm that would turn out to be a failure. This enterprise, the American Fireproofing & Cement Company of Belgian American engineer Adolphe De Man, had been established in 1901. Baekeland's concerns about its state were a major reason behind his decision not to join his wife and two children during their stay in Europe early the following year. Throughout February 1902, he occupied himself with the Loppens and De Man "affairs," being particularly concerned about the latter. For instance, he wrote to his spouse that, if she were to meet Jacobi and his family in Italy, she should by all means avoid mentioning "to them or anyone else my fears concerning our projects because that could do me much harm. These fears concern De Man, not Loppens, but in such cases the difference gets overlooked."[14]

As with De Swarte, Baekeland's acquaintance with Adolph De Man, as well as with Adolph's older brother Florimond, dated back to his years in Ghent. Baekeland's correspondence with his wife reveals that Swarts's family had stayed in touch with the De Man brothers after the latter's migration to the United States in the 1870s. When Florimond De Man temporarily returned to Ghent in 1888, for example, he and his fiancée were invited to stay at the home of Swarts's parents. On this occasion, Florimond De Man also expressed an eagerness to see Baekeland again, which suggests that he had known the then 25-year-old chemist from a young age.[15]

From the early 1880s until the end of the century, Florimond and Adolphe De Man had jointly run an engineering construction business in Detroit. Around 1901, however, Adolph sold his interest in the firm to his brother and moved from the Midwest to the Northeast to introduce what he promoted as "the De Man System of fire proofing and concrete construction" in that part of the United States.[16] By May 1901, Adolph De Man, Baekeland, and the latter's financial advisers, A. F. and F. E. Seligsberg, had incorporated the American Fireproofing & Cement Company, of which Baekeland became the treasurer as well as a director.

The little information available on the history of the American Fireproofing & Cement Company points to a business model centered on intellectual property that was notably similar to the one that Baekeland would later pursue with respect to Bakelite. In particular, just as Baekeland would grant patent licenses to various manufacturing firms to develop applications of the synthetic plastic, American Fireproofing & Cement aimed to make money by licensing individual or several components of De Man's system of concrete reinforcement structures to construction firms. Trade journals that discussed De Man's inventions (for instance, his patented tension bar) were generally positive about them, and Baekeland's personal correspondence shows that he also thought favorably of at least parts of the technology. On 20 February 1902, for example, he wrote that De Man's "latest system of block floor construction is an excellent thing and I do not doubt that it will prove useful." By this time, the firm also seemed close to receiving several orders.[17] However, Baekeland—as well as De Man himself—may have underestimated the amount of competition that the American Fireproofing & Cement Company would face from other firms in the industry that were following precisely the same business model. As historian Amy Slaton has indicated, the first decade of the twentieth century witnessed a proliferation of enterprises licensing patent-protected technological "systems" meant for concrete reinforcement. Some of these were widely adopted; the "Kahn system" of architect Albert Kahn's brother Julius, for example, had been used in well over 1,500 buildings by 1907.[18] By this time, in contrast, Baekeland had withdrawn himself from De Man's enterprise on account of its disappointing performance. Still, in mid-August 1907, he performed a final action on behalf of the firm: even though he was no longer affiliated with American Fireproofing & Cement, he agreed to sign "an affidavit…for a petition to obtain injunction on some contractors who are infringing De Man patent." According to his diary, he did this as a personal favor to De Man, who was suffering from a "hopeless illness" that had rendered him unconscious, preventing him from taking care of the paperwork.[19] As it turned out, De Man died before the end of the month. Baekeland would never again enter the reinforced concrete construction industry, one of the sectors of the American economy that had begun to employ a substantial number of academically trained scientists and engineers.[20] He and especially his wife, however, would stay in touch with the widow of their fellow Belgian American in the following months and years.

Chemical Expert and Consultant

By supporting De Swarte, Loppens, and De Man in their business ventures, Baekeland had taken on a financial risk of a kind commonly faced by early-stage technology investors. In addition, like other entrepreneurs dependent on the trust of their business associates, he had been concerned about harming his personal relationship with the Jacobi family and other professional contacts he had convinced to invest in the Continental Flax Company.[21] As these enterprises had not generated the hoped-for benefits, Baekeland's experiences with them may well have made him more cautious. There are, in any case, no indications that he participated in the creation of additional firms before the foundation of a European and an American Bakelite company in May and October 1910, respectively.

At least since 1904, however, Baekeland did advertise himself as a chemical "expert" and performed a whole range of consulting activities for industrial clients, taking on risks specifically concerning his reputation as a scientist. As several historians of science, technology, and medicine have demonstrated, by the late nineteenth century the term "expert" had become tainted because of contradictory trial testimonies of "mercenary" scientific witnesses, who were widely believed to be sacrificing truth to profit.[22] In the decades prior to World War I, Anglo-American courtrooms were one of the settings where the otherwise mostly positive public image of science was being contaminated. In England and the United States, scientific expert witnesses were engaged by either one of the opposing parties, instead of by the (supposedly impartial) court, as was common in continental Europe. And the other party's lawyers would typically leave no stone unturned in their attempts to undermine these witnesses' authority and credibility in cross-examination. Dissatisfaction with this adversary system was—and still is—widespread. However, as Tal Golan has indicated, nineteenth-century American scientists were not powerful enough a community to challenge it.[23] Nor were they unified. For instance, some preferred not to testify altogether, particularly in cases where, by scientific standards, the available evidence was inconclusive. (By legal standards, judgment always had to be passed at the end of a suit.)[24] But others, including various chemists, regarded such lawsuits precisely as an opportunity to gain a better insight into problems relevant to their discipline. In addition, there were chemists and other scientists who saw performances as expert witnesses as chances to highlight their knowledge to a broader audience or simply to supplement their income.

Some of the same trade-offs also applied to other consulting services that a rapidly growing number of scientists and engineers offered to industrial firms in the late nineteenth and early twentieth centuries, including research advice and product evaluations. For example, in the late nineteenth-century United States, a "gilded" society shaped by concerns about financial and political corruption, advocates for "pure" science did not just denounce seemingly unscrupulous courtroom expert witnesses. Rather, physicist Henry Rowland and others who defended the value of pursuing knowledge for its own sake more generally condemned professional colleagues whose work appeared to be driven by for-profit considerations.[25]

Baekeland, who had arrived in the United States during the Gilded Age and whose diaries include condemnations of various railway speculators and other "robber barons," on several occasions expressed himself critically on this issue. In 1915, as an example, he deplored chemists and other scientists who seemed "ready to sell their services to anybody who has a temporary use for them, irrespective of the underlying motives or purposes."[26] Sometimes he specifically referred to the problem of expert testimony. As a case in point, in a 1928 letter to patent lawyer Edwin Prindle, he predicted that there would be "little improvement possible as long as experts are readily to be found who, for defending their clients, are willing to twist the truth, not by direct lying, but by confusing qualitative with quantitative arguments and thus create confusion and misunderstanding instead of enlightening the Judge."[27] But in other instances he more generally highlighted tensions associated with the emergence of a science-industry nexus. In a 1907 speech, for example, Baekeland repeated a concern previously expressed by Rowland and other "pure science" campaigners. In his own words, "the fact that scientific learning has found unceasing applications in the production of wealth has fostered the constantly increasing tendency for finding in scientific education or scientific pursuits a mere means of earning a living or making money. This has led into scientific vocations a large number of persons who in their profession expected to find a quick and easier means for making a living, but outside of this stimulus possessed few if any of the qualifications of the true scientist."[28]

In light of these concerns, it is pertinent to ask how Baekeland himself sought to maintain his professional integrity in the context of his industrial consulting work. His main client was the Development & Funding Company (D&F), one of several electrochemical firms in the Niagara Falls area that were created as independent research and development (R&D) entities.[29] When established in 1903, the aim of D&F had been to invest in the development of promising new technologies up to the point where other firms could start introducing them on a large commercial scale. As it turned out, however, D&F ended up financing only one invention: a new electrolytic process for manufacturing two basic chemicals, sodium hydroxide (i.e., caustic soda) and chlorine, out of salt. This technology had been created by Clinton P. Townsend, later Baekeland's patent agent, in collaboration with Elmer A. Sperry, one of the country's most prolific inventors. After having made arrangements with Townsend and Sperry, D&F established a small plant in Brooklyn, New York, to conduct initial experiments with two model copies of the "Townsend cell," the basic unit of the electrolytic process. In 1905, it also constructed a larger factory in the city of Niagara Falls, close to the hydroelectric power station that provided multiple chemical firms in the area with a relatively cheap source of energy. D&F ran this plant until fall 1909, when the commercial value of Townsend and Sperry's process was considered sufficiently demonstrated. Accordingly, the organization was dismantled and its Niagara factory became the property of a successor firm, the Hooker Electrochemical Company.[30]

Baekeland's consulting activities for Elon H. Hooker, the politically well-connected engineer who led both D&F and Hooker Electrochemical, started in early 1904 and continued for over a decade. In other words, Baekeland would maintain his affiliation with Hooker even after his own work on the development and commercialization of Bakelite plastic started taking up much of his time.

What explains why Baekeland's professional relationship with Hooker lasted much longer than his association with the firms of Loppens and De Man? In the following paragraphs, I argue that this was because his consulting work was of a mutually beneficial nature, despite the reputational risks involved. I also show that Baekeland's largely positive experiences with Hooker's firms contrasted with those of their blue-collar factory workers, who had to labor under what even a company-friendly history characterizes as "appalling" conditions.[31] Baekeland's productive partnership with Hooker also differentiated him from local residents living close to the industrialists' Niagara plant, as several of them took Hooker to court to get compensated for damages that they claimed to have suffered because of repeated releases of chlorine gas.

A Science-Industry Partnership

The need to ascertain the commercial viability of Townsend and Sperry's process was Hooker's most immediate reason for adding a practically oriented chemist such as Baekeland to his team of consultants. In 1904 it was, after all, far from certain that the Townsend cell would be able to compete with the mercury and diaphragm cells that had been designed by multiple earlier inventors and firms who had taken up the electrolytic production of caustic soda and chlorine since relatively cheap electricity first became available.[32] To make this possible, Townsend and Sperry had focused on improving the energy efficiency of their process and on preventing the recombination of the caustic soda and hydrogen formed at the "cathode" electrode on one side of the cell with the chlorine created at the "anode" electrode on the other side.[33] However, as with Baekeland's photographic inventions, meeting these challenges at the laboratory level did not guarantee that they would not reappear in the process of scaling up the technology. In this respect, as Sperry explained to Elon Hooker in February 1904, it was Baekeland who had helped identify "the great question before us": "what will our commercial cell do under long continued action—what los[s]es are there if any in efficiency, what is the cost of upkeep"?[34] The trials in the Brooklyn plant seemed to suggest a negative answer: the anodes of the model Townsend cells, for example, suffered from strong corrosion. However, like Townsend and Sperry, Baekeland believed that this was not a reason for giving up, as the problems that had come up seemed solvable by means of further technical adjustments, informed by the systematic study of all relevant variables during a prolonged pilot-plant stage.[35] In this regard, Baekeland did not just advise D&F; in the following years, he also personally contributed several inventions, including a number of changes to the design and composition of the percolating diaphragm that separated the two compartments of the Townsend cell.[36] He thus helped Hooker and the other investors recover the costs of financing what he would later characterize as the initial, small-scale "blunders." Moreover, D&F and especially the Hooker Electrochemical Company could now start making "profits on a large scale," as they had no difficulty finding outlets for their competitively priced caustic soda and chlorine-based bleaching powder.[37] Hooker Electrochemical would indeed become the country's largest manufacturer of both these products.

When engaging Baekeland, Elon Hooker may also have anticipated that the chemist could serve as a mediator between academia and industry in a field where academic knowledge and credentials were becoming increasingly important. While American and especially European industrial electrochemists benefited from advances in physical chemistry, they did not draw as heavily on scientific theory as some manufacturers of synthetic organic chemicals were doing in the same years. However, in and beyond the US context, electrochemistry was a domain where new approaches to the organization of R&D work had been developed in small academic start-up firms as well as in giant corporations such as General Electric, which decided to establish a new kind of electrochemical research laboratory in 1900.[38]

The study of the relationships between electrical energy and chemical change was, to be sure, not a subject area that had been covered comprehensively in Baekeland's vocational and academic training in Ghent in the late 1870s and the 1880s. Shortly after the sale of Nepera, however, Baekeland had decided to spend part of a long and otherwise leisurely stay in Europe at the electrochemical laboratory of the Technische Hochschule Berlin-Charlottenburg. Unfortunately, there are no records available that reveal which topics he studied during his stint at this internationally acclaimed higher education institution, which had begun to offer separate courses in applied electrochemistry several years before the first American universities started doing the same. It is, however, clear that the electrolysis of sodium chloride (i.e., salt) was among the areas of expertise of the head of the Elektrochemisches Laboratorium, Georg Karl von Knorre.[39] Moreover, diary notes that Baekeland took before von Knorre's premature death in 1910 demonstrate that he had been personally acquainted not just with this chemist and several other members of his lab but also with personnel of various other scientific and technological institutes at the Technische Hochschule.[40] This included professors who had not yet been appointed when Baekeland first visited the establishment in the late 1880s, therefore suggesting that he may have gotten to know them in either 1900–1901 or during the applied chemistry conference of 1903. In any case, his contacts with German academics would prove valuable not just in relation to Bakelite, as discussed in chapter 7. They were also useful to Elon Hooker's enterprises as both D&F and the Hooker Electrochemical Company tried to gain additional revenues by selling patent licenses covering the Townsend cell to German and other international firms.[41]

Partly thanks to his role in New York–based associations such as the Chemists' Club, which he led in 1904, Baekeland also had close informal ties with professors at Columbia University and nearby higher education establishments such as the College of the City of New York. This is significant because Columbia was part of a still limited number of US universities that had become intimately involved in the development of the electrochemical industry. Unlike MIT and Cleveland's Case School of Applied Science, which had close ties with General Electric's research laboratory and the Dow Chemical Company, respectively, Columbia's presence was especially visible in smaller enterprises in the Niagara Falls industrial district and elsewhere in New York State. There, for example, former students of

Columbia had made major contributions to the development of electrolytic manufacturing processes for products ranging from aluminum to zinc.[42] Moreover, even before the university introduced several applied electrochemistry courses in the years after 1900, various faculty members were maintaining direct links with some of these firms. Several of the university's chemists and electrical engineers were, for instance, affiliated with the Ampere Electrochemical Company, an independent R&D entity that tried to sell or license new electrochemical products and processes to manufacturing firms.[43] (Nathaniel Thurlow, who would later make important contributions to the creation of the phenolic resin industry, also worked for Ampere.) Here again, Baekeland would rely on these academic connections with respect to both Bakelite and his consulting activities for Hooker. Around early 1908, for example, he convinced chemistry professor Charles Baskerville to testify on behalf of D&F in a nuisance suit regarding the emission of chlorine gas. In the previous year, he had introduced Elmer Sperry, who had also remained affiliated with Hooker's firm as a consultant, to another academic chemist, Charles Pellew of Columbia.[44]

Interactions between academic and industrial electrochemists, engineers, and entrepreneurs were also fostered by the American Electrochemical Society, which had been created in 1902. In the words of Sperry, the association's gatherings brought together "the brightest minds we have in this country in this line, and those at the same time that represent the heaviest [economic] interests in this field."[45] Baekeland joined the society in 1903, the year before he became affiliated with D&F. He then participated in the discussion of six research presentations at the association's annual meeting, including papers by Sperry and prominent academics Wilder Bancroft and Walther Nernst.[46] His reputation among the society's membership would grow rapidly in the following years: as early as 1909, he was elected president, without having previously served the association in a lower function. In his diary, he described his nomination for the presidency as "a misguided effort of my friends," given that "there are so many more deserving members who are entitled to the honor."[47] However, notwithstanding his initial reluctance, he ended up accepting the honor, which also increased his value as a consultant to D&F and the Hooker Electrochemical Company.

What did Baekeland, for his part, get out of the science-industry partnership with Hooker? Some benefits are only evident with the benefit of hindsight. In particular, when he began consulting for Hooker, Baekeland could not yet have anticipated that his electrochemical experiences and contacts would shape his approach to controlling the reaction between phenol and formaldehyde and to commercializing the resulting synthetic plastic. Baekeland must have been aware, however, that a position as a consultant would give him more independence than a job as a salaried employee, while also providing him with a relatively stable source of income. Moreover, the position enabled him to examine large-scale electrochemical and chemical engineering problems that he could not possibly study in his own private laboratory and gave him access to industrial knowledge and know-how that would have been hard, if not impossible, to derive from the available published literature.

"A Scientist of Undoubtful Reputation"

On the face of it, the mutually beneficial nature of Baekeland's partnership with Hooker is also evident in the expert testimony that he gave in *Gasse v. Development & Funding Co.*, a case that started in 1907. *Gasse v. D&F* was one of multiple lawsuits initiated by residents of the Niagara Falls area who charged that local electrochemical firms, most frequently the Acker Process Company, had caused various nuisances by emitting chlorine gas. As part of his testimony, Baekeland indicated that he held some stock of D&F, acknowledging his financial stake in the outcome of the litigation. However, despite his previously quoted comments on the tension between "true science" and money making, he does not appear to have had any moral reservations about defending Hooker's interests in the courtroom. In fact, he seems to have considered his legal expert testimony on behalf of D&F as a logical extension of his earlier consulting services.[48]

The original transcripts of Baekeland's testimony in *Gasse v. D&F* have not been preserved, but a later copy is available, as D&F's counsel Morris Cohn Jr. was allowed to reread relevant passages in a closely related nuisance suit initiated by another local property owner in 1909.[49] The fact that D&F reused much of the testimony is in itself significant, as it suggests that the firm considered Baekeland's statements to be an important part of its defense. The content of his remarks seems to justify this strategy: he insisted on three different points that were central to his client's argumentation in both lawsuits. First, he questioned George Gasse's evidence that D&F, rather than any of the other electrochemical firms located in the same area, was responsible for the vegetation damage and other nuisances on which the case hinged. In this respect, it is clear that Baekeland was not just following the advice of the firm's attorneys, whom he had repeatedly consulted before the start of the trial. He was also relying on independent observations that he had made in the context of his visits and inspections of D&F's Niagara factory. In July 1907, as an example, he happened to be staying close to the plant during a sequence of days when the firm was alleged to have released chlorine into the open air. Baekeland, however, felt confident that "no serious amounts of gas could possibly [have] escape[d] without I being aware of it." Moreover, he instead smelled a chlorine-like odor that seemed to stem from the works of the Roberts Chemical Company, a producer of hydrochloric acid and caustic potash located opposite D&F.[50] Based on this experience, he argued more generally that Gasse—and, by implication, other local residents—may have been aiming at the wrong target.

A second and related line of defense developed in Baekeland's testimony was that it was hard, if feasible at all, to determine that the plaintiff had actually been exposed to chlorine. "Even for the best trained chemist," Baekeland emphasized, it is almost impossible to differentiate the smell of chlorine from the one of bleaching powder or hydrochloric acid gas, if both chlorine and hydrochloric acid "are in very dilute form."[51] Moreover, hydrochloric acid fumes could also cause vegetation damage, corrosion, and irritating health effects of the kind described by Gasse. They had indeed been a major target of earlier antipollution efforts in both Britain and continental Europe.[52] In this connection, Baekeland also stressed that it was in D&F's own economic interest to prevent any release of chlorine, as it wanted to produce its bleaching powder in the most efficient way and at the lowest possible cost. While acknowledging that occasional accidental discharges had occurred in the plant's initial phase of operation, he explained how D&F had improved its methods for absorbing the chlorine by letting it react with calcium hydroxide in separate bleach chambers.

Finally, as part of an alternative legal strategy pursued by D&F, Baekeland argued that if D&F's chlorine gas had reached the plaintiff's property at all, it would have been too diluted to cause the alleged harm. Among other things, Baekeland described how he had personally tested the impact of very-low-dose chlorine by sleeping one night in a room that contained "a six-thousand part of one grain of chlorine diluted in one cubic foot of air."[53] In this experiment, which he had conducted in his own house, he had experienced neither any harm nor any of the therapeutic effects that some chemists and doctors would attribute to low-concentration chlorine exposure in the aftermath of World War I (see figure 6.2). Baekeland also critically evaluated a test described by one of Gasse's expert witnesses, who had found that guinea pigs could only survive for about an hour and a half in a room filled with chlorine gas. In Baekeland's view, the conditions of this experiment had been too extreme, as this witness, physician Chester Campbell, had used a high concentration of gas in a closed space. The chemist thus implied that humans would only experience serious harm if they were treated in a similarly drastic way. As he put it himself, "In order to kill a man by chlorine gas I think you would have to lock him up in a room and make his escape impossible and treat him with chlorine about the same way you would treat a cat if you want to drown him, and make escape impossible."[54]

As discussed in the next chapter, these types of claims would prove highly controversial in the wake of World War I, when the American chemical profession had to justify its contribution to the creation of poison gases such as chlorine and mustard gas. In contrast, Baekeland's testimony for D&F does not seem to have attracted any criticism outside the courtroom. In fact, even in the court, his integrity, contrary to that of D&F's other witnesses, apparently was not questioned. According to an April 1908 diary entry, the prosecuting attorney

> used inflated language, yelling and striking and all artifices of Rhetoric[,] called Baskerville [Charles Baskerville, the chemistry professor Baekeland had engaged on behalf of D&F] "the Young wise man from New York who knew a lot and seemed to be aware of it." In the same way he did up all our witnesses referring sneeringly to them and I was prepared to be treated in the same way but when he came to my testimony he made several very complimentary remarks[,] called me "an honest, straightforward witness and a scientist of undoubtful reputation" and similar things[. H]e only said against me that even I may be mistaken on some particulars which I called impossible, that things which had been called impossible in the past had more than once proved possible with the advance of knowledge.[55]

Even before he received these compliments, Baekeland had learned that D&F's lawyers were so satisfied with his performance that they had decided "not to dispel good effect of my testimony by new witnesses and therefore announced that they were going to abandon the right of presenting further witnesses." As he continued, "This took the prosecution by surprise because they expected we were going to bring forth many more witnesses."[56]

Despite all this, Baekeland's legal testimony also points to the limitations of his otherwise mutually helpful partnership with D&F. For example,

Baekeland experienced the proceedings as a sometimes frustrating waste of time rather than a stimulating exchange of ideas and perspectives. More specifically, he expressed his disgust at attorneys who "do not want the truth. They only want to mix up witnesses, never mind how, as long as it advances their own vanity."[57] D&F, moreover, appears to have overestimated the commercial value of Baekeland's scientific knowledge and authority in this legal context, as the chemist's main claims were accepted in neither *Gasse v. D&F* nor *Simmons-Baker v. D&F*, the second case in which his testimony had been used. As it turned out, the Niagara County branch of the New York State Supreme Court concluded that D&F had in fact frequently discharged chlorine gas "into the open air in large quantities," creating a nuisance to George Gasse. Accordingly, D&F lost this suit and was ordered to pay Gasse the relatively small sum of $702 in damages. Hooker's firm was, moreover, also unsuccessful in appealing a more punitive ruling in the case initiated by James Simmons of the Simmons-Baker real estate company, to whom it had to give more than $5,000.[58]

Finally, and most significantly, D&F's involvement in nuisance litigation points to the environmental and health costs associated with the rise of one of the most cutting-edge branches of the pre–World War I American chemical industry. Electrochemical manufacturers did not deny that their industry had contributed to the transformation of the city of Niagara Falls into a "modern Manchester," marked by trees and grass destroyed by profitable but toxic chemicals (see figure 5.1).[59] As Baekeland's testimony illustrates, they simply tried to shift the blame to other firms in the same industrial district. Moreover, they resisted attempts to introduce legislation modeled after the British Alkali Acts (1863–1906), an early but limited regime of pollution regulation created in response to nuisances similar to those experienced by Niagara Falls residents.[60]

As regards the health consequences of being exposed to chlorine and related chemicals such as hydrochloric acid, the situation of the blue-collar workers employed by firms such as D&F, the Roberts Chemical Company, and the Acker Process Company was particularly bad. Their experiences, as documented in legal records and in a 1913 report of the Factory Investigating Commission of New York State, implicitly challenge the claim of D&F's leadership that inhaling chlorine gas was "hardly ... a danger in the manufacture of bleaching powder because I have never known of any men injured by it." Most relevant, D&F's bleach packers recounted their

Figure 5.1
Factory of the Acker Process Company in 1903 or 1904. Acker was the first firm in the city of Niagara Falls to electrochemically produce caustic soda and chlorine from sodium chloride and was flooded with lawsuits by local property owners exposed to chlorine and lime dust emissions. In addition to paying damage fees, Acker was legally forced to adopt technological solutions meant to reduce the nuisances it had been causing, including the construction of a chimney of 240 feet.
Photograph reproduced from American Institute of Electrical Engineers, *The Niagara Falls Electrical Handbook.*

memories of getting "gassed." They used this term to refer to instances where they or some of their co-workers had been overwhelmed by coughing fits resulting from chlorine's irritating effect on their throat and lungs. In such cases, they had to temporarily discontinue their work and run into the open air to recover. Clearly, the rudimentary muzzles and goggles on which they relied to protect their faces against the chlorine gas present in the bleach chambers did not suffice to prevent this from happening. It was not until shortly before World War I that the Hooker Electrochemical Company provided them with proper gas masks and better sanitary facilities.[61]

In his courtroom testimony, Baekeland supported D&F's claim that there was no evidence of any permanent damage to the health of these workers. Yet, he privately expressed his disapproval of the working conditions in the bleach chambers and especially in the room where the Townsend cells produced the initial chlorine gas. In early 1908, the latter space was so filled with gas that Baekeland wondered about "the use of having cheap bleach and cheap caustic at such a [human] price." He also noted that this was "undoubtedly a case where the labor inspector should intervene and compel to obtain proper conditions or shut down."[62] According to his diary, Baekeland also defended this opinion in a report to the leadership of D&F. However, like other chemical manufacturers consulted by the Factory Investigating Commission, Elon Hooker and his brother Albert strongly opposed the introduction of workplace safety and health regulations, arguing that government interventions of this kind were unnecessary and would hinder the "personal liberty" of their employees.[63] There are, moreover, no indications that Baekeland brought up this issue again once the worst excesses had been addressed. American workers' exposure to chlorine would, in any case, soon be overshadowed by the use of poison gas in World War I. As we will see next, Baekeland also sided with chemical industrialists and other members of the US chemical profession in the controversy about chemical warfare. The "Great War," moreover, also helped accelerate his rise to prominence in this community in several other ways.

6 A Technocrat on the Public Stage

It is important not to overstate the significance of World War I in the development of a comprehensive science-industry nexus. As previous chapters of this book have shown, the decades before 1914–1918 had witnessed the growth of industrial research facilities, the intensification of interactions between academia and industry, and the creation of various science-related fields of manufacturing in both Europe and the United States. At the same time, the accelerating effects that the war had in these areas were another crucial factor shaping Baekeland's options for commercializing Bakelite and for being recognized as the inventor of this synthetic plastic. Most notably, as a member of two newly established advisory bodies, the Naval Consulting Board (NCB) and the National Research Council (NRC), Baekeland broadened his network of academic, industrial, and government contacts and was exposed to different approaches to research and development work. Partly thanks to his participation in these wartime organizations as a representative of the chemical profession, he also became one of the few US chemists known to a broader audience. As Thomas Edison, the head of the NCB, understood so well, media attention and public fame could be exploited for commercial and intellectual property purposes. Overenthusiastic press accounts, for example, had contributed to the erroneous perception that "the wizard of Menlo Park" had almost single-handedly invented a whole system of electrical technologies.[1] Baekeland, however, was more ambivalent about what he wittily described as being "placed among the celebrities and criminals nowadays" in a diary entry from spring 1915.[2] On the one hand, he was eager to correct misconceptions about the state of chemistry in the United States and, more generally, to promote his sometimes controversial views on the value of applied science and technology to a larger public. On the other hand, he remained wary of reaching out to a mass audience in a way that might seem sensationalist to other scientists and engineers.

As this chapter's next section discusses, the opinions of the scientific—and particularly chemical—community and the general public diverged the most strongly on the evaluation of chemical weapons. Although Baekeland had previously expressed pacifist views, as the 1924 president of the American Chemical Society (ACS) he joined his professional peers in supporting the peacetime continuation of the Chemical Warfare Service. On other issues, Baekeland defended views that also proved to be more divisive among chemists and other scientists. In particular, his continued questioning of the practical value of scientific theorizing may have contributed to the failure of his initial nominations to the US National Academy of Sciences (NAS), despite his presence on the research council established within the academy in 1917. In addition, Baekeland's claims regarding the superior sociocultural value of science and technology over literature and art aroused strong objections in a German chemistry journal. None of these opinions, however, appears to have hindered his further rise to prominence within the US chemical community. Indeed, his colleagues at Columbia University's Chemical Engineering Department, which he joined toward the end of the war, seem to have been amused by his repeated warnings against "starry-eyed theorists." Moreover, if anything, Baekeland's public fame and occasional eccentricity also fostered his election as president of the ACS at a time when this society was more often than not led by chemists with a "purer" profile.[3]

Chemical Warfare

As we saw in chapter 5, the problem of expert testimony continued to be discussed in the decades after World War I. However, in the context of the war and its aftermath, the relationship between science and politics had become a more pressing issue than that between science and the law. More specifically, on both sides of the Atlantic, the unprecedented level of involvement of scientists and engineers in wartime policy making and preparedness efforts had given a boost to technocratic ideas about government by experts. Along with other "rational reformers," Baekeland held that scientific methods could be extended relatively straightforwardly to the realm of governance, doing away with the partisan "sophistries" of "lawyer politicians."[4] Accordingly, he was unambiguous in his support for the likes of Herbert J. Hoover, the

"Great Engineer," who had been put in charge of the Commission for Relief in German-occupied Belgium. In November 1915, for instance, Baekeland approvingly "referred to that job of engineers, the Belgian Relief Commission[,] which was exclusively managed by engineers and was making a monument to the country."[5] And in October 1919 he wrote that he considered Hoover, whom he would later come to revere for his "apolitical," "practical" idealism, "the greatest single civilian individual in the history of this war."[6] (Conversely, in 1925, Hoover, then Secretary of Commerce, appointed Baekeland to the Chemical Advisory Committee to the Department of Commerce. Baekeland would serve on this committee for eight years.)[7]

For obvious reasons, World War I also witnessed a militarization of science and technology, and chemistry was at the forefront of this development. On 22 April 1915, for example, the German Army deliberately released chlorine gas on a large scale in the Belgian city of Ypres, crossing a new threshold in a millennia-long history of poison warfare. In notable contrast to the controversy that followed the end of the war, the event initially went largely unnoticed in the United States. Baekeland, moreover, was no exception in this regard, despite his earlier criticism of the exposure to chlorine of Hooker Electrochemical Company workers. However, this is not to say that Baekeland was not concerned with how chemistry might help break the stalemate of trench warfare in his native country and other parts of Europe. In fact, between the beginning of the war and America's entrance in April 1917, he was lecturing throughout his adopted country on the subject of chemical warfare, among other aspects of scientific and industrial "preparedness." In his "Applied Chemistry" address of autumn 1915, he even stated that "the present war has been aptly called a 'chemical war,'" with the qualification that earlier wars could have received the same label.[8] But it was not until late June 1917 that he became impressed with the significance of poison gas in this respect. After a French scientist's briefing in a meeting of the chemical committee of the National Academy of Sciences' newly established National Research Council, Baekeland wrote in his diary that "gas warfare seems much more extended and more important than what we supposed. Phosgene or carbonylchlorid[e] $CoCl_2$ in mixture of liquid chlorine seems to be used in considerable quantities; also prussic acid HCN [hydrogen cyanide] in conjunction with amine chloride; also brom derivatives of methyl ethylketone."[9] A day later, George E. Hale, Robert A. Millikan, and Arthur A. Noyes—three leading members of the NRC—asked Baekeland whether Elon Hooker, the president of the Hooker Electrochemical Company, "was qualified to take charge of gas warfare."[10] Baekeland was already aware of his consulting client's ambition in this direction and recommended him for the position.[11] Moreover, when he was queried about possible alternative candidates, he simply indicated that "I know of no other men who are not employed more usefully at this moment." This suggests that he still considered poison gas to be a topic of relatively low priority, an attitude that was widespread in the United States at the time (and arguably justified because of the limited effect that chemical weapons would have on the outcome of the war).[12]

In July 1917, however, the Germans introduced mustard gas, a more deadly successor to chlorine. In view of its own war casualties, the US Army regarded it unrealistic not to follow suit. An accelerated process of catching up was consequently set in motion. By the end of the war, the US government's poison gas research program had even become the world's largest.[13] Baekeland himself continued to give priority to other wartime projects, ranging from patent issues and the erection of a Naval Research Laboratory to the production of synthetic nitrogen, optical glass, and airplane fuel. Still, he did personally witness the rapid expansion of US gas research. In early November 1917, for example, he was shown around American University in Washington, DC, where "well equipped [chemical warfare research] laboratories have been rapidly improvised and in each room [are] as many chemists as it can contain."[14] At this stage, approximately 150 chemists seems to have been the maximum, but when Baekeland returned on 15 February 1918, there were "now over 700 chemists and physiologists engaged in this work," some of whom personally tested the newly created gas masks, in addition to trials conducted on dogs, mice, and rats. By this time, moreover, he had also been consulted regarding Hooker Electrochemical's efforts to produce ten tons per day of chloropicrin, a tear gas that would later be used as a pesticide, besides the firm's plans to start manufacturing the explosive picric acid in "enormous quantities."[15]

How did Baekeland evaluate the development and deployment of poison gases and other chemical arms? And how did his views on this topic evolve? In a 1908 address, the chemist had sought to provocatively establish himself as one of "these visionaries, these cranks, these unorthodox, unrespectable people" who believed in a warless future. As during the Cold War, the view that war could best be prevented by outcompeting one's adversaries in terms of weaponry was popular at that time. But Baekeland countered that the United States would probably have spared itself a "Philippine Problem"—its colonization of the isles had been riddled with setbacks and disillusionments[16]—if it had lacked a navy in 1898, when the Spanish-American War broke out. He also blamed classical authors, whom he referred to as the "braggarts of antiquity," for having glorified warfare.[17]

Baekeland's speech served to introduce chemist Hudson Maxim (1853–1922), a renowned inventor of explosives, at a "very well attended meeting" of the New York Section of the ACS, to which Maxim was to present a paper on "The Warfare of the Future." According to Baekeland's diary, his introductory remarks "acted as a detonating cap to Maxim who started using brilliant rhetoric defending war etc and gave us a very interesting and very well delivered lecture."[18] In Baekeland's opinion, Maxim had especially been triggered by his comments about explosives. This seems reasonable, given that the New York ACS's chair had stated, among other things, that these technologies had found "their most valuable applications in the arts of peace," whereas Maxim was about to advocate the development of more powerful torpedoes. But other comments by Maxim's professional colleague also ran completely counter to the spirit of the speech of the explosives inventor. For example, while Maxim emphasized the supremacy of scientific warfare over the exertion of "brute force," Baekeland claimed that "the intervention of science and engineering has not only made war more horrible than before, but has shorn away its picturesqueness which used to inspire the Don Quixotes whenever worms killed other worms."[19]

After the outbreak of World War I, Baekeland appears to have never again expressed similar reservations about scientific (especially chemical) weaponry. He was not, to be sure, blind to its cruelty. For instance, when he heard that French gas masks had become "entirely insufficient" because of the adoption of mustard gas, he wrote in his diary that "anybody who tries one of these masks in peaceful comfortable surroundings realizes what a hellish warfare those Huns have introduced."[20] Yet, in collaboration with his professional peers, Baekeland promoted Allied chemists' wartime work as a patriotic service (figure 6.1) and supported the continuation of the US Army's Chemical Warfare Service in peacetime, even though more than 99.9 percent of the American population favored the abolition of gas warfare according to a 1922 survey. Together with several groups of army veterans, the ACS successfully lobbied against the ratification of the Geneva

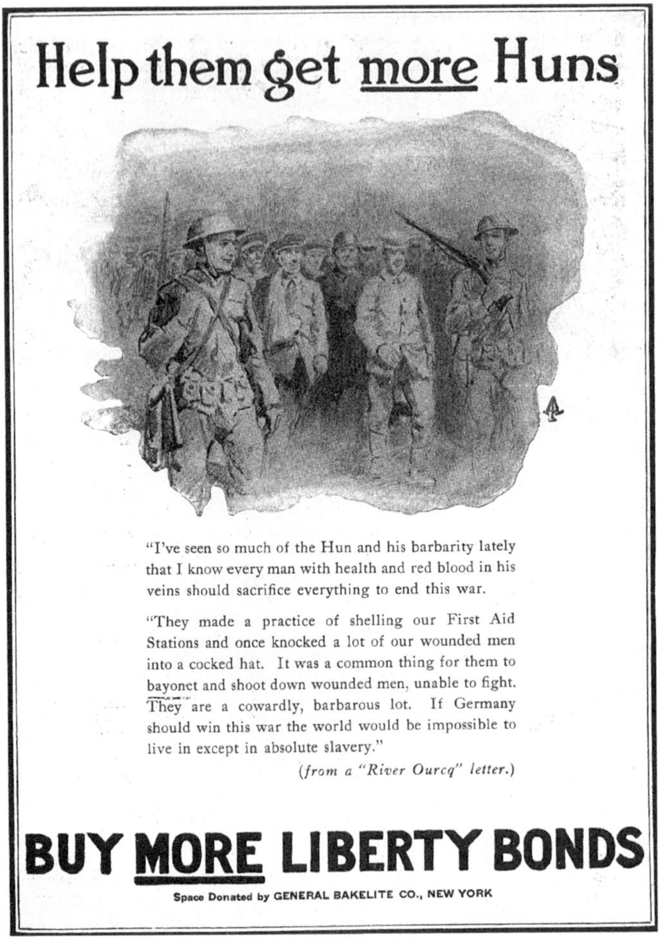

Figure 6.1
Support for the American war effort from Baekeland's General Bakelite Co. World War I dramatically changed Baekeland's attitude toward anything German.
Courtesy of Reindert Groot (amsterdambakelitecollection.com).

Protocol on Chemical Warfare in 1925–1926. As historian Daniel P. Jones has demonstrated, one of their arguments was that quantitative evidence on the relative deadliness of chemical and conventional weapons showed that poison gas was a relatively "humane" tool of war. They also made the case that its prohibition by means of an international treaty would hinder the development of the US chemical industry and, most importantly, render the country ill prepared for future wars. Consequently, as suggested by the 1925 president and secretary of the ACS, opposition could only stem from "hysterical" and "unscientific" (i.e., "feminine") activists.[21]

Ironically, Baekeland, the self-proclaimed pacifist, thus found himself on the side of a minority of scientific experts who defended the legitimacy of chemical warfare against ethical charges by a "lay" audience. The latter group, moreover, included a particularly vocal set of humanist intellectuals[22]—that is, the type of men whom Baekeland had reproached for having glorified warfare in the past. The chemist's position became particularly clear in 1924, the year in which he led the ACS (figure 6.2). On 12 September 1924, for instance, he endorsed national "Defense Day," an initiative for sustaining military preparedness. Significantly, his telegram was "promptly transmitted ... to the Chief of Staff of the Army and embodied ... in a letter setting forth certain needs of the Chemical Warfare Service" by General Amos A. Fries, the Chemical Warfare Service's second chief and its most ardent defendant. The ACS's then president also received Fries's praise for his "Prospects and Retrospects" address, delivered at Cornell University that same month.[23] In this speech, Baekeland had underscored that chemists ought not to be blamed for the Great War's atrocities, as they "certainly did not start nor encourage that war." In addition, while he correctly anticipated that even more effective "means of scientific destruction" would be developed in the near future, he argued that

> the remedy to these horrors lies not in stopping the use of chemistry in warfare. Whether it was under the shape of the stinkpots or gunpowder of the Chinese against sword or bow and arrows, or whether it was the twentieth century chemical methods, in every instance the soldiers who were subjected to it at first indignantly protested, until pretty soon they used the same or even more drastic means in return. ... [D]o not blame the chemist for what will happen if irresponsible, tactless politicians or writers continue needlessly to arouse the worst feelings in other nations. ... [A]fter the harm is committed the chemist as well as the soldier and the sailor has no choice left but to do his part and to help straighten the mess into which they have been drawn by the silliness or boorishness of others.[24]

On the face of it, these remarks were part of a debate on the integration of chemistry into war, with poison gas being the most notorious offspring. But at a deeper level, the controversy was also about scientists' status and role in society, including the sociocultural authority to which they could reasonably lay claim. As did others from the ACS's establishment, Baekeland sensed that the fundamental transformations that the chemical profession had experienced since the late nineteenth century had escaped the

Figure 6.2

As the 1924 head of the ACS, Baekeland led a delegation of chemists to the White House on 24 April, where President Calvin Coolidge applauded chemistry's civilian and military contributions to society. Baekeland (fourth from the left) is standing between his wife and Coolidge. In the following month, Coolidge famously participated in a publicity stunt of the Chemical Warfare Service, meant to illustrate the alleged therapeutic value of poison gases deployed during World War I. Eager to get rid of a cold, Coolidge agreed to be exposed to low doses of chlorine on three consecutive days in a gas chamber designed specifically for this purpose. One of the first US chemists to recommend the use of low-concentration chlorine as a prophylactic had been Charles Baskerville, whom Baekeland had engaged as a scientific expert witness for the Development & Funding Company. For more context on Coolidge's treatment, see Nader and Marcinko, "Rise and Fall of 'Chlorine Chambers.'"
Source: National Photo Company Collection, Library of Congress.

notice of the public at large, which seemed to be amazed by the existence of chemists "without a drug store."[25] In consequence, he participated in a chemical "crusade" for gaining popular support, as well as government protection (for instance, by writing popularizing articles in *Harper's* and *Scribner's* magazines).[26] However, critics alleged that chemists' support for as "sinister" a cause as the Chemical Warfare Service demonstrated that the profession could not even offer guidance on a topic that concerned its own realm of expertise, let alone on any other. Moreover, as historian Hugh Slotten has described, they highlighted chemists' perceived ethical misconduct in order to challenge the place of science in educational curricula and elsewhere.[27] Significantly, according to Baekeland's presidential address to the ACS, some of the "clergymen and ... moralists [who] now reproach us for our science ... have gone as far as to suggest that we should stop teaching science and research in our schools and that we should find our mental food exclusively in the classic lore of the past." In short, the expert services delivered by chemists during the Great War provided verbal ammunition to critics of the rise of what Baekeland claimed to be "the two most powerful men of our generation," namely the scientist and the engineer.[28]

Art No More

These spirited discussions about the merits and limitations of science and technology had roots that predated World War I. An almost religious reverence for science had, after all, been only one response to the "disenchantment of the world" resulting from the proclaimed "death of God" and the Darwinian blow to mankind's sense of specialness. Another influential fin-de-siècle view was instead that only art could keep life livable.[29] As far as Baekeland is concerned, the prevalence of this latter philosophy is especially evident from his occasional visits to Europe. In a 1913 return to Ghent, for example, he unmistakably connected a perceived reverence for "[a]rt—art!! Art!!!" with "stupidity and ignorance and arrogance!"[30] (His wife, who became a relatively successful oil painter in the second half of her life, is unlikely to have shared this assessment.) Yet the following 1889 confession to his Belgian friend Edouard Remouchamps suggests that he had noticed similar attitudes in the United States and elsewhere: "I have no sympathy for Chinese or European or American barbarians who decorate themselves with cut crystallized C[arbon] (i.e., diamonds) or other minerals and worship art to the detriment of intellectual[,] political or moral development. This sounds so unusual that I only dare say it every now and then, and solely to people like you who make the effort to listen open-mindedly."[31]

After the turn-of-the-century years, Baekeland evidently grew more self-confident, for he repeatedly denounced the "pettiness" of art, as well as artists' "short-sightedness," in public during the decade up to 1914. For instance, in a 1907 paper on "The Danger of Overspecialization," he joined Andrew Carnegie in favorably comparing "the better class of business men" to artists, especially those to whom "exact knowledge, which might help to broaden their views, is almost repulsive."[32] In May 1910, moreover, he returned to this subject in "Science and Industry," his presidential address to the American Electrochemical Society. In this speech, Baekeland temporarily put aside the "modesty" that he and many others attributed to scientists and engineers, maintaining that "the last hundred years, under the influence of the modern engineer and the scientist, have done more for the betterment of the race than all art, all the civilizing efforts, all the so-called classical literature of past ages, for which some respectable people want us to have such an exaggerated reverence." Further in the address, he clarified:

> It has been asserted so often by respectable people that science and industry cater only to our material welfare, and have little in common with culture, refinement or moral development; therefore I feel compelled to put special emphasis on this side of the question and to insist on the enormity of this error; on the contrary, the development of our industries, of our material prosperity, as well as the study and application of science are the surest and most immediate forerunners of higher civic ideals, of an improved society, of a better race.
>
> A clean, well-nourished and well-housed individual, who can enjoy the comforts and advantages of modern surroundings, and leads an active, intelligent, productive, self-supporting and self-respecting life, is certainly more of a man and a credit to his race than were some ancient saints, who lived from alms and spent their life in prayer and inaction, or who, for further education of their followers, vowed never to change their clothes, nor wash nor shave nor comb themselves; he is more of a blessing to his fellow-men than the useless drone, who lives on the work of others and gives nothing in return but arrogant presumption, based on fortune, rank or title inherited from his father.[33]

One of the people whom Baekeland may have had in mind when phrasing these comments was Eugen Kühnemann, a German philosopher and historical philologist. Baekeland and Kühnemann had met in February 1909, when the latter was in the midst of a one-year term at Harvard University as a visiting professor. In his diary, Baekeland had expressed his disapproval of the academic's scientific illiteracy, as a result of which he "not only is not aware of what he is short but thinks like all pseudo-educated people that natural Science does not contribute to culture while I believe on the contrary that no true culture is possible without it." By the same token, Baekeland had contrasted Kühnemann's knowledge of Goethe, Schiller, and Kant to his (presumed) inability to "plant a garden; or use a hand saw or a hammer." In brief, in the chemist's opinion, men like Kühnemann sadly overlooked "that culture implies not only reading & philosophizing but implies action and creation and utilizing the forces of nature and the accumulation of knowledge."[34]

The attitude against which Baekeland reacted was not monopolized by humanistic researchers with little or no background in science. Nor was it the trademark of reactionaries who experienced the rational pursuit of material ends in modern industrial societies as a threat to what they perceived as a more heroic and authentic life. As a matter of fact, critics of modernity tended to accept technoindustrial change as a value-neutral given that could and should—in view of the then widely anticipated ethical progress—be subordinated to immaterial, nontechnological goals.[35] Still, as far as the pre–World War I United States was concerned, Baekeland probably overestimated the vitality of his ideological opponents.

This at least is what the reception of "Science and Industry" and "The Danger of Overspecialization" suggests. Both of these addresses were primarily targeted at American audiences and, according to Baekeland's diary, a major aim of the latter was "to awake some of us from our onesided torpid life." Baekeland presented "The Danger" to the New York ACS on 5 April 1907, and it certainly did not go unnoticed: the speech triggered an animated debate and, as its author observed the following week at a dinner at the Chemists' Club, was even judged worthy of a satire.[36] But, to Baekeland's disappointment, the discussion on 5 April had mostly deviated "on side matters as for instance, What is the soul? Have we right to kill animals for pleasure or food?" Moreover, the paper's later appearance in *Science* does not seem to have aroused controversy of the kind that Baekeland had hoped to ignite, if any at all.[37]

"Science and Industry," in contrast, inspired a series of critical "letters to the editor." These comments, however, appeared neither in the *Transactions of the American Electrochemical Society* nor in *Science*—the two American publications in which Baekeland's presidential address had been printed entirely.[38] Instead, they concerned reactions to a one-paragraph summary of the speech in Germany, where the cultural value of technology was a much-debated issue.[39] In the *Chemiker-Zeitung*'s 16 July 1910 issue, chemist Richard Kissling urged the journal's editors to dissociate themselves from Baekeland's "inferior" claim as to science and engineering's belittling of art and literature during the century up to 1910. It is notable that the editorial staff, whose head was otherwise supportive of Baekeland's work and interests, immediately complied with this request.[40]

In response to Kissling's comment, it first admitted that a case could be made that advances in the realm of technology (*Technik*)—a term that neither Baekeland nor Kissling had used—had increased human well-being and, by democratizing access to higher learning, elevated the general cultural level. But it then countered that such a standpoint would be "superficial," for "in reality the significance of technology to the betterment of the human race is only about the same as that of the automobile to the sick person who is being driven to the doctor. As a means technology can certainly offer services valuable to human development. But to relinquish the truth, the good, and the beauty contained in the works of our poets, musicians, sculptors and scholars would almost be to lower our culture to the level of an ant colony."[41]

"Superficiality" was one of the traits associated with the German term *Zivilisation*—others being materialism and instrumental reason—when it was being opposed to an inwardly, ethical *Kultur*.[42] In the *Chemiker-Zeitung*, a journal in which industrial chemistry was a main focus, the relegation of technology to the realm of *Zivilisation* was challenged by Dr. O. Makowka, who ran to Baekeland's defense with respect to both the discussion's content and the basis on which it rested. On the latter, Makowka suggested that Baekeland's words had been taken out of their context—an interpretation that was to be verified by reading the entire address instead of a brief synopsis. Yet, this notwithstanding, it was crystal clear to Baekeland's supporter that technology did, in effect, contribute to cultural and moral progress, and that it ought to get credit for so doing. Moreover, according to Makowka's letter, a more promising way to prevent human societies from turning into ants' nests was to better guard chemists' interests against the "ignorant but cunning [business]men" employing them. This was a point that had also been included in the German abstract of Baekeland's presidential address but had not yet been commented on.[43]

To the journal's editor-in-chief, however, Makowka's counterarguments were unconvincing. In the name of the editorial staff, he reemphasized that technology, unlike science, could not in itself ever be *Kultur*, but only a means toward its creation. Strikingly, under the justification that this assertion was a self-evident truth, he did not even deem it necessary to further clarify it.[44]

Technocratic Leadership

In the US context, German-style debates about the cultural value of technology hardly took place in the decades around 1900, not least because of the absence of a precise English equivalent to the concept of *Technik*.[45] Yet the pages of the ACS's *Journal of Industrial and Engineering Chemistry*, an American equivalent to the *Chemiker-Zeitung* to which Baekeland regularly contributed, did feature somewhat related discussions that are relevant for understanding why the chemist's speeches were not similarly criticized in the United States. Like Baekeland, two of this journal's first three editors—Milton C. Whitaker and Charles Herty—played leading roles in the self-glorifying "crusade" that US industrial chemists carried out during the twentieth century's first decades.[46] To this purpose, they zealously spread, as Baekeland put it in a 1929 letter to Arthur D. Little, another influential participant, the "gospel of service through Science."[47] More to the point, even before World War I, they did not refrain from championing the supremacy of "applied" chemistry over its "pure" counterpart. As historian David J. Rhees has written, for example, in the campaign, Little implied that "pure science was self-indulgent, while applied science represented idealistic service to society"—a view with which Baekeland more often than not concurred.[48]

The leadership of the ACS, the main society of what probably was America's most "industrialized" scientific community, was also remarkably successful in unifying rank-and-file members in support of its agenda.[49] This was the case not only when it came to resisting the introduction of restrictions on the development and use of chemical weapons but also applied to various other issues.[50] Most relevant, there are no indications that US chemists jeopardized the association's promotional efforts by putting into perspective the importance of (especially chemical) inventions and industries in terms as strong as the *Chemiker-Zeitung*'s editorial staff had used.

Likewise, while industrial chemists' advancement within the ACS did not go uncontested, no serious "revolt" against the society's proximity to commercial interests occurred.[51] In this the society differed from America's major engineering associations; as historian Edwin Layton Jr. has demonstrated, the latter's traditional, pro-business leadership was challenged during the World War I period. Around this time, "patrician reformers" such as Morris L. Cooke maintained that their profession's mission ought to consist in serving the public at large, instead of electrical utilities, railway companies, and steel manufacturers, among other industries.[52]

Unlike Cooke, who would direct the Rural Electrification Administration during the New Deal era, Baekeland did not judge the interests of the public to be incompatible with those of industry,[53] but he did share Cooke's view that engineers were well equipped to solve sociopolitical problems and ought to play a significantly larger role in this realm. In November 1915 and March 1916, Baekeland expressed his opinion on this topic in reaction to a statement by Gano S. Dunn, a prominent electrical engineer. Notably, Dunn and Baekeland seemed to draw different conclusions from a similar experience. Both men belonged to their professions' establishment, a group that tended to have a natural interest in preserving the status quo.[54] Moreover, during the war and its aftermath, either one of them was being consulted on matters of national security because of their expert knowledge. Yet, while Dunn approved of this increased participation of engineers in government affairs, he cautioned that "the engineer should never go so far as to think that he can take the place of the statesman. The function is different and requires a different kind of man.... I feel that we should stay in our own province as technical experts, and advise about the things we know."[55]

Baekeland, on the other hand, argued that scientists and engineers would run the country more efficiently than "lawyer politicians," a standpoint for which, according to his diary, he was heartily applauded at a banquet of the Washington Society of Engineers in late 1915.[56] Then, during a series of lectures in February 1916 on the role of science and engineering in warfare, he perceived "more and more that socalled [*sic*] preparedness with many people is only a pretext for preparedness for [the] next election."[57] Baekeland contrasted this political partisanship with the (alleged) impartiality of exact science, which, in his view, could and should also be "applied" in other domains, including government. For example, according to his "refutation" of Dunn's opinion,

> This country [the United States], as it grows larger and larger cannot keep on being run in the easy-going ways of a mining camp. We have to come to a point where we have to change our methods, and in this we have to do some experimenting, just as we are doing in our chemical industries.... These problems cannot be solved by a general formula, inflexible and everlasting. Furthermore, whenever we discuss these problems, we are apt to be carried away by one single point of view or another and on this account, we frequently exaggerate one single

> feature by looking at it from one standpoint. This is the mistake of most of our politicians. When we talk about these matters, let us discuss them like engineers. Let us use quantitative argumentation and let us impress those of our friends in Congress or the Senate who are really willing to learn that we, at least, can look at the situation with a due sense of proportion by using quantitative arguments and not be overawed by merely qualitative considerations.[58]

Baekeland's advocacy of quantitative reasoning can be contrasted with his skepticism as to the utility of mathematics to chemistry; significantly, on one occasion he equated the latter with a coffee mill, which, "however good the mill is ... will ... not convert bad coffee beans into good coffee." Yet he genuinely believed that society would benefit from being "pushed ahead" by a technocratic elite, "just like a herd of unthinking sheep is unwillingly driven forward by the shepherd and his dogs."[59] His argument was, moreover, in accordance with the efficiency-speak of the contemporary "progressives" who wanted their government to be run more like an industrial corporation.[60]

Baekeland left little doubt as to what types of scientists and engineers should be part of his wished-for aristocracy of talent. Like Edison, he preferred "practical men who are accustomed to doing things, and not talking about it" above scientists of the "ivory tower" type.[61] Academically oriented scientists, however, also regarded World War I as a unique opportunity for furthering their own interests. In particular, a group of activists led by George E. Hale, an astronomer depressed by the perceived underrepresentation of scientists on the NCB, set out to create broader support for fundamental scientific research by stressing its (alleged) practical utility. They, moreover, received President Wilson's permission to establish a research council within the National Academy of Sciences.[62]

As one of only two people who were elected into both the NCB and this National Research Council (NRC), Baekeland was well positioned to compare the two advisory bodies' performances. He had been particularly active in the NCB, where he "jumped from one topic to another" but nevertheless played a leading role in initiating and coordinating comprehensive R&D projects related to the domestic production of nitric acid—a source of both fertilizers and explosives—and optical glass.[63] Baekeland's diaries and personal correspondence also make clear that he more strongly identified himself with this organization. In fact, given that several historians have characterized the NRC's science-centered approach to R&D as more apt than the NCB's reliance on engineers and "heroic" inventors such as Edison, Elmer Sperry, and Hudson Maxim, it is striking how critical Baekeland was of both the NRC's methods of working and its underlying philosophy.[64] For instance, after having attended an April 1917 meeting led by a fellow chemist, Marston Bogert of Columbia University, he concluded that NRC members such as Bogert, Hale, and Robert A. Millikan, later a Nobel laureate in physics, were "all men whose time is not very valuable and who on this account drag out the proceedings." Moreover, he pitied the Navy commanders who had agreed to attend the "dreary" gathering.[65] Baekeland noted this criticism not only in his diary; more diplomatically worded, he had also expressed it at the meeting itself. In addition, in letters to the NRC's leadership, he proposed to organize the council's internal communication more efficiently—or "businesslike," as he put it in a comparable situation.[66] In a letter to Bogert on 1 May 1917, he also indicated that the NRC's meetings were "strikingly different" from those of the NCB. In his view, this was because "many of the members of the Naval Consulting Board have won their spurs by constructive work," including the development of "very important devices which are now used in the Navy, while the men of the National Research Council have mostly all earned their reputation by work of an analytical nature. All these tendencies are necessary but when it comes to ... [solving] a constructive problem, engineers, manufacturers and inventors are at their best. ... The situation here is very similar to what one encounters in a practical manufacturing establishment where a scientific research laboratory is a very useful addition, but where the real constructive work has to be devised and carried out by the engineering staff."[67]

In the following month, Baekeland returned to the subject in the University Club in New York City, one of the exclusive social organizations where he frequently spent his evenings. With Columbia's electrical engineering professors Michael Pupin and Francis B. Crocker, both of whom had been affiliated with the Ampere Electrochemical Company, he then "discussed limitations and restricted views of pure scientists and how in applied Science one has to be more thorough and careful in the details of a problem." Baekeland seems to have especially enjoyed one example given by Pupin, an academic inventor and patentee who "achieved a financial success which has seldom, if ever, been equaled by a full-time engineering professor."[68] The story concerned Albert A. Michelson, the first US scientist to win a Nobel Prize. According to Baekeland's diary, Michelson had "confessed" his ignorance about "amplificators" and "oscillators"—two recent innovations in the area of communication technology—to Pupin after a joint meeting of the NRC and the NCB on 9 June, at which telephony and wireless telegraphy had been the main focus. In other words, "Michelson in his isolated laboratory and professor's occupations had failed to follow some of the most important technical advances in this important branch of physics."[69]

How did less technology- and business-savvy scientists respond to Baekeland's verbal assaults? Was his career hindered because of them? In chemical circles, this does not seem to have been the case. Although Baekeland was "looked upon ... as rather eccentric" by many of his professional colleagues, his experience as an industrial executive was appreciated.[70] Similarly, although one prominent scholar later found Baekeland "too loquacious" to invite him to a conference on academic patent policies, the chemist's denunciation of "intellectual freaks" also did not preclude a return to academia. According to Baekeland's diary, for example, Johns Hopkins University repeatedly, but to no avail, offered him a professorship in applied chemistry.[71] Columbia's president Nicholas M. Butler, on the other hand, could convince Baekeland in May 1917 to join his university's newly established Department of Chemical Engineering (figure 6.3). Baekeland accepted Butler's proposal under the condition "that I should have absolute free disposal of my time and any other obligations according to my own judgment."[72] In a similar vein, he minimized his responsibilities at Columbia in a letter to the university's longtime chemistry professor Charles F. Chandler, with whom he had developed a supportive relationship during the previous decades. Baekeland noted to Chandler that "I shall need no desk, nor office room, nor any laboratory. My intervention ... will mainly consist in giving, now and then, a lecture on some subject with which I am more or less familiar, and in trying to confer with the other professors on anything which may improve the teaching of applied chemistry, and to turn out chemical engineers."[73]

Still, from other sources it is evident that Baekeland took his tasks seriously and was able to leave his mark. For instance, he had a major say in a restructuring of his department, including the hiring of new personnel. In this connection, it is noteworthy that Baekeland stressed to George Pegram, the dean of Columbia's Faculty of Applied Sciences, that "a true University" should steer clear of industrially funded research and development work that was not scientifically challenging.[74] Yet, at the same time, he took steps

Figure 6.3
Baekeland as Doctor Honoris Causa of Columbia University.
Courtesy of Ghent University Museum.

to strengthen the practical orientation of Columbia's chemical engineering program, therefore suggesting that his reentry into academia did not substantially alter his previously discussed educational views. This is also evident from a 1922 meeting on the facilities required for science instruction at the home of Butler, with whom Baekeland was on friendly terms. The chemist then agreed that the availability of large laboratories was desirable, but cautioned that the work conducted in them should be "worthy [of] the effort." In his opinion, "it is a common mistake of selecting uninteresting research topics in University laboratories. Industrial laboratories as for instance Schenectady [of General Electric] etc. chose true and fascinating results. Same could be done in our Universities. Each problem of applied Science had certain factors of Pure Scientific research which make splendid subjects because they inspire the investigator with the possibility of great applications."[75] What is more, according to a colleague at Columbia, each year Baekeland closed his final lecture, entitled "Dreams and Realities," by warning his students "in his thundering sonorous voice" against the "devious ways" of not just bankers, lawyers, ministers, and politicians but also "all starry eyed theorists who live by the sweat of the other fellow's brow and to whom the scientist, inventor[,] engineer and honest workers often fall a prey"![76]

"Dreams and Realities" was also the title of a lecture that Baekeland delivered at Johns Hopkins in fall 1931 and was published in the *Journal of Chemical Education* the following year. As with several of his earlier addresses, Baekeland sent reprints to a long list of colleagues, friends, and acquaintances. On this occasion, the reactions contained not merely the customary praise but also a sobering observation by a fellow chemist, James B. Conant of Harvard University. In a brief response, Conant, who would become Harvard's president in 1933, indicated that he had been "interested" in Baekeland's lecture "and particularly in the sad prophecy which you there made of my fate."[77] The comment probably referred to Baekeland's prediction that the minds of academics who had taken on administrative responsibilities "gradually … will be compelled to work along business channels." In addition, it is questionable whether Conant, who held industrial research in relatively low regard, agreed with Baekeland's contrast between such research and investigations "undertaken for purely scientific purposes." Among other things, Baekeland had included a critical note on the incentive for speedy publication, which had become more central to academic reward systems since the advent of the research university and the modern scientific journal.[78] Probably drawing on his own experiences, he had stated that "the pure scientist may choose to rush into print as fast as he has any sketchy results, and publish some half-baked theories and reap the glory thereof. Even if other workers later contradict the accuracy of his observations, the worst that can befall him is to be involved in a polemic. In industry, superficial or incomplete work spells financial disaster, besides recriminations of the victims."[79]

Of all the reactions to "Dreams and Realities," as well as to Baekeland's other nonresearch addresses and publications documented in his personal papers, Conant's was about the least positive. This clearly suggests that Baekeland's outspoken assertions were less contentious than he seemed to believe. Still, together with a dramatic shift in his evaluation of German science after the beginning of World War I,[80] Baekeland's scientific "impurity" and his anti-intellectualist views may well have delayed his admission into the US National Academy of Sciences (NAS), in critics' eyes a bastion of the "self-satisfied" scientific elite.[81] While Baekeland had been involved in the NAS's research council from 1917—the year after its foundation—until the early 1930s, his election did not take place until 1936. In comparison, Sperry and Edison, who neither held an academic degree in science nor primarily conceived of themselves as scientists, had been voted into the NAS about a decade earlier.[82] (According to Baekeland's diary, Edison joked, "I know you—you are the man of the molecules—This goes way above my head!," when he and Baekeland were introduced to one another by Josephus Daniels, the secretary of the Navy, in October 1915.)[83]

Edison's election had previously been blocked by NAS members who considered his work too "applied," and the available evidence suggests that Baekeland's name had likewise aroused resistance within this organization. Most notably, a diary entry that Baekeland wrote in December 1930 reveals that he had been nominated to no avail at the academy's annual meeting of that year, as well as on at least one earlier occasion.[84] In addition, some of the professional colleagues who congratulated the chemist in 1936 hinted at the relatively late date of his admission. As an example, Gano Dunn, who had presided over the NRC from 1923 until 1928 and had been elected to the NAS in 1919, claimed to have "been persistent for a long time past in urging a recognition that has long been your due." Jerome Alexander, an authority on colloid chemistry, likewise opined that Baekeland "should have been elected a member long ago, but perhaps in this, as in other matters of similar nature, other factors than merit alone play some role."[85]

In his 1930 diary entry, Baekeland had held "some German born [NAS] members" who had remained supportive of their native country after the outbreak of World War I responsible for the failure of an earlier nomination effort. Yet the May 1936 announcements of the academy's latest entrants in *Nature* and *Science* seem to point to other factors. Significantly, neither journal mentioned Bakelite or Velox, nor Baekeland's expert services during and after the war. Instead, Baekeland was referred to as an "honorary professor of chemical engineering in Columbia University."[86] On the one hand, this suggests that his affiliation with Columbia increased his respectability, even though his responsibilities at the institution were limited. On the other hand, it is noteworthy that Baekeland's well-known contributions to the synthetic plastics industry apparently were considered no match for an academic title.[87]

In any case, Baekeland did not depend on the NAS to get wide recognition for his work on Bakelite. As the remainder of this book documents in more detail, he had already succeeded in doing so by the mid-1930s. More specifically, the following chapters argue that a particular combination of intellectual property and entrepreneurial strategies had enabled him to further increase his wealth and his technological legacy while also gaining scientific and popular acclaim for the invention and commercialization of the first fully synthetic plastic.

7 Bakelite: Transatlantic Patenting

In summer 1905, part of which Baekeland spent away from his family, his wife sent him an emotional letter that casts further light on his motives as an entrepreneur, as well as on his weaknesses as a father and husband. By this time, Baekeland had become closely involved in the scale-up of Townsend and Sperry's electrolytic process, among other consulting assignments, while also conducting various independent experiments in his personal laboratory in Yonkers. In addition, as we have seen, he had started to play a prominent role in several professional chemical and chemical engineering societies. Still, none of these pursuits—let alone his responsibilities as the father of a ten-year-old son and a nine-year-old daughter—could prevent him from feeling that he had lost his "ambition in life." According to Céline Swarts's letter, Baekeland was blaming her for his reduced appetite for work, presumably because of her efforts to have him spend more time with his family. Swarts, in contrast, pointed out that he had "taken back [his] freedom and [his] independence long ago" and instead associated her husband's complaints with his difficulties adjusting to a rich man's life. "Your lack of ambition is the result of a material point reached, and I am astonished that your moral or fame ambition does not satisfy you. But I am not the cause of it," she argued. At the same time, she encouraged him not to "allow your disappointment in me to hurt your life; have your ambition, your career, and above all your self respect, for yourself, for your children." Moreover, in a sarcastic comment that revealed that she had not yet found peace with their less than perfect marriage, she added that, "I shall never again bother you with my love or my affection."[1]

What neither Swarts nor Baekeland realized at this time was that the project that would eventually provide their family with much additional wealth, fame, and moral satisfaction had already begun. According to later courtroom testimony, Baekeland had turned his attention to the reaction between phenol and formaldehyde after having observed a market opportunity: the steep rise in the price of camphor, which was primarily obtained from camphor trees in the Japanese colony of Formosa (Taiwan), after the outbreak of the Russo-Japanese War in early 1904.[2] The demand for camphor had increased substantially because of the expansion of the celluloid plastic trade, one of a range of new industries in the United States and Europe that depended on raw materials imported from elsewhere. As one of his first steps, Baekeland hired Nathaniel Thurlow, an electrochemist and inventor of a synthetic camphor process that, although complex and ultimately unsuccessful, had seemed sufficiently promising to warrant expensive development and commercialization efforts in the two preceding years. However, while none of the compounds created in the course of Thurlow and Baekeland's initial experiments on the phenol-formaldehyde condensation reaction seemed suitable for the aim of finding "a better and cheaper product than camphor for celluloid," one appeared to hold potential as an alternative for shellac resin. Shellac was another natural substance imported from Asia, and its price had gone up also, partly from its use as an insulating coating in the rapidly growing electrical industry. Therefore, Baekeland and Thurlow had both technical and commercial reasons for abandoning their original goal and instead starting to focus on the development of a shellac-like varnish or lacquer.[3] The earliest preserved accounts resulting from their laboratory research project show that, by February 1907, Thurlow had effectively come up with a new process for creating a soluble and fusible phenol-formaldehyde condensation product, which his employer chose to call "Novolak" (later spelled "novolac"), a neologism. Moreover, by April of that year—when Baekeland started keeping a diary—Thurlow was working with a Detroit-based manufacturer of varnishes to commercialize this invention.[4]

At the same time, Thurlow and Baekeland had not neglected the insoluble, heat-hardened "spongy mass" that had long puzzled students of the phenol-formaldehyde reaction.[5] This type of condensation product had been less relevant to Baekeland and Thurlow's research program, as its properties differed notably from those of camphor and shellac. Still, a patent application of 18 February 1907 reveals that they had experimented with a new and original way of trying to control its production and turn it into a commercially useful substance. In its original formulation, this application outlined a method for reacting phenols with aldehydes within soft wood or other fibrous materials, with the aim of indurating these products and improving their technical qualities. Importantly, the document explicitly stated that both sorts of condensation products that could be obtained—that is, not just the alcohol-soluble substance in which Thurlow and Baekeland had primarily been interested but also the "very hard and insoluble gum"—were suitable for this purpose.[6]

In June 1907, while Thurlow was in Detroit, Baekeland personally continued this line of work on several subsequent days. During one of his wood impregnation experiments, however, his boiling of a phenol-formaldehyde mixture unexpectedly led to the creation of a viscous condensation product that was relatively easy to heat-harden in isolation.[7] This substance seemed to have enormous potential not only as an improved version of—or supplement to—Thurlow's novolac but also as a molding material that could substitute for hard rubber and celluloid (see chapter 8 for a discussion of the initial applications that Baekeland envisioned for Bakelite). Accordingly, he felt prompted to secure far broader patent protection at the earliest possible date, thus initiating a dynamic and highly contested process that would ultimately give his contributions to the creation of a new industry much more visibility than the accomplishments of other chemists and inventors.

The intellectual property strategy that Baekeland pursued in the following months and years was most definitely not the approach of someone who lacked ambition. As an example, besides his US patent applications, he and his patent agents also filed for IP protection in Belgium, Canada, Denmark, Hungary, Japan, Mexico, Russia, Spain, and several other countries. Moreover, Baekeland later strengthened his legal position by concluding a new contract with Thurlow, who abandoned his ownership interest in patents resulting from his work in exchange for substantial financial compensation.[8]

One of the notable outcomes of Baekeland's desire to obtain an almost global legal monopoly was that he ended up participating in the formation of a European firm before having settled on a longer-term strategy for the US market. In fact, as this chapter demonstrates, Baekeland's interactions with European associates, competitors, and patent examiners affected his entrepreneurial decisions and IP record in the United States in several ways.

These experiences also influenced his related efforts to reform the US patent system, which culminated in his chairing of the Committee on the Patent Office of the National Research Council from 1917 until 1922. Most relevant, Baekeland emphasized that the US patent regime favored affluent inventors and corporations, and he advocated the adoption of northwest European, especially German, institutions and practices to level the playing field.

The Transfer of Bakelite to Europe

By 1907, as we have seen, Baekeland had accumulated substantial experience in dealing with patents in both Europe and the United States. However, partly because he had relied on trade secret law to protect Velox photographic paper, he was not yet fully aware of the difficulties that he would face with regard to Bakelite. The available literature suggests that both corporate and individual US innovators faced considerable hurdles when attempting to obtain and enforce IP rights across the Atlantic in the late nineteenth and early twentieth centuries. This period, which was characterized by an intense wave of globalization, saw the introduction of international agreements prohibiting discrimination against nonnational patent applicants, most notably the Paris Convention for the Protection of Industrial Property of 1883. Still, as economist Ha-Joon Chang has observed, "Even the most advanced nations were still routinely violating the IPR [intellectual property rights] of other countries' citizens well into the twentieth century."[9]

There is, moreover, some evidence that the situation of independent innovators such as Baekeland was particularly challenging at a time when large multinational enterprises' share in international patenting was becoming disproportionately high.[10] After all, as economist Edith Penrose noted when reviewing inventors' criticisms of the international patent regime initiated by the Paris Convention, "The greater the administrative difficulties in obtaining patents in different countries due to language differences, methods of drawing up applications, forms of samples and models, enumeration of claims etc., the more the large, well-financed and legally equipped firms have an advantage over the smaller ones."[11] Recent studies have empirically documented some of the setbacks experienced by US inventors. Legal historian Christopher Beauchamp, for example, has demonstrated that the British IP position of Alexander Bell was weakened because a description of his invention had entered the public domain before the issuance of the patent.[12] For this same reason of prior disclosure, the German airplane patent of the Wrights was invalidated around the end of 1911.[13] About four and a half years earlier, moreover, Baekeland's fellow electrochemical inventors Elmer Sperry and Anson Betts had shared with each other their skepticism regarding the value of international patents.[14] To some extent, these problems and concerns could be offset through collaborations with internationally active investors and patent agents.[15] Based on the available evidence, however, it remains largely unclear whether and how these arrangements benefited the inventors involved.

Besides preparing US and international patent applications, Baekeland also took detailed notes documenting the progress of his work and showed samples to visitors, whom he then requested to sign his laboratory notebooks to prove his priority of invention.[16] As long as it remained unclear whether his patents would be granted, however, he preferred not to draw too much attention to his research and therefore limited his commitment to improving the technology. After having amended several of his US patent applications in the middle of March 1908, he noted in his diary that:

> If I only knew that my patents will be granted and that there will be no interference cases which may drag me into long and costly litigation the whole matter would be very simple and I would not hesitate making the necessary expenses to forge ahead and engage one or two more assistants and equip my laboratory so as to be able to work on a larger scale and investigate several possible industrial applications.... I am precluded from doing much as long as my American and foreign patents are not allowed. This delay should be used to the best advantage to do as much work as possible, yet without going in heavy expenses.[17]

Not until February 1909, more than a year and a half after his laboratory breakthroughs of June 1907, did he feel sufficiently secure to publicly announce the invention of Bakelite in a paper presentation to the New York Section of the ACS.[18] Even at this time, however, most of Baekeland's patents, including all the US ones, had not yet been granted. His decision to publish some of his research findings therefore primarily reflected increased confidence in the chances of his IP claims being allowed. In fact, Baekeland and his patent agent would be quick to provide the US Patent Office with a copy of the paper, hoping that the review of prior research included in it would conclusively demonstrate the newness of the chemist's work. An early exchange with the German Reichspatentamt may have contributed to these tactical choices. Like its US counterpart, the German patent office examined whether submitted inventions were sufficiently novel and ingenious; Baekeland had been excited about its first judgment and gave priority to responding to the objections raised in it.[19] Other countries (for instance, Belgium, France, and Spain), in contrast, registered patent applications without investigating the prior art. It may therefore have been difficult to convince investors that these patents would not be invalidated in court on account of their lack of novelty. The shortening of the administrative procedure could, however, be exploited strategically. As an example, Baekeland seems to have consulted Belgian patents to anticipate potentially conflicting IP claims pending in countries with examination systems.[20]

Baekeland's first presentation on Bakelite not only inspired members of the New York ACS, who would award him its Nichols Medal, which recognized the best paper delivered at its meetings.[21] Importantly, the paper also garnered international attention. For instance, Baekeland was contacted by Eduard von Vietinghoff-Scheel, editor-in-chief of the *Chemiker-Zeitung*, who desired to publish an authorized translation of it in the journal. Published in late March 1909, this German translation by a correspondent of the *Chemiker-Zeitung* aroused interest for the invention and the European IP rights. Furthermore, Baekeland also personally prepared a French version of his paper, its style being later improved by his wife, who also recorded accounts of early sales of Bakelite samples.[22]

When Baekeland traveled to Europe later that year, his fellow chemists and engineers were eager to hear additional details about Bakelite. As Sperry, who had crossed the Atlantic on the same steamer, commented: "Arriving in London May 24th, Dr. L. H. Baekland [*sic*] was immediately surrounded by people clamorous to know all about 'Baklite' [*sic*]; it soon became apparent that the time that he had at his disposal for outside matters would be considerably limited."[23] Baekeland would remain in Europe until the second week of August. Although he did not neglect his other professional obligations, Sperry rightly felt that his priority was to conclude manufacturing arrangements for Bakelite.[24] But how did he go about this?

To secure industrial property protection, Baekeland pursued a transnational, language-based strategy. For instance, he made use of the services of New York–based patent agents to file applications in the United States, Canada, and the United Kingdom; patent agents in Berlin for applications in Germany, Austria, and Hungary; and native French speakers for Belgium and France.[25] He did not, however, subsequently initiate the formation of

(a) (b)

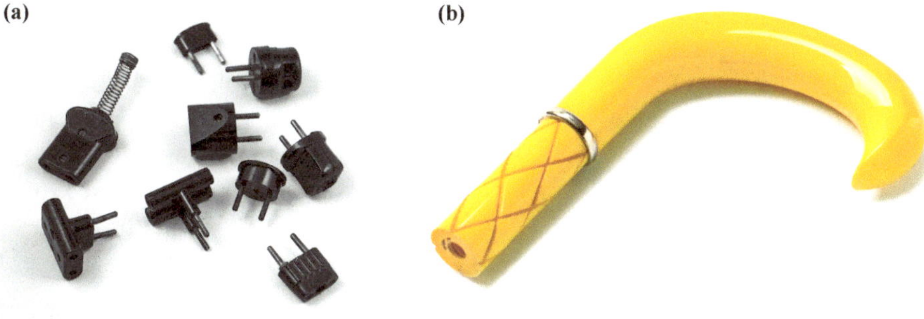

Figure 7.1
Early applications of Bakelite: (a) electrical plug-ins (molded Bakelite) and (b) umbrella handle (cast resin).
(a) Courtesy of Ghent University Museum; photo by Ivo Verhaeghe.
(b) Courtesy of Karel Julien Cole (juliensart.be).

various companies for exploiting the respective national rights but instead opted to sell these rights on all his European patents (including those for the European colonies except Canada) to a syndicate of investors based in Berlin and dominated by German chemical corporation Rütgerswerke A.G., which had large supplies of phenol at its disposal. Why Berlin? Aside from the local demand and raw-material supplies, the available expertise concerning a most promising market for Bakelite—the material's use in electrical applications (figure 7.1)—may have influenced his decision. Baekeland and Hermann Hecht, a German industrialist who had offered to bring the chemist "in relation with any manufacturer or bank I desire," wanted to include Siemens & Halske and Allgemeine Elektrizitätsgesellschaft (AEG), the two companies that were largely responsible for Berlin's status as "the electrical metropolis" of the world, in the syndicate.[26]

Baekeland's negotiations with Siemens & Halske and AEG illustrate the role of patents in attracting funding and expertise. Although he may have felt sufficiently confident about his IP position, potential investors had to be convinced as well. Indeed, Siemens made the granting of Baekeland's patents a condition for its participation in the planned syndicate. The chemist wrote in his diary that one of Siemens's directors expected that it would "take many more months to have [the] patents straightened out" and therefore considered the duration of the presented option too short.[27] Similarly, one of AEG's patent attorneys had queried Baekeland about the

IP situation. His company's interest in Bakelite appeared to be less than that of Siemens, and by the end of June Baekeland decided to no longer involve it in the negotiations. On 7 July, Baekeland, Hecht, and Rütgerswerke's president, Sally Segall, agreed to drop Siemens as well. They had probably concluded that the costs of keeping it in had become too high and consequently, to use one of Baekeland's favorite expressions, the game was not worth the candle anymore (meaning that the game's stakes would not even cover the cost of the candle burned to light game play).[28]

Before his arrival in Berlin, Baekeland seemed not to be particularly concerned about differences between the respective US and German IP legislation.[29] Once he started to work out the details of his entrepreneurial strategy, however, he was forced to face the reality of the situation. Tellingly, he spent back-to-back days in the offices of Meffert und Dr Sell, his German patent agents, to deal with unanticipated complications. For instance, to circumvent restrictions for foreigners, he authorized Hecht to secure trademark protection in the name of his own company, the export firm Hecht, Pfeiffer & Company. Moreover, Baekeland described in his diary how he got "a creepy feeling" when he was informed about the total costs of German patents up until their expiration after 15 years.[30] (Because of annual renewal fees, these were much higher than in the United States.) As he had previously done in the United States, Baekeland also attempted to influence German patent examiners by further clarifying his claims in personal interviews, in which he and Meffert inquired about the status of his applications. While this gambit at the Reichspatentamt had not been completely unsatisfactory, even Meffert remarked the following day that the Bakelite IP rights would drive him crazy if the application process lasted much longer![31]

Baekeland may have anticipated that Germany would become an important testing ground for his IP claims. After his first lecture to the New York ACS, for example, Hugo Schweitzer, a well-connected chemist and patent expert of Bayer's American branch, had already questioned the value of Baekeland's work, and he feared that Schweitzer would use his clout in Germany to damage his interests. While in Berlin, Baekeland was confronted with the first major contender—chemist Hans Lebach of Knoll & Company, who had also invented a phenol-formaldehyde resin and claimed that his work anticipated Baekeland's.[32] Baekeland first heard about a 22 May lecture by Lebach on 7 June from Vietinghoff-Scheel, whose *Chemiker-Zeitung* had briefly reported on it. Baekeland's association with Vietinghoff-Scheel, whom he had asked to join the European Bakelite syndicate and to defend his personal interests after his return to the United States, proved helpful in a further stage of the dispute as well when Vietinghoff-Scheel forwarded a follow-up clarification by Lebach to Baekeland prior to its publication. Nevertheless, the controversy with Lebach also put IP issues at the heart of Baekeland's negotiations with Rütgerswerke.[33] The chemist even came to believe that he "had made a mistake by starting all these negotiations before patents were out" and before he "had made enough progress in the U.S.," fearing that the following months "will be used to pluck down my invention like a fresh hatched chicken until it looks like thirty cents. … Better should I wait a few months and then throw before them [Rütgerswerke's negotiators] what I have done and treat the whole matter as a subject of record and not of doubt."[34] A "few months" would prove an optimistic estimate: not until July 1910 was Baekeland informed that Knoll, Lebach's employer, had revised an "absurdly broad" patent claim that, if accepted, would have curtailed Baekeland's IP rights.[35] The next month, a settlement was reached giving Knoll "a finger in the Bakelite pie" while uniting Baekeland's and Lebach's inventions in what would prove to be "a prosperous marriage."[36] However, Rütgerswerke, Vietinghoff-Scheel, and Hecht, the latter two of whom represented Baekeland's interests after his return to the United States in August 1909, had not awaited the result of the dispute with Knoll before forging ahead: the Bakelite Gesellschaft mbH was established on 31 May 1910, the terms of which satisfied Baekeland (figure 7.2).[37]

Entrepreneurial inventors and scientists, as several historians have pointed out, may be as committed to the adoption of their technologies as to the financial performance of the firms established for exploiting these inventions.[38] Baekeland's diaries and correspondence suggest that criteria of success other than (direct) financial returns may have been equally important for understanding foreign investments. Because of World War I and its aftermath, Bakelite Gesellschaft would not become a significant source of income to Baekeland and his family until the late 1920s, when dividends on his German stock generally exceeded the highest yearly salary paid by the American Bakelite company.[39] Prior to the war, Baekeland probably benefited more from the technology-transfer stipulations and export restrictions included in the European contract, which facilitated the development of a

Figure 7.2
Baekeland (second from left) with a 1910 delegation of German visitors to the United States, including several employees of Siemens. Although Siemens was not part of the consortium of investors that established Bakelite Gesellschaft, it remained eager to keep abreast of the latest technological and legal developments concerning phenolic resins.
Courtesy of Leo Baekeland Papers, Archives Center, National Museum of American History, Smithsonian Institution, AC-0005-0000001.

domestic Bakelite business in the United States. By 1936, when Baekeland made another trip to Europe, that US business had proved so profitable that the extra income derived from European equity had probably become of secondary importance when compared with "the thrill of [technological] progress and achievement"—"a greater reward than any dividends"—that he experienced when reflecting on the growth of the industry "since my humble beginnings in my laboratory in Yonkers" (figure 7.3).[40] Above all, however, these different factors suggest that Baekeland's motives for taking out European patents and the success of his international entrepreneurship should be evaluated in relation to his accomplishments in the United States.

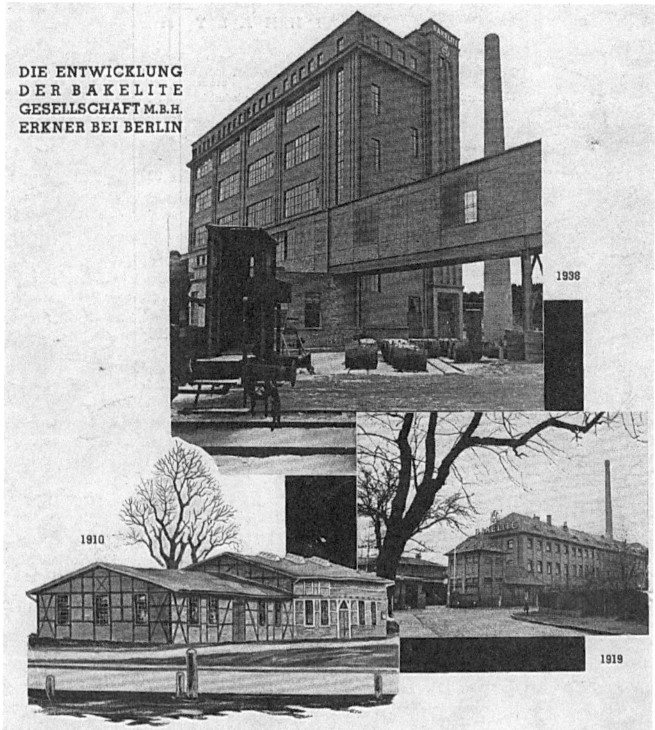

Figure 7.3
Bakelite Gesellschaft's manufacturing facilities in Erkner, Berlin.
Courtesy of Reindert Groot (amsterdambakelitecollection.com).

From Licensing to Litigation?

In America, Baekeland purposefully introduced Bakelite only gradually. Rather than establishing a Bakelite company of his own, he collaborated with a limited number of firms with whom he proposed IP licensing arrangements regarding either broad (e.g., as electrical insulator) or very specific (e.g., for grinding wheels) uses of the material. In doing so, he effectively outsourced much of the development work, thereby allowing himself and the small staff of chemists and engineers that he had started to employ in his private laboratory to focus on preparing different grades of phenolic resin and providing technical support.[41] During his stay in Berlin, Baekeland had found some time to reflect on a future business plan for the United States. After another round of negotiations at Rütgerswerke, he had confided to Sperry his "idea of in the U.S. trying Bakelite with several then finding out by experience who are really the wide-awake firms, then afterwards grouping them all in one big Bakelite consolidated Co. thus proceeding by natural selection, eliminating the non fit and non desirable and thus building up an enormous enterprise."[42]

The General Bakelite Company was formed at the end of September 1910, but it did not consolidate parties emerging from a social Darwinian struggle. Instead, the firm's board of directors was controlled by executives of a single corporation, the Roessler & Hasslacher Chemical Company (R&H), which produced formaldehyde at its works in Perth Amboy, New Jersey, and had "ramifications with all important industrial countries," among other assets.[43] In September 1909, Baekeland had queried Roessler & Hasslacher's Hugo DuBois and Jacob Hasslacher about whether they would be interested in joining a Bakelite company, but plans did not become more concrete until the following May. By then it was clear to Baekeland that his association with the Protal Company, an early Bakelite licensee with whom he had closely connected himself, seriously risked damaging his interests.[44]

Contrary to Baekeland's negotiations in Berlin, the IP situation was not much of an issue during the talks leading up to the formation of General Bakelite. On 9 December 1909, the chemist had been granted five US patents concerning various products and processes, including his so-called heat-and-pressure method for controlling the phenol-formaldehyde reaction.[45] About two months earlier, he had filed additional patent applications to prevent competitors from being able to "invent around" his IP.[46] By summer 1910, he could clearly substantiate his claim to potential investors that "the patents are so arranged that it is almost impossible to carry out anything practical without infringing at least three or four patents at the same time."[47]

In his prospectus for forming a Bakelite company, Baekeland also explicitly referred to his international patents, claiming that his IP position was even more satisfactory because of "the fact that the same patents have been granted in other foreign countries where rigid preliminary examinations are involved."[48] The next year, he would even state that, "One of the main reasons why I took out foreign patents was to have the benefit of as much criticism as possible, so as to know better the value of my claims to priority."[49] Both statements may well have been exaggerations, but the suggested connection between Baekeland's foreign and domestic IP records was very real. Especially relevant in this regard were the public-opposition proceedings included in the Austrian, German, and British patent systems, among other countries', but not in the US system.[50] In the German system, with which Baekeland became most familiar, the claims of patent applications accepted by one or several examiners were published prior to the patents' issuance. The interested public could then file objections, which were presented to the inventor. In this process, the introduction of prior research that had been overlooked by the examiner could lead to the invalidation of a patent. Baekeland's German heat-and-pressure patent application, among others, had reached the public-opposition stage between the chemist's return from Europe in August 1909 and the establishment of General Bakelite in September 1910, and he had successfully countered additional objections.[51] The experience and knowledge thus gained could be invaluable for managing his US Bakelite patents, which, because of the worldwide novelty requirement, could be invalidated by means of prior research anywhere. This included facts disclosed in documents that were unavailable at the US Patent Office, whose collections of Belgian, Canadian, and French patents were, for instance, incomplete.[52] Through the combined scrutiny of European patent examiners and interested parties, however, Baekeland significantly reduced the risk of being confronted with prior research of which he was not aware.

The strength of Baekeland's US IP rights would be tested throughout the following decade in patent fights with two major challengers: the Condensite Company of America and the Redmanol Chemical Products Company. As previous historians have described the outcome of these contestations, I will only briefly summarize it here.[53] Condensite formed in 1910 as a spin-off of Thomas Edison's National Phonograph Company in order to introduce its own phenolic resin, invented by chemist Jonas Aylsworth, for applications other than phonograph records. The conflict with Condensite, which started around early 1911, was resolved when General Bakelite and Condensite signed a patent cross-licensing agreement in May 1913, although their competition for market share and further IP rights continued. Bakelite's patent litigation with Redmanol, on the other hand, resulted in a major consolidation in the phenolic resin industry. In 1921, Redmanol's patents were held to infringe General Bakelite's, and Baekeland, the company's president, was not inclined to compromise. However, because of the licensing arrangements between General Bakelite and Condensite, Redmanol could indirectly acquire access to the Bakelite patents by purchasing Condensite stock. The ensuing negotiations led to a merger among General Bakelite, Condensite, and Redmanol during 1922–1923, the successor firm being named the Bakelite Corporation. Contrasting his German and American experiences up to 1910, Baekeland had written, "It is gratifying to know that I have the field so very clear here in the United States." Although this turned out to be overly optimistic, his patents proved to be a major force in the industry.[54]

The advent of an effective alternative to Bakelite resonated throughout the industry and put Baekeland's gradual approach under pressure. His earlier business model, which was centered on providing raw materials and technical advice, had not been revised after the formation of General Bakelite. As before, Baekeland and his associates preferred to pass on the development of specific applications to licensees, to whom they offered either nonexclusive or exclusive rights. Through these licensing arrangements, firms other than General Bakelite consequently depended on the chemist's IP rights. As one customer noted, "Through your patents, outside concerns have no way of protecting themselves in connection with bakelite, except through you, and we naturally look to you for any protection we might require through formulas developed by us and known to you."[55] This was not just a domestic problem, for Aylsworth had also applied for European patents despite the warning of his associate Frank Dyer, a patent expert, that this might give Baekeland "a very good opportunity of seeing just what we are doing."[56] (Consistent with the concerns about international patenting mentioned at the beginning of this chapter, Dyer also indicated in this letter that "foreign patents are expensive to maintain and we have never made any money out of them.") Accordingly, perhaps prodded by their licensees, the owners of the Bakelite patents took legal actions against the Condensite Company on both sides of the Atlantic.

At the same time, though, General Bakelite was reluctant to directly challenge its rival in court. Why was this so? Above all, a clash with the Condensite Company risked being long and costly. As the firm seemed "disposed to fight the case vigorously," Baekeland estimated that a final decision would only be made after three or four years, with annual costs of at least $10,000.[57] Consequently, he and his associates expected the best outcome to be no more than a Pyrrhic victory. Condensite was similarly discouraged by the prospect of prolonged litigation. In particular, its negotiators emphasized that while the two companies would be exhausting their resources fighting each other, a third party might appear to snatch the prize.[58] Both sides thus had incentives for avoiding an open confrontation and reaching a settlement. This point of view was also suggested to Baekeland by renowned patent lawyer Frederick Fish, who had indicated that a consolidation had "happened most of the time at the beginning of [a] new art when two or more parties seemed to control the patent situation."[59]

However, what would be the cost of making such a compromise? So as to impress the other side with the strength of its position, both General Bakelite and Condensite exchanged some of their central arguments in negotiations that were paralleled by talks with Bakelite Gesellschaft regarding a consolidation in Europe. For instance, Condensite indicated that it would present a patent by an individual named Allen in order to demonstrate the (alleged) invalidity of Baekeland's claims. Baekeland, on the other hand, could emphasize that the value of his patents had already been tested. His reaction once more points to the transatlantic nature of the dispute:

> I told him [Dyer] furthermore that his arguments were not new[,] that his arguments together with others had all be[en] presented by my opponents before the German Patent Office which had decided finally in my favor and that just recently my patent on the use of bases after a lengthy and energetic objection by [Fritz] Raschig had finally been adjudged in appeal in my favor. That of the four patents on which we were suing them only one[,] the method of molding[,] had been lost on contestation and then only in a technicality because my patent had been filed a few days after one of my [other] patents was already established. Dyer said we were here in the U.S. and not in Germany[,] to which I remarked that just the same the german [*sic*] patent office had the reputation of being very thorough and careful in chemical patent matters and that I consider the German Patent office at least as good judges as Mr. Dyer.[60]

Patent Reform

Baekeland's management of his European patents was not only useful for defending his interests in America, it also influenced his judgment concerning the US patent regime. The chemist gave his first public lecture on patent reform in January 1909, one month before his official announcement of Bakelite. At a symposium of the New York ACS, organized on his own initiative, he presented "the inventor's standpoint."[61] In an introductory note, US Patent Commissioner Edward Moore wrote that it was a most appropriate time for discussing IP issues, as reports on the new British patent law of 1907 were becoming available and the US Patent Office's rules of practice were undergoing revision. Baekeland's involvement grew in 1912, when President

Taft and congressional members of both parties pushed for a bill that would annul the Supreme Court's pro-patentee decision in *Henry v. Dick*, which interfered with the government's antitrust agenda. Baekeland criticized the result of their efforts—the Oldfield Bill—as being a threat to technological progress and suggested that the initiative for less "incongruous" legislation should be left to a committee of experts, as had been done in Germany. Around the end of World War I, he would serve on such a committee of scientists, inventors, and patent lawyers organized by the National Research Council at the request of the Patent Office Society, an interest group created in 1917. After committee chairman William Durand's departure for Europe, Baekeland was even appointed as acting chairman.[62] General Bakelite's president thus gradually moved to the center of the policy debate.

Baekeland's perspective was probably influenced by patent lawyer Louis Raegener. In the 1909 symposium, Raegener had criticized the 1891 reforms that had delegated the Supreme Court's jurisdiction over patent appeals to nine regional circuit courts of appeal. While confessing that lawyers had benefited from the reforms, "the star of the evening" (according to Baekeland's diary) deplored the fact that the reforms' main effect had been that "now there is never an end in sight for the unfortunate patentee." For instance, patent infringers in New York who had been convicted by the United States Court of Appeals for the Second Circuit could reincorporate in New Jersey or Pennsylvania, which were within the jurisdiction of the Third Circuit Court, "and everything, every detail ... must be done all over again at similar expense of time and money."[63] As a remedy, Raegener suggested that jurisdiction be conferred to a single tribunal, either the Supreme Court or a specialized Court of Patent Appeals. Baekeland agreed wholeheartedly. While considering the United States' patent law an unsurpassed piece of legislation, he regretted that ineffective enforcement procedures seriously diminished its practical value to independent inventors and small firms, which were "at an inexpressible disadvantage" against large corporations with the resources to engage in prolonged litigation. Like Raegener, who had privately explained his opinion to him in May 1908, Baekeland repeatedly stressed that the establishment of a single Court of Patent Appeals would be an important step toward a more just system.[64] As the head of the NRC's patent committee, he also supported the demand to reorganize the US Patent Office, suggesting that its separation from the Department of Interior and higher salaries for staff members would improve its standing and efficiency. In return for better services, Baekeland and his constituents were prepared to pay higher fees.[65]

In addition to stricter examinations by the Patent Office, Baekeland suggested that the interested public should be given the opportunity to object to patents before they were granted.[66] Furthermore, he supported the Inventors' Guild's resolution favoring patent protection for the first applicant, rather than the first and true inventor, in order to hasten the disclosure of inventions and diminish legal uncertainty.[67] In addition, he believed that infringement suits could be shortened by having testimony made in open court, and he advocated proper compensation for patentees whose rights had been violated. Finally, Baekeland suggested that judgment on the validity of challenged patents should be left to a court of technical experts at the Patent Office instead of nontechnical external courts.[68]

As Baekeland and other campaigners pointed out, most of these proposals were already in force on the other side of the Atlantic, in Northwest and Central Europe; consequently, their practical feasibility had already been tested. Rather than designing new institutions, Baekeland wished that, "If only we could borrow some of the more efficient methods with which the German patent law is administered, and enforced, we might succeed in making an American patent real property for poor and rich alike, instead of a pretext for expensive and endless litigation with all the advantages it gives to the richer litigant, to the detriment of the consumer who, in the end, pays the bill."[69] Whereas many other reformers presumably relied on English-language reports available in America, Baekeland had an additional opportunity to assess the German patent regime on the ground. In March 1913, he attended two infringement suits in Berlin to improve his understanding of the procedure and queried German experts about their opinions on the country's patent system. For instance, he discussed its functioning with patent agent Julius Ephraim, whose services had been used by Bakelite Gesellschaft, and reported on the US reform campaign to Bayer's Carl Duisberg and BASF's August Bernthsen, among other German chemists.[70]

Consensus Lost

When scientist Bernhard Hesse suggested at the 1909 symposium that the adoption of another feature of the German patent system—its annual renewal fees—might ensure the actual exploitation of patents, he said,

"Everybody (so it seemed) jumped on me and so hard that I thought I hadn't a friend in the room."[71] Ten years later, however, he felt that the time was ripe for reconsidering the proposal. He elaborated it in an open letter to *Industrial and Engineering Chemistry*, which would be extensively discussed at a meeting of the ACS in September 1919. Already skeptical about the idea, Baekeland feared above all that it, together with other minor reform proposals, would distract attention away from "the much more important fundamental reforms, such as the reorganization of the Patent Office and the reorganization of our system of court procedure through the creation of one single Court of Patent Appeals."[72] Not coincidentally, Baekeland, patent lawyers Edwin Prindle, Frederick Fish, and Thomas Ewing, and physicist Robert Millikan had agreed to focus the efforts of the National Research Council's Patent Committee on those two points.[73]

Although previous historians have argued that the committee generally accomplished the goals of its most active members and the interest group that had pushed for its creation, Baekeland was disappointed with the results.[74] As he commented to his friend Milton Whitaker in 1932:

> Our Committee, with the help of various associations, had succeeded in creating a "miracle"—a bill which had the endorsement of the engineering and chemical societies as well as the approval of patent law associations and even a large number of inventors. The Congressional patent committees were so much in favor of that bill that it was given priority for immediate action by unanimous consent. Then came the wreck! Just when action was to be taken, some of our well meaning but impractical friends in the scientific departments of the Government thought it the proper occasion to get through Congress some new laws concerning patents taken out by men in Government Service. So they introduced a rider into the pending bill. This rider was worded in such a way as to arouse violent protestations from the engineering societies, the manufacturers, the patent lawyers[,] as well as the independent inventors.—So that bill got lost.—This was the first time after many decades of endeavor that we came so near to introducing some needed reforms and to bring our patent system more up to date.[75]

The Nolan Bill, providing for the restructuring of the US Patent Office, passed the House of Representatives almost unanimously on 5 March 1920. Senator George Norris of Nebraska, however, then added the amendment to which Baekeland was referring, which authorized the Federal Trade Commission to manage the patents of government employees. Despite intense lobbying by the ACS, among others, the rider remained part of the Senate version of the bill, and ultimately Congress failed to reach a compromise.[76]

In several respects, this setback would be only temporary. Not least thanks to the tireless efforts of Prindle, who had replaced Baekeland as chair of the NRC committee in April 1922, salaries of patent examiners were raised, and the Patent Office was separated from the Department of Interior in 1925. Still, the demand to which Baekeland had attached special importance—the creation of a single Court of Patent Appeals—would not be met until 1982. Significantly, this court has since received much criticism on account of its apparent bias in favor of patentees.[77]

Private Gain and the Public Good

How closely were Baekeland's patent reform proposals related to his efforts to obtain broad IP protection regarding Bakelite? Whether or not they agreed with his views, contemporaries did not fail to notice that his personal stakes were high. In IP reform discussions within professional societies of chemists and chemical engineers, he was repeatedly brought up as an example of an independent inventor-patentee whose interests risked being harmed by the Oldfield Bill or similar legislation. In 1913, for instance, chemical consultant William M. Grosvenor mentioned Baekeland's name, among other American chemists', to challenge the claim that it was mainly large German producers of synthetic dyes that were taking advantage of the US patent system to protect chemical inventions.[78] Another acquaintance of Baekeland, who evidently was not aware that Baekeland's success in the photographic industry had owed nothing to the patent system, likewise praised the inventor of Bakelite as an important "practical example of a man who has climbed his own ladder, you might say, on the [patent] protection that the country has given him, and but for that protection he never would have had the resources to operate his private laboratory and produce empirical results in various chemical lines."[79]

There are no signs that Baekeland was uncomfortable with the use of inaccurate references to his career in attempts to justify the importance of strong patent protection. Indeed, as discussed in chapter 4, in his Perkin Medal address in 1916, he personally gave an explanation for his decision not to patent Velox that was misleading but consistent with his agenda for patent reform. There is no doubt that he also approved of the patent-centered account of the invention of Bakelite that his own son George gave in the late 1930s, when the US patent system was under renewed political scrutiny. George Baekeland, who had joined his father's company around the end of 1922, was summoned as a witness by a congressional committee appointed to address the dangers of economic concentration.[80] In his testimony, he spoke in defense of comprehensive patent protection, emphasizing its significance for the development of Bakelite from 1907 onward. In making his case, he succinctly encapsulated the modern economic justification of patents as tools for encouraging invention and technology diffusion through, respectively, the prospect of (temporarily) obtaining exclusive rights on an invention and the requirement of disclosing its technical details in a manner intelligible to anyone "skilled in the art." George Baekeland stated that the patent system "enabled him [his father Leo] to build up a glass wall around himself, behind which he could work in security without being rushed and through which the public on the outside could peer in and see what he was doing, and wait until [the patents expired after] 17 years when they knocked the walls down and everybody could then come in."[81] Thus George suggested that the societal costs resulting from the reduced competition during the lifetimes of the "glass walls" from approximately 1909 until 1926 had been a "necessary evil," without which the American synthetic plastics industry would not have been established as rapidly, if at all.

George Baekeland was right to emphasize that his father's patents and research papers had drawn others to the field of synthetic plastics. As chapter 8 demonstrates, he was also correct that the US phenolic resin industry had been opened to competition after the expiration of the Bakelite Corporation's main patents from 1926 onward. His suggestion that this industry would not have been created in the absence of patent protection is, however, impossible to verify, because of its hypothetical nature. True, among the members of the NRC patent committee on which his father had served, it was a common view that investors would not be willing to finance promising inventions without being able to exclude others from also commercializing the resulting technologies after the completion of the development work.[82] Leo Baekeland's own experiences regarding Bakelite, moreover, generally supported this assumption, as his patents played a crucial role in the establishment of the Bakelite companies and in these firms' strategies for securing market dominance. Yet, as discussed in chapter 4, historical and contemporary studies have also documented many cases where, contrary to this supposition, inventors, scientists, and firms attached relatively little importance to patents or even regarded them as harmful.

Scientists' perspectives on this issue are particularly pertinent because, as Leo Baekeland wrote to Whitaker, the involvement of some of them in patent policy making had contributed to the failure of the congressional bill that Baekeland himself had promoted as chair of the NRC committee. To some extent, this debacle can be traced back to an experiment in patent management that Berkeley chemist Frederick Cottrell had initiated in the early 1910s. Like Baekeland, Cottrell had obtained his first patents in the course of his employment at a public university and was eager to see his invention, an electrostatic air pollution control device, implemented in practice. By 1912 at the latest, Cottrell also shared Baekeland's belief that the provision of exclusionary IP rights was crucial for securing funding because, as Cottrell repeatedly put it, "what is everybody's business is nobody's business."[83] Cottrell's commercialization strategy, however, was very different from the approaches that Baekeland had followed, as he had no interest in becoming a business entrepreneur and was hoping to use the money that his invention might generate for philanthropic purposes. After having first approached his own university and the Smithsonian Institution, Cottrell's commitment to public service eventually led to the creation of Research Corporation in 1912. This was a nonprofit organization to which academic (and other) inventors could assign their patent rights and whose earnings would be used to fund scientific research.[84]

Cottrell's charitable actions generated considerable enthusiasm within (and beyond) the US scientific community. In 1919, for instance, Chicago chemistry professor Julius Stieglitz argued that the "patent difficulty" could be solved by emulating the example of his professional colleague, who had moved from the University of California at Berkeley to the US Bureau of Mines. More specifically, in Stieglitz's opinion, the use of Cottrell's patents as a means for ensuring the broad availability of his pollution control device showed that patenting could be a tool for maintaining ideals of academic openness. Accordingly, Stieglitz ended his paper by encouraging universities, government laboratories, and research institutes to disseminate chemical and pharmaceutical inventions "free[ly] to the world" by licensing them at a fair price to all interested parties.[85] He must have been aware that Cottrell had taken the lead in drafting a bill meant to make it easier for government researchers to handle their IP rights in such a public-spirited way. However, both Stieglitz and Cottrell could hardly have anticipated that the addition of a revised version of this proposal to the Nolan Bill of 1920–1921 would prevent the latter from being turned into law.

Stieglitz expounded his opinion on patent reform because he felt that experts such as Baekeland "may be speaking for the people who take out and own patents" rather than for society at large.[86] Baekeland himself, as we have seen, claimed to be mainly protecting the interests of inventors and small firms, which he did not consider to be in conflict with those of the general public. Indeed, he defended the view that protecting patents and other forms of "intellectual property" at least as strongly as chattels or real estate was a much better way of serving the public interest than including provisions meant to foster technology diffusion by restricting the rights of patent holders.

Baekeland's disappointment about the last-minute failure of the Nolan Bill must have been tempered by the victorious completion of his own firm's patent suit against the Redmanol Company. When he resigned his chairmanship of the NRC patent committee in spring 1922, he was in the midst of merger negotiations with Redmanol and Condensite. As it turned out, the merger did not just consolidate the monopoly over the production of phenolic resins held by Baekeland and his now enlarged group of associates. Importantly, it also strengthened the power of the "Bakelite" trademark, as the former rivals would start promoting all their phenolic resin varieties under this brand from the mid-1920s onward.

Partly thanks to Baekeland's international publication strategy and the efforts of the Bakelite Gesellschaft mbH, the Bakelite name was also widely adopted outside the United States. In addition to the US and German factories, by 1916 Bakelite plants had been established in Bulgaria, Canada, England, France, and Russia. After World War I, moreover, further factories were built in Italy and Japan, the latter becoming the East Asian center of production. Traveling in Latin America in 1920, Baekeland opposed "propaganda that had been started against Bakelite cigar and cigarette holders as being of a poisonous nature."[87] For reasons that are explored in chapter 8, Bakelite smoking accessories were a particularly widespread application. In Spain in 1923, for instance, Baekeland commented that, "When I see a man driving a Ford car and smoking a bakelite cigarette the world seems small after all."[88]

Baekeland did not just have plenty of reason to be satisfied about the outcome of his own patent suits. At least to some extent, he had also enjoyed the challenge of asserting his claims in court in support of his economic interests and of what his wife had referred to as his "moral or fame ambition." As mentioned in chapter 5, previous studies have portrayed Baekeland as a "reluctant innovator ... drawn away from his lab" to ensure that Bakelite would enter the marketplace.[89] At first sight, this interpretation is supported by the archival record, which offers numerous statements in which he expressed frustration over his prolonged involvement in business. Of all his distractions, patent litigation was among the biggest: "Of all the irritating, time-robbing experiences in my life," he claimed in 1932, "I consider patent suits as some of the worst."[90] However, a previously unknown letter from Baekeland to history professor Paul Fredericq, a friend from his hometown of Ghent, puts his vexation into perspective. In it, Baekeland admitted that he had actually enjoyed his involvement in *General Bakelite Co. v. George J. Nikolas*, a suit about which his complaints had been particularly vocal. After pointing out that patent suits are "unusually expensive," he confessed in the letter:

> I like a first-class fight of the kind, as long as I can feel that I can hold my own. I suppose if I were born one or two centuries sooner, I would be fighting something else.... The past patent suit was more interesting to me than a first-class baseball game. Sometimes I have fun in thinking that people who are not acquainted with an inventor's life imagine that all that he has to do is to get a lucky inspiration in his laboratory and forthwith the dollars are pouring in. They do not know all the work of development, all the struggles, all the difficulties, all the responsibilities, steady fight against infringers and imitators, the commercial difficulties besides the technical problems which have to be solved. And yet a struggle of the kind is very beneficial to the one who undergoes it. It strengthens his character and urges him to further efforts.[91]

8 Fancy Consumer Goods in an Industrial Age

When Leo Baekeland publicly announced the invention of "Bakelite" in February 1909, as we have seen, he was acutely aware of the material's commercial potential. He maintained that applications "in more than forty different industries" were being studied and, although not numerous enough to justify that claim, the examples he gave did cover a remarkably broad territory. On the one hand, the chemist presented his synthetic plastic resin as a substitute for celluloid and for "natural" plastics such as ivory, which had primarily been used in the manufacturing of decorative and small consumer items such as collars, combs, knife handles, and billiard balls. On the other hand, he predicted that Bakelite, thanks to its technical qualities, would be of great value in industry as well. "As an insulator," he claimed, "and for any purposes where it has to resist heat, friction, dampness, steam or chemicals it is far superior to hard rubber, casein, celluloid, shellac and in fact all plastics."[1] Such a stress on versatility may have been a sign of weakness rather than strength, for the development and commercialization of early plastics and other new materials could require innovators to identify a single focus instead of casting too wide a net.[2] Yet Baekeland seemed to have clear priorities. After referring to possible Bakelite consumer items, he continued by stating that "its use for such fancy articles has not much appealed to my efforts as long as there are so many more important applications for engineering purposes."[3]

The prioritization of industrial products over decorative ones was indicative not only of Baekeland's personal preferences but also of his times. The very notion of "technology," for example, was closely connected with industrial manufacturing when it began to enter into popular usage in the 1930s. More than the earlier keyword of "the useful arts," *technology* became the province of white middle-class men, notably corporate engineers, who had a professional interest in associating themselves with science.[4] According to historian Leo Marx, the rise to prominence of the knowledge-intensive chemical and electrical industries in the late nineteenth and early twentieth centuries contributed to the creation of the "semantic void" that the revised technology concept was about to occupy.[5] As discussed in chapter 5, Baekeland regarded these industries as markers of societal progress. Thanks to his consulting work for chemical and engineering companies, he also had a better understanding of industrial markets in which aesthetic considerations were largely irrelevant.[6] He could, moreover, fall back on an extensive professional network in this area, in part because of his involvement in industry-oriented scientific associations such as the American Institute of Chemical Engineers and the American Electrochemical Society.

But to what extent could Baekeland control the technology's development? In his acclaimed cultural history of plastics in the twentieth-century United States, Jeffrey Meikle emphasized Baekeland's loss of influence after his February 1909 announcement, which brought his work to the attention of competitors and new customers, who would attribute various meanings to "Bakelite" during the ensuing years.[7] Some of the resulting challenges were highlighted and analyzed in chapter 7, yet we have also seen that Baekeland's basic patents on Bakelite were nevertheless powerful tools of control. Building on that, the first part of this chapter investigates the extent to which Baekeland and his business associates could effectively translate the legal monopoly to which a "first and true" inventor in America was entitled into market dominance. To this purpose, I pay special attention to *General Bakelite Co. v. George J. Nikolas*—the patent suit that the chemist found more interesting "than a first-class baseball game"—as well as to changes in the phenolic resin industry around the end of 1926, when several of his basic patents expired. This chapter then returns to Baekeland's stated preference for industrial plastic over novelty items. This distinction was reflected in General Bakelite's organizational structure: Bakelite variants sold to producers of "fancy goods" were predominantly manufactured by a single department. This was the cast resin unit—known as the "transparent" department among the firm's employees because of the relative translucency of this type of Bakelite—about which separate data are available. Baekeland's declared preference hence can be compared to quantitative evidence regarding the evolution of his company's sales. In addition, qualitative information from his diaries, among other sources, helps explain the tendencies revealed by these data.

This chapter also continues to explore the perceived moral legitimacy of science-for-profit. In his public role as a professional spokesperson, Baekeland countered the allegation that industrial chemistry was merely "the chemistry of dollars and cents."[8] Yet, as president of the American Bakelite companies, he risked facing the potentially damaging impression that his firms' expansive IP management and monopolistic pricing had slowed the growth of the phenolic resin industry, impeding technoeconomic development. A prioritization of industrial Bakelite would, moreover, turn out to be less attractive from a financial point of view than Baekeland figured in early 1909.

A Powerful Monopoly

Baekeland's potentially precarious position as a scientific monopolist is perhaps best illustrated through the example of his follow-up paper, "The Use of Bakelite for Electrical and Electrochemical Purposes," which he presented to the American Electrochemical Society on 8 May 1909. As suggested by his title, Baekeland's talk focused on practical applications of Bakelite, discussing its use in, respectively, block working, impregnating, coating, and molding processes. Thereby, he strategically disclosed technological details of his production processes, including his use of a "Bakelizer," the vessel in which he simultaneously applied heat and pressure. He also described the properties of Bakelite "A," the initial condensation product of phenol and formaldehyde; Bakelite "B," the intermediate substance; and Bakelite "C," the final condensation product, which was practically insoluble and infusible. On the whole, the content of the presentation suggests that Baekeland had primarily intended to further clarify Bakelite's potential for technical purposes. Nevertheless, he turned reticent when Samuel A. Tucker, the head of Columbia's electrochemical laboratory, queried about a parameter that was more than relevant in this connection. The episode, included in the professional society's *Transactions*, is worth quoting at some length:

> Prof. S. A. Tucker: I would like to ask Dr. Baekeland the probable comparative costs of finished Bakelite and viscose. Apparently viscose has gone out, due to the expense of its manufacture, but I should think that that was rather a cheap product.
>
> Dr. Baekeland: This is rather a commercial question, which it is hard to answer before a scientific body. I am amazed at the commercial instincts of dear Prof.

> Tucker. But I may say this that it depends, of course, on the price at which [Bakelite] A is sold, which is a matter that I am not prepared to present here in public. But I can state this, that the ultimate possibility is that Bakelite will ultimately be available at such a startlingly low price that a large number of applications will be possible, even beyond these I have mentioned.

Was it really considered inappropriate to discuss the price of a new technology at a scientific meeting? Given the industrial orientation of the American Electrochemical Society and the topic of Baekeland's talk, Tucker's question was certainly not far-fetched. It moreover reflected a worldliness that, in Baekeland's own opinion, other academics unfortunately lacked. The fact that a university professor raised the issue of cost at a public scientific gathering, however, allowed Baekeland to appeal to an ideal of disinterestedness with which much of his audience must have been familiar. In this specific instance, he clearly did this as an excuse for not disclosing confidential information, which further illustrates his skills in navigating the boundaries between the worlds of science and industry.

All the same, it is revealing to compare Baekeland's quoted prediction about the eventual lowering of Bakelite's cost to the price evolution of the phenolic resins sold by his companies (see figure 8.1). Starting at a level of about $0.40 per pound in 1911, a drastic price increase occurred after 1917, leading up to a peak of $0.65 per pound in 1920. This rise was largely caused by raw material shortages during and after World War I, which prompted large chemical firms such as Dow to invest in the development of synthetic phenol processes.[9] Then a remarkable decrease took place between 1923 and 1933, by which year the average sale price of the Bakelite Corporation's resins had dropped to about $0.17 per pound, less than half its initial level. The declines were especially pronounced in 1926–1927 and 1927–1928, when Bakelite on average cut its prices by about 20–22 percent. Industry watchers predicted that this decrease would convince several manufacturers to start using Bakelite instead of metal or wood. Similarly, in this period, "the old belief that molded products [of phenolic resin] were limited practically to the electrical insulation field" was vanishing fast.[10]

As with the rise before 1920, prices of phenol, formaldehyde, and their equivalents influenced the pattern shown in figure 8.1. Partly thanks to the introduction of synthetic processes, these raw materials had become cheaper and their American consumers no longer depended on European imports. Manufacturers' increased ability to exploit economies of scale and

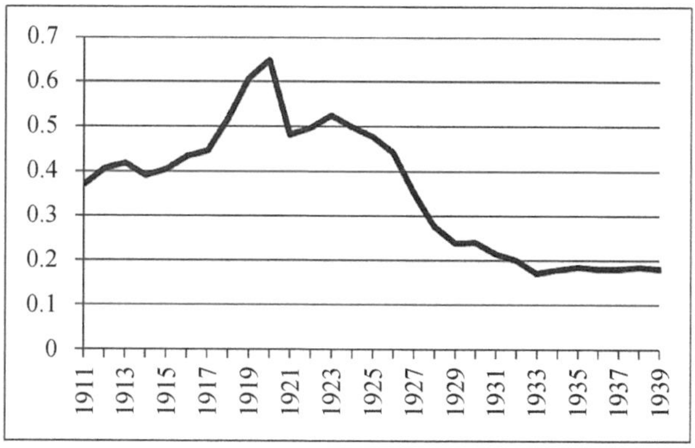

Figure 8.1
Average net sales price of phenolic resin for Bakelite companies (in dollars per pound). Based on data from Lawrence Byck, "A Survey of the Bakelite Thermosetting Business," in LBP, series 1, box 2, folder 5.

scope also contributed to the price decline.[11] Yet the exact chronology of events suggests that another factor may have been more important. Precisely at the end of 1926, several of Baekeland's basic patents had expired, seriously undermining his company's monopolistic position in the sector.

The Bakelite Patents

Baekeland's most renowned patent probably was US Patent 942,699, entitled "Method of making insoluble products of phenol and formaldehyde" but commonly referred to as the "heat-and-pressure patent."[12] The history of this patent, from Baekeland's initial application of 13 July 1907 until its expiration at the end of 1926, provides further evidence of the clever ways in which he and his legal advisers managed his IP rights. In addition, it provides a new perspective on how Baekeland's own understanding of his invention changed between his laboratory breakthroughs in mid-1907 and his first public research presentations on Bakelite almost two years later. At the latter time, Baekeland pointed out in his diary that some of his rivals seemed not yet aware of "the great importance of my broad claim of heat and pressure" in his pending patent application.[13] Yet, the amendments made by Baekeland personally and by his patent agent, Clinton P. Townsend, reveal that both of them had only belatedly recognized its strategic significance themselves.

Only one out of nine IP claims presented in the original application concerned the combined use of heat and pressure, and the accompanying description of the invention also did not emphasize this feature. Instead, the patent focused on Baekeland's process for obtaining his initial phenol-formaldehyde condensation product and for hardening this viscous or oily substance—either alone or in combination with other materials such as wood—by heating it at a temperature of approximately 110–140 degrees Celsius. It was not until March 1908, after Baekeland and Townsend had extensively discussed this and other pending patents with each other and had visited the examiner who had dismissed the original submission, that they first highlighted Baekeland's joint application of heat and pressure in the production of Bakelite C in an amended application (figure 8.2).[14] Such changes in emphasis were not unusual; as a case in point, the lawyers of Baekeland's childhood hero Alexander Bell essentially reinvented the meaning and significance of Bell's original telephone patent in the context of high-stakes litigation.[15] Still, notwithstanding his personal meeting with Baekeland and Townsend, himself a former Patent Office employee, the responsible examiner rejected all heat-and-pressure claims, asserting that "the original specification ... nowhere discloses pressure." Moreover, even after Baekeland and Townsend had successfully countered this allegation, they had to make additional amendments to sufficiently differentiate Baekeland's revised claims from those of earlier inventors such as Adolf Luft and Henry Story, who had previously patented phenol-formaldehyde molding materials. To some extent, the discussion of prior research in Baekeland's February 1909 paper on "The Synthesis, Constitution, and Uses of Bakelite" can be seen as a response to these objections. It took an extra personal interview with a senior patent examiner, however, before Baekeland's now fundamentally revised heat-and-pressure patent was finally granted in November of the same year.[16]

As we have seen, Baekeland also secured several other US and international patents. They covered not only different steps of his production processes, such as the use of bases as condensing agents, but also specific end products. As an example, one of his patents concerned cardboard created by compressing alternating layers of paper and Bakelite—an important application for electrical insulation purposes.[17] To prevent competitors from "inventing around" his IP rights, Baekeland had even filed for patent protection of alternative processes that he considered impractical.[18]

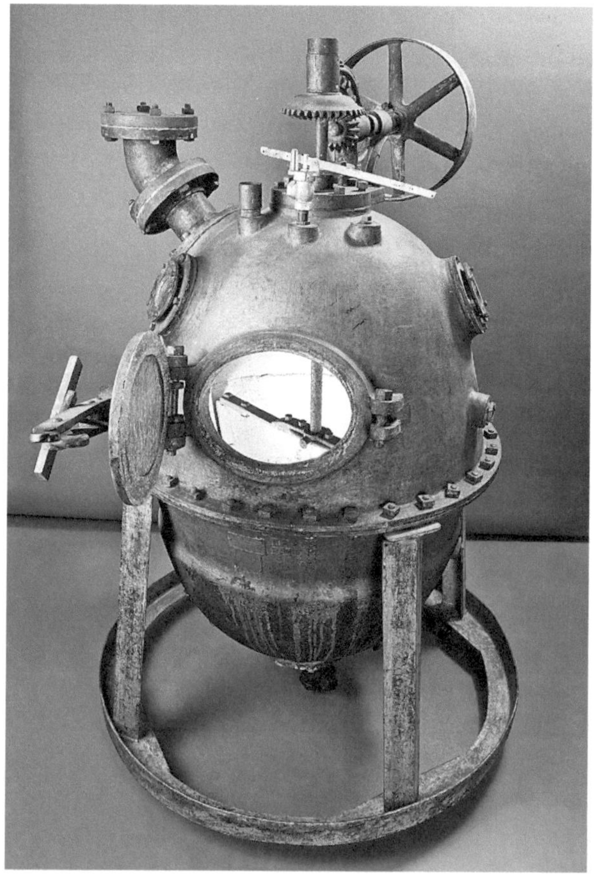

Figure 8.2
Baekeland's original "Bakelizer," a steam-pressure vessel used to convert early- or intermediate-stage Bakelite into the final condensation polymer. In the initial application of what would become his crucial "heat-and-pressure" patent, Baekeland recommended the use of closed rather than open vessels to avoid the vaporization of formaldehyde. He did not, however, explicitly discuss the importance of applying a countervailing pressure while heating Bakelite A or B at high temperatures. In contrast, a related patent application shows that, by December 1907, Baekeland and Townsend had started to put more emphasis on the combined use of heat and pressure as a key technological feature of the Bakelite process that might also serve the legal purpose of differentiating Baekeland's IP claims from rights granted to earlier inventors. In December 1907, Baekeland also designed a "big Bakaliser"—possibly the one shown in this photograph. See diary 4 (18 December 1907), laboratory notebook BKL IV (10–22 December 1907), and the files of US Patents 942,699 and 942,700.
Courtesy of National Museum of American History, Smithsonian Institution; photo by Harold Dorwin, 2006-28230-000001.

Arguably, however, the value of any patent portfolio remained open to question until it had been tested in court, and, as highlighted in chapter 7, patent litigation could be long and costly. In an apparent "eureka moment" during the preparations for an early infringement suit, Baekeland foresaw the desirability of expounding to the judge "the manysided results of my research" before turning to the three specific patents that his opponent, George J. Nikolas & Company, a Chicago-based manufacturer of varnishes, was alleged to have infringed.[19] Even if Baekeland and his legal advisers had desired to keep the basic Bakelite patents out of the proceedings, their scope and validity would probably have been reexamined. Nikolas & Company's strategy of defense, after all, partly consisted in arguing that the inventions claimed in these patents were insufficiently novel to be eligible for IP protection. In any case, Baekeland's sense that it was desirable to stress the multifaceted nature of his invention proved justified. After the suit, he enthusiastically commented to Columbia's Charles F. Chandler that *General Bakelite Co. v. George J. Nikolas* had been "started on three of my patents covering the manufacture of varnishes or lacquers" but had been "conducted in such a way that the status of most of my fundamental patents was examined."[20] Consequently, the ramifications of the verdict against Nikolas were broader than Baekeland's firm had anticipated when submitting its bill of complaint. The decision would, in effect, strengthen General Bakelite's position in later lawsuits that had nothing to do with the varnish industry.

The preceding analysis of the Nikolas case is consistent with what historian Wiebe Bijker has characterized as the "interpretative flexibility" of technological artifacts such as Bakelite. Rather than attempting to attribute to the invention a fixed meaning, Baekeland pragmatically adapted his descriptions of it so as to ensure the broadest possible level of protection. When Nikolas & Company's counsel William R. Rummler queried whether Baekeland considered varnishes or lacquers to be major applications of Bakelite, for example, he responded that, "Just now I would say no, but that changes from day to day; what is not important today becomes important tomorrow." Moreover, he went on to claim that not merely varnishes but "every thing was a side issue [at the time of his first paper presentation to the ACS] because they were all applications of one principle."[21] In this connection, it is important to note that Baekeland repeatedly reconsidered not only the nature of his invention but also the meaning of specific passages from his research papers on Bakelite. He had also provided the US Patent Office with copies of his earliest papers to justify amendments that he desired to make to previously filed IP applications.[22] For this reason, among others, it seems misleading to consider solely the loss of control that Baekeland's public disclosures caused by bringing his work to the attention of challengers. In effect, from an IP point of view, these information releases at least occasionally helped strengthen his grip on the phenolic resin industry.[23]

Strained Relations

Rummler did not dispute Baekeland's IP entitlement to solid variants of Bakelite but reproached him with having "invaded" the field of varnishes, which he considered "our territory."[24] He would not be the only person to resent the business policies of Baekeland and his associates. Perhaps most notably, in November 1926, Richard W. Seabury of the Boonton Rubber Manufacturing Company suggested to five other licensed Bakelite molders to consolidate their companies into a General Molding Company. A major reason for doing so was to strengthen their bargaining position vis-à-vis Baekeland's firm, their supplier of phenolic resin, for which Seabury claimed to have been "shoveling sand" during the lifetime of Baekeland's basic patents.[25] In consultation with Baekeland and Nathaniel Thurlow, who continued to work for Baekeland until the late 1910s, Boonton had started experimenting with Bakelite molding mixtures around the beginning of 1908. The firm then negotiated an IP licensing arrangement for the use of Bakelite for electrical insulation purposes and, to enter this branch of manufacturing, erected a new plant in 1909. However, when the Bakelite company was losing business because of the success of Condensite, the rival phenolic resin that had entered the market around the end of 1910, it started delivering its products directly to important customers such as the Remy Electric Company and the Dayton Engineering Laboratories Company (Delco), which had previously purchased Bakelite from Boonton Rubber. By doing so, Seabury charged in summer 1912, Baekeland's company had violated at least the spirit of its agreement with his firm.[26] Although Boonton did find new customers and would gradually increase its sales volume during the following decade, it could not put its Bakelite molding department on a profitable basis (see figure 8.3). In contrast, General Bakelite's small circle of shareholders received the first of a long series of dividends at the beginning of 1914, while cash reserves could simultaneously be built up.[27] Unfortunately, no profit and loss accounts of the firm have

Figure 8.3
The unprofitability of Boonton Rubber's Bakelite sales. These profit/loss calculations included costs for labor and R&D. Based on data in J. Harry DuBois Collection on the History of Plastics, Archives Center, National Museum of American History, Smithsonian Institution, series 3, box 3.

been preserved. Yet, in the words of an early employee, it was "common knowledge that the profits were decidedly substantial," not least thanks to the legal protection offered by Baekeland's patents.[28]

In August 1921, Baekeland opined that "it is a well known fact that the public at large is always siding with the infringers—the pirates[,] and seems to bear a grudge against the inventor because he has a limited monopoly of 17 years."[29] Broader US audiences, however, were mostly unaware of the resentment caused by the assertive, comprehensive exploitation of the Bakelite patents. Complaints from enterprises like Boonton certainly pale when compared with those faced by, for example, airplane inventor Orville Wright and the Wright Company, whose irreconcilable stance toward patent infringers had developed into a public relations catastrophe by 1914.[30] How could Baekeland avoid a similar fate? One difference was that synthetic phenolic resin could be effectively publicized as an "entirely American," "science-based" technology, whereas the United States was clearly lagging behind European, notably French, competitors in the field of heavier-than-air aviation by the onset of World War I.[31] Without a doubt, Baekeland's own public relations efforts also played their part. As early as 1910, he had faced the objection that Bakelite had not yet stood the test of the market, despite the high expectations surrounding it. During the negotiations concerning the formation of an American Bakelite company, Jacob Hasslacher had confronted him with the question "Why is it that nothing has been accomplished thus far [although] we have heard since so many years about Bakelite?"[32] Besides putting the facts straight—Bakelite had only been publicly announced the year before, and sales had been increasing slowly but steadily—Baekeland pointed to the step-by-step approach described in chapter 7. This was an explanation that seems to have satisfied Hasslacher, who was about to become General Bakelite's first vice president and treasurer and one of the firm's main financiers (see chapter 10).

On other occasions, Baekeland blamed collaborators and licensees for causing unnecessary delays by mishandling the phenolic resin. In this respect, too, his Perkin Medal address of 1916 was particularly significant because of its wide circulation and its uncritical use as a source by contemporaries and later historians. In this autobiographical speech, Baekeland described how he had been astonished to find "that people who were proficient in the manipulation of rubber, celluloid or other plastics were the least disposed to master the new methods which I tried to teach them or to appreciate their advantages." Similar to his explanation for why hobbyists had experienced fewer problems than professional photographers in using Velox, he believed that this was because of the reluctance of such Bakelite users to abandon familiar working methods and routines and simply "listen patiently to suggestions from newcomers in the field."[33] As mentioned, however, techniques for molding Bakelite had been pioneered at Boonton *Rubber*. Moreover, along with other evidence, other companies' eagerness to learn from Boonton suggests that the accomplishments of Seabury's enterprise had by no means been limited to implementing instructions from Baekeland and Thurlow.[34]

Indeed, in reality, Baekeland also depended on the specialized knowledge, skills, and equipment of firms such as Boonton to determine the suitability of Bakelite for several possible applications. With regard to Bakelite phonograph records, for example—a use that was eventually abandoned—Baekeland honestly admitted to a visiting expert that he had little understanding of the mechanical requirements of commercial sound records.[35] At this time, furthermore, experiments at Baekeland's private laboratory had not been advanced to a stage where pecuniary payoffs were within easy reach; nor did he consider it his duty to continue them to this point. The technological uncertainty inherent in his proposed division between chemical and mechanical tasks, however, probably contributed to at least some customers' sense that his financial demands were excessive. The National Phonograph Company's manager, Frank Dyer, for instance, claimed that the patent royalty figures proposed by Baekeland were "prohibitively large."[36]

Similarly, Bakelite's comparatively high price partly explains the initial reluctance of General Electric to purchase the material. GE may have had little choice, though; according to its leading engineer, Charles Steinmetz, the electrical giant was "badly in need of an insulating material practically regardless of its cost."[37]

Prices Down

General Electric was one of the companies that started manufacturing Bakelite themselves as soon as Baekeland's relevant patents had expired in and after the end of 1926. In addition to the loss of business to these customers-turned-producers, new competitors also entered the market in this period. Significantly, according to data from the US Tariff Commission, the number of phenolic resin manufacturers rose from two to seven between 1926 and 1927 and grew further to fourteen before the end of 1930.[38] Baekeland had anticipated this increase in competition and had felt nervous about it.[39] In response, he and his coexecutives agreed to drastically lower prices of the Bakelite Corporation's molding mixtures, a core product line, which would even be temporarily sold at a loss around September 1928. More generally, in this period of transition, the Bakelite Corporation's average profit per pound of sales dropped from almost seven cents to two cents.[40] Even before Bakelite's price reductions had been fully implemented, industry watchers had been amazed by the extent to which phenolic molding powders had cheapened.

The rapid expansion of output in the phenolic resin industry after 1926 was perhaps equally remarkable. By 1930, the total sales volume had more than doubled. While it was widely acknowledged that the expiration of his basic patents played a key part in this acceleration, Baekeland tried to counter the impression that his companies' market dominance had delayed the development of the industry. Even to the carefully formulated comment that the industry's "spectacular growth" had not begun until around 1927, for example, he objected by highlighting earlier accomplishments.[41]

Despite the increase in competition, the Bakelite Corporation did succeed in retaining a high market share after the expiration of Baekeland's basic patents. Until 1939, when the firm was sold to Union Carbide, it accounted for up to 45 percent of all phenolic resin sold, according to the best available estimate.[42] What happened in the 1930s is, however, less relevant to the goals of this chapter, for it would be far-fetched to study the influence of Baekeland's personal preferences and choices on technological trends in the industry at this stage of its development. Even putting aside the sector's rapidly increasing size and diversity, the fact remains that crucial initiatives at the Bakelite Corporation were now being taken by a new generation of executives, including three sons of one of the founders of the Condensite Company of America.[43] In this decade, Bakelite also became a relatively large corporation: because of rapid growth in the years before 1936, it employed about 1,300 workers by the end of the decade.[44] General Bakelite, in contrast, had been a small company by any standard. It had also been financially controlled by Baekeland, its president and founding entrepreneur. Together with Baekeland's strong patent position and the newness of the industry, that explains why priorities indicated in his seminal 1909 paper deserve to be taken seriously.

Industrial Plastic Revisited

Baekeland expressed his preference for industrial Bakelite not only in his 1909 announcement but also on later occasions. In 1924, for example, industrial chemist Arthur D. Little had agreed to give a lecture on "The Relation of Research to Industry" at Princeton University. As "the development of Bakelite naturally comes to mind as an outstanding example of applied research," Little desired to discuss it in his talk and asked his professional colleague to provide him with some data on the size of the phenolic resin industry. Apparently eager to nudge Little's interpretation of those statistics, Baekeland first claimed that "as far as I know there is not a single synthetic coal tar product which, even not excluding the famous indigo synthesis, has in the same number of years after its birth, found such a volume and diversity of industrial applications." Then, he added that "I should mention that transparent Bakelite and ornaments are mere trifles

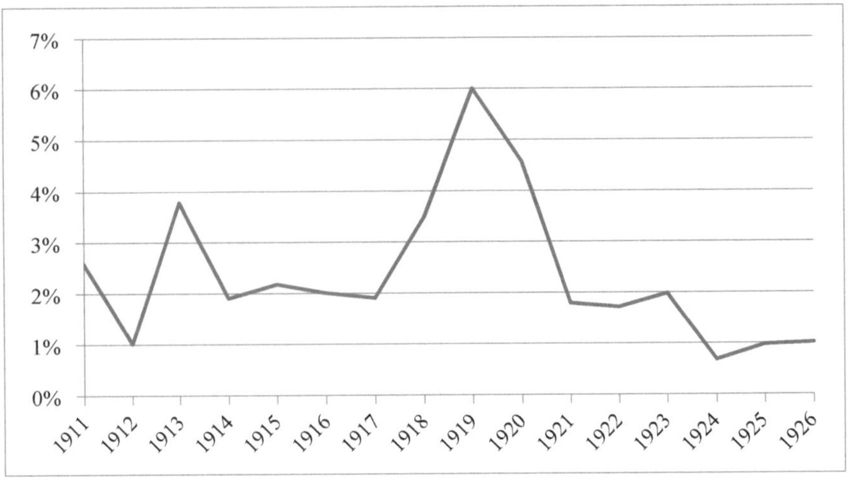

Figure 8.4
Bakelite cast resin sales as percentage of total sales volume. Based on data from Byck, "A Survey of the Bakelite Thermosetting Business."

in volume compared to the incomparably more important and technical application in the electrical and mechanical arts."[45]

"Bakelite transparent and ornaments" were predominantly cast and, as figure 8.4 illustrates, Baekeland was right about the relatively negligible volume: cast resin sales of General Bakelite (until 1921) and the Bakelite Corporation (from 1922) never accounted for more than 6 percent of these companies' total sales. The electric, automotive, and radio industries, on the other hand, on average consumed 85 percent of General Bakelite's output between 1911 and 1921 (see table 8.1). Instead of cast resin, these customers primarily purchased laminating varnishes and, to a smaller extent, molding mixtures.

It would, however, be misleading to take into consideration only the amount of Bakelite sold, for there is strong evidence that the average profits on cast resin considerably exceeded those on other product lines up to about 1920. One of General Bakelite's first employees even claimed that "money from the Transparent business built the Company."[46] Moreover, contrary to what Baekeland's public statements might suggest, the dominance of industrial Bakelite did not result from preferential treatment by the material's inventor. Rather, even before "technical" product lines were being operated on a relatively large-scale, profitable basis, Baekeland put

Table 8.1
General Bakelite, sales volume by use classification (in pounds and as percentage of total sales).

Year	Automotive		Electric and radio		Cast resin (smokers' articles)	
1911	8,524	8.53%	77,516	77.60%	2,576	2.58%
1912	103,145	22.47%	231,679	50.47%	4,661	1.02%
1913	181,163	25.60%	420,616	59.43%	26,725	3.78%
1914	203,521	25.97%	481,336	61.42%	14,968	1.91%
1915	276,683	23.15%	726,370	60.78%	26,022	2.18%
1916	533,952	26.64%	1,201,770	59.95%	40,297	2.01%
1917	698,314	24.78%	1,849,398	65.64%	53,696	1.91%
1918	467,374	16.01%	2,154,159	73.77%	101,601	3.48%
1919	663,060	28.32%	1,336,432	57.09%	140,508	6.00%
1920	762,224	24.33%	2,005,597	64.01%	143,449	4.58%
1921	379,515	27.93%	772,615	56.86%	24,346	1.79%

Source: Byck, "A Survey of the Bakelite Thermosetting Business."

great hope and effort in Bakelite's use for smokers' articles and, to a lesser degree, buttons.

Anticipated profit margins go a long way in explaining this contradiction. As early as March 1908, Baekeland had expressed his intention to provisionally limit himself "to such applications of Bakelite as can pay 25c per lb."[47] Two years later, moreover, he realized "that people are willing to pay [a] high price for fancy buttons and that one must strive to get a special effect with Bakelite more than to make Buttons cheaply" (see figure 8.5).[48] In the short term, the business case for this decision was sound. In 1911, for instance, General Bakelite sold Transparent C—a variant used in, among other things, the production of pipe stems—for $2.50 per pound; that is, more than six times as much as its average selling price around this time (figure 8.1). Accordingly, Baekeland also required his Canadian licensees, Wallace Cohoe and Harold Van der Linde, to pay larger royalties on products incorporating "transparent" cast resin.[49]

General Bakelite's largest consumer of this type of Bakelite became William DeMuth & Company, a New York–based manufacturer of pipes. Baekeland had first met Leopold DeMuth, the company's president, on 20 April 1910.

Figure 8.5
Buttons, pipe stem, and cigar and cigarette holders made of transparent Bakelite. Courtesy of Karel Julien Cole (juliensart.be).

Baekeland's diary notes regarding their following meetings confirm that he was less acquainted with DeMuth's trade than with the chemical and electrical industries, among other capital- and knowledge-intensive sectors of the second industrial revolution. For instance, Baekeland was disappointed about the low level of mechanization at DeMuth's plant in Brooklyn.[50] On the other hand, the size of the industry positively impressed him, and DeMuth promised large orders if Baekeland could deliver phenolic resin of satisfactory quality.

At this early stage, there was much discussion between Baekeland and DeMuth as to whether Bakelite should be presented as a substitute for amber or as a new and distinct material of superior quality. According to his notes, Baekeland had convinced DeMuth of the benefits of the latter strategy by the middle of June 1910. However, at least until November 1911, customers of DeMuth objected when informed that Bakelite pipe stems, unlike those made from amber, were not color-permanent. When hearing this from a DeMuth salesman, Baekeland repeated that Bakelite should not be offered "as an apology for amber" but as a better alternative that, among other qualities, "in time ... acquires a pretty glow which amber never can attain"

Figure 8.6
William DeMuth & Company advertisement for Bakelite smoking articles. Reproduced from *The American Magazine* 88 (1919).

(figure 8.6). Despite this endorsement of Bakelite's potential for smoking accessories and General Bakelite's extensive research and development efforts in this area, the remainder of Baekeland's diary entry indicates that he felt ambivalent about his involvement in this trade. Whereas General Bakelite dubbed its own salespeople "service engineers," the chemist now grumbled that, "I have now to train pipes salesmen!"[51]

DeMuth apparently overcame its customers' objections relatively quickly, for in 1912 it committed itself to ordering phenolic resin for at least $100,000, or more than $2.5 million in 2018 US dollars.[52] Moreover, seemingly confident in its ability to maintain its early lead, DeMuth allowed General Bakelite to start selling phenolic resin to its competitors. Why, then, did General Bakelite's cast resin sales not take up a larger proportion of its total sales and, after 1919, scarcely increase in absolute numbers (see figure 8.4)?

Without a doubt, the company suffered from at least three major constraints. First, there were repeated setbacks caused by technical shortcomings of General Bakelite's cast resin. Despite Baekeland's claim, color instability was one major concern. It proved especially challenging to develop phenolic resin that properly withstood prolonged exposure to daylight. Significantly, in the months after Baekeland's meeting with the DeMuth salesman, Bakelite's researchers continued earlier attempts to retain "a permanent amber color" and, in consultation with Townsend, Baekeland prepared a patent application covering this topic.[53] This IP request by no means concluded this line of work. In fact, a research and development program on transparent Bakelite's color stability, in which the majority of General Bakelite's research scientists were involved in one way or another, ran well into the 1920s.[54] Moreover, owing to Bakelite's brittleness, cracks on pipe stems made out of the material also appeared. This problem first emerged in June 1911, as a side effect of General Bakelite's efforts to overcome color changes. In February 1919, production at General Bakelite's transparent department even had to be temporarily halted on account of it.[55]

Second, General Bakelite's lead in the cast resin field was challenged by the Redmanol Chemical Products Company, which, besides the Condensite Company, was its most important domestic competitor. Unlike Baekeland, the president of Redmanol, Lawrence V. Redman, and his research assistants Archie J. Weith and Frank P. Brock, did not hierarchically rank different possible applications for their phenolic resin in their public communications. Nor were they concerned about their invention being perceived as a substitute for amber. In 1918 and 1919, for example, their enterprise was the sole producer of "synthetic amber" listed in the *Annual Chemical Directory of the United States*.[56] Previously the firm had even been named "Amberoid Chemical Products Company." That designation was changed to "Redmanol" because of conflicting IP rights, but Redman and his associates nonetheless seem to have paid special attention to the market for smoking accessories, for which aesthetic appeal was a major sales factor.[57] This focus was only natural in view of their IP position. As will be discussed in chapter 9, Redman and his collaborators claimed that their "dry" production process differed from Baekeland's "wet" methods. Yet, as chemical inventor Carleton Ellis observed in a 1935 textbook, "On account of the expense of anhydrous aldehyde, dry processes are generally used only for the preparation of transparent, translucent or opalescent, light-colored cast or turned articles such as cigar and cigarette holders, umbrella handles, dentures, amber substitutes, beads and similar ornamental objects where the cost of raw materials is less vital than with cheaper molding compositions."[58]

Unfortunately, because of a lack of data, it is impossible to pinpoint the respective shares of General Bakelite and Redmanol in the manufacture of smokers' articles. It is, however, plausible that Redmanol's had become at least as large as Bakelite's by the early 1920s. When Baekeland visited Redmanol's plant in November 1923, for instance, he was positively impressed with the company's "shaving and finishing machine for…transparent slabs," as well as with its "new alloy for casting molds for the transparent."[59] According to a 1926 editorial in the journal *Plastics*, moreover, Redmanol had acquired an enviable reputation among smokers, more so than among other groups of users of phenolic resin products.[60] In the same year, the Bakelite Corporation alleged infringement of a Redmanol patent that particularly concerned production of "phenolic condensation product articles, such as cigar-holders, cigarette-holders, pipe and pipe parts, or other articles which are manufactured from an infusible, and for practical purposes made substantially insoluble, phenolic condensation product."[61]

Competition from abroad constituted a third and particularly heavy constraint on the development of Bakelite's cast resin business. Before the establishment of General Bakelite, Leopold DeMuth had already shown Baekeland transparent plastic resin, imported from Vienna, which was remarkably similar to his. At least by the beginning of 1923, imports from Central Europe had become a serious concern to American manufacturers of phenolic smoking accessories. At that time, five of them, including DeMuth & Company, petitioned the US Tariff Commission for stronger barriers to trade, complaining that they could not possibly compete on price with articles manufactured in Czechoslovakia under the import duties established by the Tariff Act of 1922. In Czechoslovakia, as well as in other Central European countries, small family firms relied on relatively cheap manual labor, consequently reducing production costs to less than half the American level.[62] Besides American manufacturers' dependence on higher-priced labor in the United States, the IP situation in Europe was another factor that had contributed to their comparative disadvantage. Baekeland's German patents, in particular, had effectively covered Bakelite's use for technical purposes when the phenolic resin was mixed with filling materials such as wood flour or asbestos. Yet Bakelite Gesellschaft had not been able to similarly dominate the production of cast resin. As several challengers successfully evaded Baekeland's IP rights in this area, the cast resin business had already become highly competitive during the lifetimes of the chemist's patents. Accordingly, despite some notable R&D accomplishments (for instance, creating "transparent" Bakelite that "is almost colorless and darkens only slightly in the light"), Bakelite Gesellschaft ended up dissolving its cast resin division in the early 1930s.[63]

In 1923, the tariff commission could not meet the request of Baekeland, who had presented his views in an informal conference, and the manufacturers of smoking accessories. A second try would be more successful, though. In 1926, in an unusual step hailed in the journal *Chemical and Metallurgical Engineering* as "the first attempt to use a means of defense previously overlooked by American industry," a Bakelite coalition was able to convince the tariff commission to investigate whether imported phenolic resin violated the Bakelite Corporation's American IP rights. Aside from Baekeland's firm and the same five manufacturers of smoking accessories, two producers of Bakelite jewelry and beads, Abse Brothers and the Embed Art Company, had joined in on the complainants' side.[64] The tariff commissioners evaluated their case favorably, and Presidents Coolidge and Hoover followed their advice. However, through contestations by the defendants—small importers from New York—the case was elongated until October 1930, when the US Supreme Court denied a review of the Court of Customs Appeals' decision. By that time, some damage had become irrevocable. Most relevant, Embed Art, a firm whose 1921 creation as a subsidiary of General Bakelite had been supported by Baekeland, had already been dissolved.[65]

Whereas he had personally lobbied tariff commissioners in 1923, Baekeland expressed his disapproval of the proceedings that had started in 1926. For one thing, he feared that his company would have to disclose data that he regarded confidential; for instance, on its profitability and its production costs. For another, he indicated that his collaborators' efforts to defend the Bakelite companies' interests in this line of business had begun "much too late" and were disproportionate to its commercial significance.[66] Contrary to his publicly stated priorities, there are no indications that a preference for industrial plastic contributed to this critical evaluation. Profit considerations were instead the decisive criterion.

The same "almighty dollar" doubtlessly explains why, as we have seen, Baekeland had dedicated considerable attention to making Bakelite more suitable for smoking accessories at a time when sales to the electrical and automotive industries had not yet been put on a firm basis. Ironically, the longer-term business case for that decision may have been weak. With hindsight, it seems as if Baekeland and his collaborators could have been even more successful if they had fully concentrated their efforts on non-decorative uses. With Lawrence Byck, a longtime Bakelite employee and a cofounder of the Embed Art Company, it can indeed be concluded that "possibly a disproportionate amount of time and effort were spent" on research and development work regarding cast resin "at a time when so many other, larger, more important industries were clamoring for attention."[67]

Technological Utility and Scientific Recognition

When "Bakelite" became a household term in the United States of the 1920s and 1930s, it was primarily known as the typically black or brown molding material out of which automotive, telephone, and especially radio parts were made.[68] Even among some industry insiders, moreover, there was an assumption that this type of technical application had always been prioritized in the United States. In 1930, for example, German American chemist Kurt Pelikan maintained in *Plastic and Molded Products* that since the early 1910s, "Baekland [*sic*] and his imitators on the American continent stuck to the idea of molding powders, i.e., mass production; and [European Fritz] Raschig followed his idea of [cast] block, sheet and rod production, i.e., competing with the celluloid and hard rubber industry. The interesting part of it is that both succeeded extremely well and that both continents finally added the experience of the opponents, or their co-inventors and imitators, to their own. This brought Catalin [a cast phenolic] and others to the United States and made Bakelite establish manufacturing places in almost all of the European countries."[69] While this explanation was consistent with Baekeland's publicly declared priorities, this chapter has demonstrated that it was far from accurate. In particular, the early importance of cast resin to the American phenolic plastic resin industry was overlooked.

Why is this dissonance between reality and perception relevant for understanding Baekeland's renown as the "father of plastics"? A comparison between him and celluloid inventor John W. Hyatt (1837–1920)

Figure 8.7
Bakelite billiard balls.
Courtesy of National Museum of American History, Smithsonian Institution; photo by Richard W. Strauss, RWS2016-09957.

suggests that it was not inconsequential. Baekeland and Hyatt both created a new and commercially useful "giant molecule" without fully understanding the chemistry involved (see chapter 9). Hyatt, however, only received one major scientific prize—the Society of Chemical Industry's Perkin Medal—whereas a fellow US chemist jokingly wrote Baekeland as early as 1916 about his surprise that "the government has not yet prosecuted you for a monopoly in [scientific] medal taking."[70]

Hyatt had personally summarized his career to Baekeland in March 1909, when he had come to inquire about the prospects for manufacturing billiard balls from Bakelite (figure 8.7) instead of celluloid.[71] The fact that Bakelite would prove superior for this application is ironic, as the experiments that eventually led to Hyatt's invention are famously said to have been driven by the demand for a substitute for ivory in billiard balls.[72] More important, Hyatt's acknowledgment of this incentive sharply contrasts with Baekeland's public emphasis on industrial plastic. As historian Robert Friedel has shown, the promoters of celluloid had been aware that the material's use for technical purposes might have raised its symbolic status.[73] Unlike Bakelite, however, celluloid had primarily remained a substitute for so-called natural plastics in small consumer articles, therefore limiting the material's value to scientists and engineers.

In addition, while Hyatt is said to have been stimulated by a monetary prize of $10,000, nonpecuniary incentives were emphasized in congratulatory letters to Baekeland. The president of a California company that distributed Bakelite varnishes and molding materials, for example, complimented Baekeland in 1932 by saying: "You are one of the few who could combine science with its necessities of application, and in your determination to give the benefits of your researches to the world, without the possibility of their relegation to the forgotten things, you were strong enough to subordinate at the proper time your idealism in scientific work by promoting and disseminating your works for the benefit of mankind. I am fully convinced that your own material success was but an incident in your mind at the time, yet it is a powerful endorsement of your sacrifices."[74]

Remarks of this type were in accordance with the promotion of industrial chemistry as a lofty "service to society" and suggest that Baekeland, as befitted a "true scientist," was widely perceived to be financially disinterested. As with other professionals whose honorability and reputation for integrity rested in part on the belief that they were not primarily driven by profit motives, it was in Baekeland's own interest to maintain such an image of selflessness.[75] Accordingly, he systematically veiled his monetary incentives and concerns regarding the Bakelite companies, which will be analyzed in more detail in chapter 10. Moreover, as we will see next, similar considerations occasionally made him engage in seemingly pointless theoretical speculation.

9 A Mysterious Macromolecule: Speculating about the Structure of Bakelite

Whether considered in terms of output, sales value, or number of employees, the production of phenolic plastic resins constituted only a minor industry until the mid-1920s. Its relative smallness encouraged historian Ludwig Haber to leave it out of his classic study on the international development of the chemical industry between 1900 and 1930. In line with the evidence discussed in this book's two previous chapters, Haber attributed the ostensibly slow diffusion of these synthetic compounds to patent disputes and monopolistic prices.[1] Other scholars' views, however, are more supportive of the 1931 claim of Bakelite research director Lawrence Redman that it was more valuable to have "a monopoly of new knowledge" than to have "monopolies of existing patent rights."[2] According to economic historian Joel Mokyr, for example, the limited knowledge about the chemistry of Bakelite was a serious constraint. Mokyr writes that Bakelite, and celluloid, were "developed on a very narrow epistemic base." In his opinion, the advance of the synthetic materials industry "into large-scale manufacturing of mass produced commodities such as nylon and polyester had to await the establishment of its epistemic base by [German chemist] Hermann Staudinger, who discovered the chemical structure of large polymers in the 1920s."[3]

Consistent with this interpretation, in 1927, Charles Stine, a top executive of E. I. du Pont de Nemours and Company, proposed to conduct basic research on polymers because barely anything was known "about the actual mechanism of the change which takes place, so that the methods used are based almost solely on experience."[4] Moreover, within a decade, the theoretically inspired research program that was launched to close this gap in understanding had led to the invention of nylon and neoprene synthetic rubber.

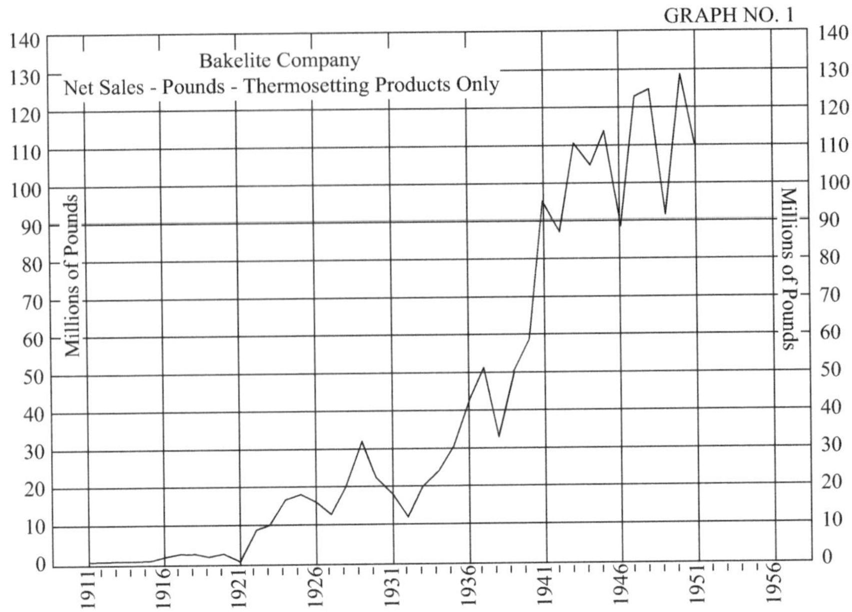

Figure 9.1
Evolution of net sales of General Bakelite and Bakelite Corporation (excluding thermoplastics such as polystyrene, which Bakelite started producing in the late 1930s). Given that Bakelite's market share in the phenolic resin industry dropped from almost 100 percent in 1926 to about 45 percent by 1939, the sharp increases in sales that occurred during the 1930s surely underrepresent the expansion of the entire industry. Reproduced from Byck, "A Survey of the Bakelite Thermosetting Business," by courtesy of Leo Baekeland Papers, Archives Center, National Museum of American History, Smithsonian Institution, AC-0005-0000125.

With respect to the phenolic resin industry, however, Mokyr's argument needs to be qualified. By the mid-1930s, this sector was expanding rapidly (figure 9.1), while macromolecular theory had begun to gain broad acceptance, including in the Bakelite Corporation's research laboratories. Yet the structures of the different varieties of Bakelite, and the reactions occurring during their formation, remained more poorly understood than the chemistry of other synthetic plastics. This was partly because C-stage Bakelite is a relatively complex example of what came to be known as a "thermoset" plastic—that is, a macromolecule or polymer resulting from an irreversible hardening process—as opposed to "thermoplastics" created via reversible polymerization reactions. In addition, the economic rationale for supporting academically oriented studies on this topic had simply been weak. After all, as one of the participants in a historic 1935 conference on polymers put it, "The immediate requirements of the [phenolic resin] industry have been satisfied without the necessity of a great deal of fundamental research."[5]

It was not until the second half of the 1940s that the growth of this sector started to depend more strongly on studies that were categorized as "basic" or "fundamental," two terms that had gained popularity as alternatives for the higher-minded concept of "pure" science.[6] At the same time, as late as 1950, it was obvious to industry insiders such as German chemist Kurt Hultzsch that "the *theoretical-chemical* study of phenolic resins has not by any means kept pace with their technical development."[7] This observation, which had been made by many others during the previous decades, stands in tension with Mokyr's argument on the importance of a comprehensive knowledge base as an enabler of sustained growth. It is indeed ironic that, precisely when technological breakthroughs with regard to phenolic plastic resins started being more firmly grounded in science, the industry was being overshadowed by new and even more versatile plastics such as polyethylene and polyvinylchloride (PVC).

Still, before World War I, chemists associated with the phenolic resin industry had already advanced various interpretations concerning the molecular structures of these synthetic compounds, even though it was widely acknowledged that the empirical basis on which their "speculations" rested was weak. Characteristically, Baekeland minimized the significance of these theoretical deductions in strong terms, among others by dismissing them as "flights of fancy … of very subordinate interest" and by stating that "it does not matter very much" whether C-stage Bakelite was a chemical compound or a mixture.[8] Yet he did personally take part in this discussion. Moreover, the interpretations that he proposed were more extensively cited and discussed in the scientific literature than his earlier photochemical ones.

This chapter analyzes Baekeland's motives for participating in these seemingly academic exchanges, despite his long-standing reservations about largely theoretical arguments. Four possible explanations are taken into consideration, all of which are known to have mattered in other cases and help throw further light on the connections and interdependencies between science and industry. First, I investigate whether legal intellectual property (IP) issues provided an impetus, keeping in view that in order to obtain strong patent protection, the ability to offer convincing scientific explanations regarding the inventions concerned could be of great importance. Mobilizing scientific authority as an economic resource, however, was not without risk: in the process, the scientists in question might damage their reputations and their dignity. Second, I connect Baekeland's occasional digressions into chemical theory with his familiarity with another type of IP, associated with scientific reward systems to which peer evaluation and intellectual credit were central. One way of comparing early twentieth-century academic and technoindustrial payoff structures is to have a closer look at Baekeland's responsibilities at Columbia University, where the General Bakelite Company began funding research in 1919. In this connection, one hypothesis could be that the theoretically driven publications resulting from the company's research fellowship served to associate it with fundamental science. As is evident from studies on Du Pont and other corporations, investments in basic research, whether in-house or via industrial scholarships, could also be aimed at facilitating the recruitment of PhD scientists.[9] Finally, I explore whether Baekeland regarded his theoretical reflections as compensation for a strategic silence about other kinds of information that he or his business associates were not willing to reveal. In this regard, I take my cue from earlier research on publication restrictions and censorship in knowledge-intensive industrial settings.

For the purposes of this chapter, Baekeland's main research articles concerning Bakelite (see table 9.1) can be divided into three groups. To begin with, the series of papers that he presented from 1909 until 1912, I argue, must be interpreted in the context of the patent management efforts that he was making around this time. His predominantly theoretical Willard Gibbs Medal Address of Acceptance from 1913, in contrast, mainly took into account the expectations of academic colleagues, including self-styled "pure" chemists within the ACS. Finally, Baekeland's articles from the 1920s differed from the previous ones in that they were coauthored with Mortimer Harvey and Howard Bender, both of whom had been granted Bakelite fellowships at Columbia. The latter of these two publications has been identified as Baekeland's main contribution to the understanding of the chemical structure and formation of synthetic plastics. However, as with several of his earlier research papers, Baekeland's diaries and the other sources examined here contain evidence that requires us to reconsider its meaning and significance.

Table 9.1
Baekeland's research articles about Bakelite.

Year	Title of article
1909	The Synthesis, Constitution and Uses of Bakelite
	On Soluble, Fusible, Resinous Condensation Products of Phenols and Formaldehyde
	The Use of Bakelite for Electrical and Electrochemical Purposes (in *Transactions of the American Electrochemical Society*)
1910	"Bakelite," a Condensation Product of Phenols and Aldehydes, and Its Uses (in *Journal of the Franklin Institute*)
1911	Recent Developments in Bakelite
1912	Phenol-Formaldehyde Condensation Products
1913	The Chemical Constitution of Resinous Phenolic Condensation Products (Willard Gibbs Medal Address of Acceptance)
1921	(with Mortimer T. Harvey) Further Studies on Phenolic Hexamethylenetetramine Compounds
1925	(with Howard L. Bender) Phenol Resins and Resinoids

Note: All papers appeared in the *Journal of Industrial and Engineering Chemistry* unless otherwise noted.

Intellectual Property Rights

One way to evaluate Baekeland's publication record regarding Bakelite is to differentiate between his practical accomplishments and his understanding of the underlying science. In the former respect, his legacy has been highly influential. This was partly because Baekeland introduced several innovations to control the reaction between phenol and formaldehyde, ranging from his combined use of high temperatures and a countervailing pressure to his selection of base catalysts and the addition of wood flour or other fillers to the phenolic resin to strengthen its mechanical qualities. Baekeland's differentiation between an initial "A"-stage phenolic resin, an intermediate "B" resin, and the final "C"-stage polymer also continues to be adopted today, even though these steps were ill defined from a chemical point of view. As a corollary, his ability to halt the Bakelite process after either of the two initial stages, particularly after the first, has shaped the division of labor within the phenolic plastic resin industry.[10] Baekeland's theoretical deductions, by contrast, have largely become a historical curiosity. Already by the mid-1930s, for example, it had become clear that his observations on

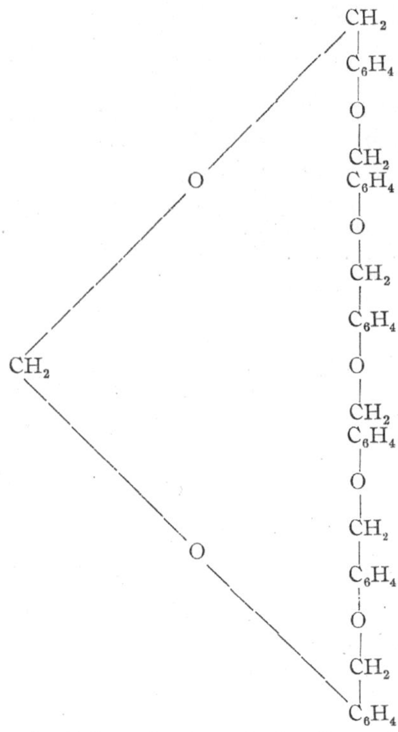

Figure 9.2
Baekeland's tentative formula for B-stage Bakelite.
Reproduced from *Journal of Industrial and Engineering Chemistry* 1 (1909).

the chemical structure of Bakelite had been more off the mark than those of early competitors such as German chemist Fritz Raschig. At any rate, Hultzsch and other mid-twentieth-century experts also referred to Raschig when concluding that "all these [early] theories ... were insufficiently based on fundamental experimental work, if at all, so that already on this account none of them could be considered as having been proven. They rather were subjective, partly even virtually speculative views about the structure of the C-stage resins, which, due to their insolubility, did not seem to be open to more thorough chemical analyses."[11]

Ironically, Baekeland probably would have agreed. In his first research paper on Bakelite, for example, he proposed the molecular ring structure depicted in figure 9.2, which quickly became a target for criticism, as the chemical formula for B-stage Bakelite "until we have anything better." In later publications, he likewise emphasized that, in view of the difficulties encountered in this "greasy" branch of chemistry, one could merely hypothesize about the precise nature of the ultimate chemical reactions occurring in the Bakelite process.[12] Hence, it is intriguing that he nonetheless opened himself to the reproach of having speculated carelessly and, as a corollary, to the risk of damaging his scientific reputation.

To explain this apparent paradox, it is helpful to consider to whom Baekeland's early theoretical reflections and hypotheses were targeted. Importantly, the view that he was primarily seeking the recognition of the chemical profession could be misleading. Given that he also referred to his research papers in his attempts to obtain and enforce patent rights, the appraisal of the patent examiners and judges who had to evaluate his IP claims should certainly be taken into account, too. As discussed in chapter 7, in the United States, the former were underequipped and overburdened, and the latter did not necessarily have the educational background required for mastering all the relevant principles and details. Hence, while contentions about the chemistry of the Bakelite process could make a difference in patent offices as well as in courtrooms, they would not be subjected to conventional methods of academic peer review.

The linkage between chemical theory and IP rights was particularly obvious in Baekeland's disputes with Lawrence Redman and his collaborators, which preceded the formation of Bakelite Corporation in the early 1920s. Already in their first paper on phenolic resins, Redman, Archie J. Weith, and Frank P. Brock took interpretations and suggestions by Baekeland for facts (allegedly) pointing to substantial differences between his process and their own, which was based on anhydrous phenol and hexamethylenetetramine (hereafter hexamine), the crystalline compound whose "ungodly name" Thomas Edison considered a cause for "mental fatigue and absentmindedness."[13] For instance, Redman and his coauthors argued that the chemical reactions supposedly leading to Baekeland's depicted compound could not occur in their process, as "no water or oxygen is present which could possibly form such an inner anhydride."[14] They advanced similar lines of reasoning until the patent infringement suit between General Bakelite and General Insulate was concluded in 1921.

One of the core contentions of Redman and his team was that Baekeland had only belatedly realized the possibility of directly combining phenol with hexamine at the beginning of an alternative process for manufacturing commercially viable phenolic resin. While Judge Thomas I. Chatfield would not follow them on this point in the General Insulate case, it was an argument about which Baekeland had been concerned as early as 1910. Indeed, in the final months of that year, he had worried that his IP rights might be circumvented by using hexamine instead of formaldehyde. Hence, even though he then considered this alternative method unappealing in view of its comparatively high cost, he attempted to patent it.[15] His patent titled "Reaction product of hexamethylenetetramin and phenolic bodies and method of making same," however, was not granted until June 1916. Moreover, its scope had been reduced to the "wet" hexamine process as a result of interference with other inventors, the "dry" process being granted to Byron B. Goldsmith of New York, who assigned his IP rights to the Redmanol Chemical Product Company.[16]

While his hexamine patent application was pending, in a parallel and eventually more successful IP strategy, Baekeland started making the case that his earlier patents already covered the phenol-hexamine process. In particular, he stressed that he had specified the use of ammonia as a catalyst and "any chemist knows that hexamethylentetramin, and ammonia and formaldehyde, are tweedledum and tweedledee."[17] As a corollary, Baekeland began emphasizing that his use of the term "formaldehyde" should be understood "in a rather broad sense," also comprising "mixtures of various chemical compounds which ultimately react in the same way as CH_2O."[18] He even started a September 1912 paper on "Phenol-Formaldehyde Condensation Products" by noting that this very designation was unfortunate, as "it is well-known that these products can be obtained without the use of so-called formaldehyde.... [I]t is generally accepted that other methylene compounds, for instance, methylal, trioxymethylen [or paraformaldehyde], hexamethylentetramin, etc. can replace formaldehyde in this reaction."[19] Earlier in 1912, Baekeland and his associates had received "an unexpected knock on the head" by the granting of a patent to Condensite inventor Jonas Aylsworth that claimed the use of methylene-amino compounds such as hexamine, instead of "paraformaldehyde or other polymerised formaldehyde," in forming the final, C-stage phenolic plastic resin.[20] In addition, they had been informed that Baekeland's own patent application on hexamine interfered with applications by Goldsmith, Redman, and Charles Steinmetz of General Electric. It surely had been reasonable for them to hedge their bets.

Baekeland's paper on "Phenol-Formaldehyde Condensation Products" is relevant here not only because it helps demonstrate that he and his advisers sought to rewrite the history of the invention of Bakelite by claiming that they had from the outset regarded hexamine as a viable alternative to formaldehyde and ammonia. It also matters because of the audience. Baekeland delivered his paper at the Eighth International Congress of Applied Chemistry in New York and Washington, D.C., which he had helped organize and whose section on "India Rubber and Other Plastics" he presided over. More than half the non-US chemists who attended the congress were German.[21] Moreover, this large delegation included the president of the chemical division of the German Patent Office, as well as Fritz Raschig, the chemist whom Baekeland had been "fighting...in the german patent office and we are still fighting."[22] According to Baekeland's diary, Raschig "refused to discuss altho urged repeatedly" after the end of Baekeland's lecture, illustrating that IP concerns figured prominently in his evaluation of the effectiveness of his own talk.

Conversely, Baekeland found nothing to protest against in Raschig's paper on "The Scientific and Technical Significance of the Coal Tar Phenols," delivered to the Verein Deutscher Chemiker shortly before the beginning of the international chemistry congress. Indeed, after his own presentation, Baekeland favorably referred to it to clarify his views on the chemical characteristics of C-stage Bakelite, which he plainly described as "one of the difficult problems of organic chemistry; one of those problems which have been very much neglected by the average chemist, who has the habit of kicking out of his laboratory anything which does not crystallize in [nice] crystals."[23] Moreover, in his diary, he noted how he "after many delays arrived...just in time to hear that part of Raschig's lecture which he devoted entirely to Bakelite and gave prominent importance to it. Everything he said was unobjectionable and gave my work much prominence. In fact it was so to speak the scientific endorsement of a man who is considered the principal authority on the chemistry of phenol."[24]

Hans Lebach, Baekeland's former rival at Knoll & Company, also attended the conference. Lebach was in charge of Bakelite Gesellschaft's newly founded British subsidiary. As a result of his joining Bakelite, his and Baekeland's patents could be pooled in disputes with other contestants. The two chemists regularly met up around the time of the conference and discussed the "Condensite patent contestation in England."[25] They probably also conferred about the IP situation with regard to phenol-hexamine, for Baekeland explicitly referred to work by Lebach when arguing against the novelty of this process. In his conference paper presentation, which Lebach could not attend because of illness, he stated that "the fact that hexamethylentetramin can suitably replace formaldehyde in the formation of the infusible phenolic condensation products was published as far back as December 31, 1907, by Lebach."[26]

The dispute about hexamine climaxed in the patent infringement suit between General Bakelite and General Insulate, a Redmanol customer, from 1917 until 1921. What role, if any, did theoretical arguments play at this stage? During the court trials, Insulate's counsel John H. Lee interrogated Baekeland about the chemistry of the Bakelite process along lines similar to those expressed in the earlier publications by Redman, Weith, and Brock. Yet Baekeland, who could draw on his earlier courtroom experiences, showed himself to be a combatant witness. Irritated by Lee's "boresome, monstruous, repeating, doleful crossexamination and trick questions," he contested the relevance of Lee's cross-examination in strong, anti-intellectualist terms.[27] For instance, when asked whether he was referring to "actual commercial operations" or to "the patented art or published art," Baekeland responded, "I am not referring to freaks. I mean used only for commercial purposes, goods that you can sell." In addition, in reaction to Lee's querying about the chemistry of the phenol-hexamine combination, he stated, "When I used to be a school boy, I used to talk about chemical radicals grabbing each other, and all that, but having become so old, and having accumulated experience enough, I don't want to make myself ridiculous by answering a question that way."[28]

In a later continuation of the cross-examination, Baekeland acted less obstinately and did respond to questions concerning the basic chemistry of the Bakelite process. Instead of dismissing this subject as immaterial, he now emphasized its complexity. For example, to Lee's question whether "in producing these final condensation products is it your theory that you pass through the phenol-alcohol stage," he answered,

> I don't know what you call my "theory." I have got so many theories; but since I have been making use of this complicated reaction I have come to the conclusion that a good many things all happen at the same time; anybody who tries to describe this reaction as a straight unique reaction, makes a mistake for the reason that I have found the most divergent products in the transition from the original

> materials to the final materials; and I believe that all these products are produced to a limited extent; but that ultimately the final result is the same.[29]

Moreover, with regard to the nature of the ultimate product, Baekeland once more explained that in organic chemistry of this kind "we do not have at hand the simple means with which we can determine molecular weights for crystallized bodies or simple bodies which can be volatilized, or which you could dissolve, or which you could crystallize." As Wiebe Bijker, Yasu Furukawa, and other historians have pointed out, the fact that phenolic resins could not be subjected to these methods of chemical analysis had indeed been one of the reasons why several earlier researchers had turned away from the phenol-formaldehyde reaction.[30]

The nontechnical judge to whom the case was assigned, however, obviously could not postpone his verdict until a more accurate chemical knowledge of Bakelite and "Redmanite" had become available. Did he take theoretical claims advanced during the dispute into consideration? Scientific credentials were certainly relevant to Thomas I. Chatfield, who described Baekeland as "a man of broad chemical knowledge and experience" in his 36-page verdict. Judge Chatfield also accepted Baekeland's unspecific definition of "formaldehyde," vaguely indicating that the term "embraces different compounds of methylene groups."[31] With respect to hexamine, Chatfield urged keeping in mind the exact chronology of Baekeland's patent applications and the research papers that he had presented in 1909. US patents 942,809 ("Condensation product and method of making same") and 1,038,475 ("Process of making insoluble bodies derived from phenol alcohols") were of particular importance. The latter, filed after Baekeland had become concerned about the phenol-hexamine process as a possible alternative to his own methods, specified that hexamine could be used in place of "ordinary formaldehyde." In this patent, Baekeland and his patent counsels had also referred to the former one, in which it was explained that "if a large amount of ammonia be used, hexamethylentetramin is formed, which is a crystalline body of definite chemical properties."[32] As Redman, Weith, and Brock had pointed out, from the remainder of that patent it was clear that Baekeland had regarded the formation of hexamine as an unwanted side effect.[33] To Chatfield, though, this was clearly beside the point; he simply concluded that Baekeland "evidently ... was aware of the chemical formation of hexa as early as the proceedings in the Patent Office resulting in Patent No. 942,809, which was issued Dec. 7, 1909."[34]

This interpretation surely came as a relief to Baekeland and his associates at General Bakelite. Yet, for the purposes of this chapter, Chatfield's reading of the chemist's 1909 research articles is more important, because he explicitly discussed some of Baekeland's views concerning the basic chemistry underlying the Bakelite process. In particular, Chatfield highlighted the parallels Baekeland had drawn with processes of biosynthesis, analogies that had previously been suggested by, among others, German agricultural chemist Oscar Löw. This passage of Chatfield's ruling is worth quoting at length, for it stands in sharp contrast to the later, critical evaluations set forth at the beginning of this chapter. Chatfield reasoned:

> In one of these articles, almost the last paragraph calls attention to researches by one Oscar Low, who "many years ago" called attention to the formation of formaldehyde by the chemical action of sunlight on CO_2 in the presence of water in chlorophyl as a starting point for the synthesis in plant life. Baekeland also calls attention to the fact that the willow tree produces in its cells, glucocides of saligenin. Saligenin or oxybenzyl alcohol in the presence of more formaldehyde CH_2O gives Bakelite.
>
> Baekeland calls attention to articles published several years before, in which the phenolic nature of resinous substances, such as Japanese lacquer, show some analogy with Bakelite and similarity to the exudation of the plant known as poison ivy.
>
> This paragraph throws a strong light upon the entire nature and construction of synthetic resins, and clears up much of the mystery which to the lay mind arises from the incomprehensibility of the organic reactions.[35]

Chatfield was paraphrasing the part of "The Synthesis, Constitution, and Uses of Bakelite" in which Baekeland had also proposed his erroneous chemical formula; that is, the very same passage that signaled to later polymer scientists and historians of chemistry that Baekeland's theoretical understanding of the Bakelite process was not as sophisticated as that of Fritz Raschig. But Chatfield, ironically, brought forward precisely this excerpt to illustrate Baekeland's (alleged) grasp of the more fundamental scientific side of his invention. This difference is significant in that, from a legal-economic point of view, Chatfield's decision was surely more consequential than the criticisms by scientists. Indeed, his verdict set off the chain of events that resulted in consolidations of the industry in the United States, where the Bakelite Corporation had been established, as well as in Britain and Canada (see chapters 7 and 8).

Scientific Prestige

At the same time, another early research paper by Baekeland clearly shows that he was not indifferent to real or perceived normative ideals and expectations of his academic colleagues. Indeed, I would argue that the lengthy theoretical interpretations that he gave in this paper should be read as a concession to some of them. Baekeland presented the paper when accepting the Willard Gibbs Medal of the Chicago Section of the American Chemical Society in 1913. Only two other chemists, Svante Arrhenius and Theodore W. Richards, had received this award prior to him. On both occasions, moreover, this had been on account of accomplishments in the area of basic chemical research. Hence, as Julius Stieglitz, the chairman of the Chicago ACS, pointed out in his introductory address, "Perhaps the most significant feature of our gathering here this evening... lies in the fact that for the first time in the history of the medal it is to be bestowed upon an eminent investigator in the field of *applied* chemistry."[36] As discussed in chapter 7, Stieglitz would later criticize Baekeland's agenda for patent reform, suggesting that it did not sufficiently take into account the interests of the general public. But, as a younger brother of renowned photographic artist Alfred Stieglitz, and as someone who is said to have personally "attained results of high artistic merit" as an amateur, Stieglitz did share his professional colleague's passion for photography. Consequently, it may not be surprising that Stieglitz, as Baekeland reported, "put special stress on Velox." (In fact, the invention of Velox was the only accomplishment of Baekeland explicitly referred to in the published version of Stieglitz's address.) Yet, for the most part, Stieglitz dwelt on the relationship between "pure" and "applied" chemistry, two concepts that academic scientists had been using extensively to define and negotiate their interactions with industry and the state.[37] It is notable that he particularly highlighted prejudices encountered by "pure chemists," including a questioning of their disinterestedness. Namely, when Herbert McCoy, a colleague of Stieglitz at the University of Chicago, had declined to analyze a sample of shoe polish on the grounds that it was his department's policy not to engage in this kind of commercial work, the man who had made the request countered that McCoy and Stieglitz surely would have accepted the job if he had offered them a fiver![38]

In a vein similar to that of Stieglitz, Baekeland himself wrote as late as 1933 that he was "still the only medallist [*sic*] who got this reward for industrial chemistry."[39] To him the Gibbs Medal had come as a surprise; he had been on a trip in Europe that was supposed to last until early May. But, as he wrote to his friend Paul Fredericq, the lecture that he was expected to give in Chicago "requires some preparation because all of this happens quite formally—with experiments, exhibits, etc. Therefore I will try to leave here as soon as possible, and I had to cancel most of the numerous invitations in Germany and England that I had already accepted."[40] Baekeland also explained to Fredericq that the award linked his name to those of Gibbs, Richards ("the famous professor of Harvard who was repeatedly offered a chair at german Universities"), Arrhenius, Wilhelm Ostwald, Jacobus van 't Hoff, and other seminal contributors to the field of physical chemistry. He gave special credit to Gibbs in connection with the historical breakthrough of this discipline, on account of his discovery of the so-called phase rule, a mathematical equation derived from thermodynamics. In the mid-1880s, Baekeland himself had briefly referred to Gibbs's work in his thesis on dissociation processes.

Baekeland embarked on the steamer T.S.S. *Rotterdam* on 5 April 1913 and arrived in New York eight days later; that is, more than a month before the medal presentation. His diaries contain no evidence that he struggled to finish his paper in time. On the day of the award ceremony, moreover, Baekeland received help from Hylton Swan, a General Bakelite sales engineer, who arranged a display of various Bakelite artifacts and "all chemical formulae were hung on the wall."[41] These preparations notwithstanding, Baekeland was clearly caught off guard during the proceedings. According to the *Journal of Industrial and Engineering Chemistry*, "Dr. Baekeland on receipt of the medal was for several moments rather at a loss for words, but deeply felt the honor in having his name connected with those of Arrhenius and Richards, the earlier recipients of the medal, and Willard Gibbs for whom the medal is named."[42] Moreover, at the end of his presentation, Baekeland stated that he "was just as glad as they [the audience] that my lecture was over and that I felt like a schoolboy who was just thru with his prize oration."[43] I would suggest that his uneasiness stemmed not merely from the distinguished setting in which he presented his paper but also, and in no small measure, from the content of his talk. Unlike his earlier lectures, its focus was on the "purely theoretical" side of the Bakelite process. In view of the daunting challenges in this area, he therefore found himself dangerously close to what he called "the nebulous realm of gratuitous assumptions and hypothesis."[44]

As indicated previously, during his later court testimony in *General Bakelite v. General Insulate*, Baekeland was reluctant to make theoretical digressions. Consistent with his criticism of textbook abstractions discussed in chapter 2, he then held that an explanation of the chemical reaction between hexamine and phenol advanced by the interrogating attorney was "too simple" and "good for a schoolboy, but I don't want to put myself in the position of accepting that hypothesis in that way."[45] Why, then, was he prepared to act as a prize-winning pupil during the Gibbs Medal award ceremony of 1913?

To some degree, the interpretation proposed in the previous section of this chapter also applies here. For example, in his Gibbs address, Baekeland once again singled out hexamine as a possible equivalent to formaldehyde. Yet the strengthening of his legal IP position was certainly not a major goal of this presentation. In fact, as suggested by the paper's predominantly theoretical content, Stieglitz's introductory remarks, and Baekeland's initial gratification and later discomfort and relief, it seems to have been a concession from an industrial chemist toward the academically inclined committee that had granted him this prestigious award. Put differently, I would claim that, on this occasion, Baekeland was seeking to receive credit as a scientific author within a reputation-based reward system influenced by notions of disinterestedness. In the terms introduced in chapter 1, Baekeland's publication record points to two distinct forms of intellectual property that motivated and affected the careers of scientific entrepreneurs. On the one hand, Baekeland and his business associates were incentivized to create powerful narratives about the invention of Bakelite that sought to broaden the scope of their patent rights in response to conflicting interpretations set forth by competitors. But, on the other hand, Baekeland, as well as other "applied" chemists who later received the Gibbs Award (for instance, General Electric's research director Willis Whitney and Frederick Cottrell of the Bureau of Mines), apparently deemed it inappropriate to elaborate on their personal technological accomplishments in a "pure" academic setting.[46]

Fellows in Bakelite

Baekeland's position at the crossroads of different institutional environments and reward systems can also be studied through his association with Columbia University. From a financial point of view, it was certainly a sound decision of Columbia to offer an honorary professorship to the chemical industrialist, who had long maintained informal contacts with several members of its teaching and research staff. Baekeland not only contributed to the university's endowment via private cash donations, he also established a Bakelite research scholarship in the Department of Chemical Engineering. He had been authorized to do this at a General Bakelite board meeting on 7 May 1919. The fellowship's official goal was to develop "the subject of Phenolic Condensation Products or matters relating thereto," and the maximum cost was initially set at $5,000 per year.[47] This relatively narrow focus already indicates that General Bakelite was not planning to give its research fellows (see table 9.2) as much autonomy as some larger firms were doing. Du Pont, for example, had initiated a comprehensive fellowship program the previous year, which was mainly aimed at recruiting chemical researchers, and the company did not consider it necessary to specify any problems they were supposed to examine. But, on the other hand, Bakelite's grants were clearly meant to be less restrictive than those of the corporate sponsors of chemical R&D at the Mellon Institute in Pittsburgh and schools such as the University of Michigan at Ann Arbor and the State College of Washington in Pullman. The Mellon Institute was the largest academic center of industrial research in interwar America and extensively advertised the terms of its fellowships, including in the *Journal of Industrial and Engineering Chemistry*. It therefore seems likely that it served as the main reference point for this journal's August 1919 announcement that General Bakelite's scholarship differed from "the general type of industrial fellowships" for two reasons. First, Baekeland's firm was more generous because it not only financed a scholar and his equipment but also compensated Columbia for the use of its facilities. Second, "No time or other limitation is put upon the publication of the results of the investigation." In other words, the Bakelite fellowship was also claimed to be special on account of its commitment to what is sometimes referred to as "open science."[48]

Did this liberal publication policy interfere with the exchange of information between General Bakelite's research fellows and its regular employees? And how did the projects at Columbia relate to the investigations that were being carried out within General Bakelite itself? The available evidence indicates that the research undertaken in Columbia's Chemical Engineering Department was generally of a more fundamental nature than

Table 9.2
Columbia's Bakelite Fellows (1919–1929).

Name	Year	Educational and/or professional background
Mortimer T. Harvey	1919–1920	Bachelor of Science in sanitary engineering, Union College, NY (1917) Graduate studies in chemical engineering, Columbia (1917–1920; assistant in chemical engineering lab in 1918–1919) PhD, Faculty of Pure Science, Columbia (1921)
Howard L. Bender	1922–1923	Bachelor of Arts, Marietta College, OH (1917); MA (1919) Drury Fellow, Case School of Applied Science, OH; Bachelor of Science in chemical engineering (1919) Research chemist, Dow Chemical (1919–1921) Instructor, Marietta College (1921) PhD, Dept. of Chemical Engineering, Columbia (1925)
William D. Turner	1927–1929	Bachelor of Science, University of Chicago (1912); PhD (1917) Instructor in chemical engineering, Columbia (1928–1929); assistant professor (1929–1930)

Note: No fellowships were awarded in 1929–1930 and 1932–1933. From 1930 until 1932, the Bakelite Corporation financially supported Kenneth W. Coons and Lawrence M. Debing, both of whom were advised by William Turner. The former submitted a PhD thesis on "The Demethylation of Ring Alkylated Phenols," a topic suggested by Virgil Meharg of Bakelite Corporation, around the end of this period. The latter did not complete his dissertation, entitled "Factors Influencing the Dielectric Properties of Phenolic Resins," until 1945.

General Bakelite's in-house R&D work. None of the three Bakelite fellowship holders in the 1920s—Mortimer Harvey, Howard Bender, and William Turner—were assigned to work on the final, C-stage resins, to which conventional methods of chemical analysis could not be applied. The PhD research of Harvey and Bender is better documented than Turner's postdoctoral examinations because it was disseminated in the form of dissertations and articles.[49] Their projects were aimed at developing a better theoretical understanding of the earlier phases of the Bakelite process, notably the formation of the intermediate, B-stage products and these products' chemical constitutions. As part of the goal to "abandon mere speculation," Harvey and Bender attempted to isolate some of these substances, an approach common among organic chemists.[50] To facilitate this process, they worked with substitutes for phenol and/or formaldehyde—hexamine in the case of Harvey. Turner likewise tried to obtain crystalline compounds in his Bakelite-funded investigations on phenol-formaldehyde resin of the so-called novolac type, but his research never found its way into any ACS journal and was probably never published at all. This difference is consistent with the fact that Columbia did not impose any formal publication requirements on its postdocs, whereas its PhD degrees were "awarded only on the basis of *published* results."[51]

The academically based projects of Harvey, Bender, and Turner were thus not unequivocally designed to meet a market need. At the same time, though, their work was being actively coordinated with in-house experiments at General Bakelite. For example, in November 1919, Harvey visited Baekeland's private laboratory in Yonkers—General Bakelite's main research facility—to familiarize himself with the company's laboratory methods. In September 1922, Baekeland confidentially informed Bender about examinations by Conrad Schrimpe, one of the chemists who had participated in the R&D program on color permanency discussed in chapter 8. And two months later, he noted down his intention to have various products prepared by Bender tested "as solid solvents, specially in transparent bakelite" by Schrimpe and Harvey D. Shannon, the head of the transparent department.[52] This kind of coordination and knowledge transfer also took place in the case of Turner. This postdoctoral fellow had to regularly present his findings to Baekeland or members of the Bakelite Corporation's research laboratories in Bloomfield, New Jersey, in the late 1920s. Therefore, Columbia's Bakelite scholarships obviously were not carried out in an academic "ivory tower."

How much autonomy did Baekeland leave to Harvey, Bender, and Turner in filling in the details of their projects? One indicator is provided by the journal articles in which the PhD research of Harvey and Bender was presented. Harvey appeared as the first author in his publication "Further Studies on Phenolic Hexamethylenetetramine Compounds"—an exact copy of his PhD dissertation, except for the acknowledgments. But in the article based on Bender's examinations, Baekeland's name was mentioned first. Alphabetical ordering cannot explain this difference and was, in any case, a coauthorship practice that neither Columbia's nor the Bakelite companies' chemical researchers adopted in the period covered in this book.[53] Baekeland's seniority also cannot have been the cause, as he had been Harvey's adviser, too. Therefore, this difference seems to suggest that Baekeland had more strongly shaped Bender's project and/or was more intent on getting recognition for the findings and ideas presented in this fellow's dissertation.

Whatever the reason, this authorship arrangement turned out to be not inconsequential. A recently completed "prehistory" of polymer science, for instance, even omits Bender's name in the discussion of his own PhD research as presented in his article with Baekeland. By implication, Bender's adviser is given most of the credit for a study that, thanks to its comprehensiveness and promising research approach, was internationally recognized as a major early contribution to the understanding of the structure and formation of phenolic resins.[54] This disregard of Bender's role was not without precedent. In a 1926 textbook on *The Chemistry of the Natural & Synthetic Resins*, for example, schemes from Baekeland and Bender's publication repeatedly were attributed solely to the former. Likewise, chemist William J. Hale did not mention Bender in the letter to Baekeland in which he complimented that, "Your article on phenol condensation products is certainly very well written and replete with many interesting problems."[55]

Was Bender treated unfairly? It is clear that Baekeland did not actually co-write the paper. But he had repeatedly met up with Bender and, as with Harvey and Turner, had provided him with detailed, strict guidelines. (Significantly, in 1929, the year in which Turner advanced to the position of assistant professor in Columbia's Chemical Engineering Department, Baekeland requested that he "write me out my suggestions so as to see whether he understood my views"!)[56] Moreover, it should be taken into account that Bender was no longer based at Columbia when his PhD research was published in 1925. It turned out that Baekeland had been favorably impressed by the capabilities and personality of this Bakelite fellowship holder, whom he described as a "quick correct worker, who grasps what is wanted and knows how to get there." Accordingly, he had recruited him for his company in June 1923.[57]

The difficulties of knowledge-intensive corporations in hiring top academic scientists and in keeping them satisfied have been highlighted in several studies on the emergence of industrial research in the early twentieth-century United States. Complaints about the (allegedly) dim career prospects of research chemists in the American chemical industry, however, had become less frequent by the 1920s. Moreover, as historian

Steven Shapin has argued, mid-twentieth-century commentaries by industrial research managers did not support the assumption of contemporary sociologists that scientists transitioning from academia into industry would experience a clash of moral values.[58] How, then, given these different findings and perspectives, did Bender's career continue after his departure from one of the country's still relatively small number of research universities?

Initially Bender was assigned to work at General Bakelite's plant in Perth Amboy, New Jersey, under the direction of Louis M. Rossi, the firm's works manager. However, not Rossi but Lawrence Byck, the research chemist who cofounded the Embed Art Company, was asked to train Bender, therefore suggesting that General Bakelite's onsite R&D facilities were the newcomer's primary workplace. In any case, in Baekeland's view, "As usually he [Bender] is ready to go and has the same cheerful attitude with which he undertakes any new work."[59] Bender's further career advancement suggests that this enthusiasm did not fade. By the late 1920s, he had become a leading member of the Bakelite Corporation's independent research laboratories in Bloomfield, New Jersey, involved in several R&D projects for which Baekeland cherished big expectations. Moreover, after the 1939 takeover of Bakelite by Union Carbide and Carbon, Bender continued his investigations at the latter company, where he was promoted to the job of assistant director of research. Tellingly, in 1952 he received the John Wesley Hyatt Award for the Advancement of Plastics, honoring "achievement of wide importance to the plastics industry," in recognition of his decades-long work on the chemical structure and formation of phenolic resins.[60]

That Bender's transition from academia to industry did not imply a clear shift from "pure" to commercially oriented research is also evident from the handful of scholarly papers that he authored during his employment with Bakelite. These publications reveal that Bender, as well as several of his co-workers at the company's research laboratories, kept a close eye on theoretical developments that seemed relevant to the study of synthetic plastic resins. This is particularly evident in the physical studies that Bender conducted to explore which features such resins shared with other "colloids"; that is, gluelike substances that could not (or almost not) be crystallized and whose molecular structures were therefore very hard to determine. In the mid-1920s, colloid theory had been "one of the most popular topics in American science," so it is not unnatural that Bender's work, as well as his earlier article with Baekeland, were influenced by this intellectual framework.[61] From his 1932 paper onward, however, Bender's writings also engaged with Staudinger's argument that what most colloid scientists had regarded as physical aggregates of relatively small particles actually were chemically united "giant molecules."

In the second half of the 1930s, Bender also had the chance to discuss the polymerization theories of Staudinger and other prominent researchers at an exclusive conference series, providing further evidence that Bender's industrial position did not prevent him from pursuing scholarly activities and getting recognition for them. These summer meetings were organized at a yacht club on Gibson Island, Maryland, with the aim of fostering scientific knowledge circulation between the few academics and the much larger group of industrial scientists knowledgeable about the chemistry of synthetic plastic resins. In July 1936, Bender had the honor of being the second paper presenter on the program, which meant that there was plenty of time available to discuss his work. (The previous speaker, E. Emmet Reid of Johns Hopkins University, had been given the thankless task of replacing Du Pont's Wallace Carothers, the foremost US expert on polymerization.) In 1939, Bender was even the only paper presenter on the opening day. This may have partly reflected his role in preparing the meeting, but it also speaks to his standing in scientific circles. In fact, in both that edition and the next, conference organizer Neil Gordon had relied on Bender for putting together the program and serving as a symposium chair. Bender thus contributed to the success of these "vacation gatherings," which renowned academic chemist Herman Mark later described as "an ideal trading post for scientific and technical information and discussion in an extremely attractive, informal and relaxing atmosphere."[62]

Did all this mean that Bender also continued to experience the academic freedom of publication that, as mentioned, was claimed to make Bakelite's fellowships at Columbia different from other industrial stipends? It is crystal clear that, when it came to the dissemination and sharing of research findings, the company's commercial interests were of paramount importance. In June 1930, for instance, Baekeland reproached Bender, as well as some of his co-workers, for attaching "undue importance" to the technical term "resinoid." Baekeland himself had introduced this designation around 1925 to differentiate Bakelite, a thermoset, from thermoplastic resins of the novolac type.[63] Later, however, it had been more generally put forward as "a single word for designating synthetic resins," and Baekeland now reassured

Allan Brown, his firm's head of advertising, "that the whole discussion is merely academic and has no important bearing on the fact that 'Bakelite' is our valuable trade-mark which we must maintain in its full importance. That I have noticed lately in 'ads' and publications the absurd attitude of our emphasizing 'resinoids' etc. while avoiding the use of the word 'Bakelite': an absurd paradox."[64]

In the following month, Baekeland returned to the subject in a meeting with an in-house patent counsel, Bender, Brown, and Victor H. Turkington, a co-worker of Bender who had presented a paper on "Phenol Resinoids in Oil Varnishes" to the ACS's Paint and Varnish Chemistry division earlier that year.[65] Here again, Baekeland stressed that terminological exactitude should be subordinated to the maintenance and extension of a powerful, trademark-protected brand. In his opinion, the company "should not be pedantic here but bear in mind that any omission of the word Bakelite diminishes its trade mark value. That we are not here a chemical committee on nomenclature but member[s] of Bakelite organisation who must try to advance the interests of its trade-mark and consider all the money we have to spend yearly for publicity purposes of which our competitors profit in no small degree."[66]

Did Baekeland also keep a critical eye on the actual content of the presentations and research papers of his R&D staff? As I will show in chapter 10, he was generally more supportive than other top executives of the Bakelite Corporation with regard to R&D work. Baekeland was also proud that, notwithstanding the unanticipated absence of research director Lawrence Redman, no other firm had as many employees attending the 1936 Gibson Island symposium as Bakelite (see figure 9.3).[67]

Still, the Gibson Island meetings also illustrate the censorship policies restricting these corporate scientists' autonomy and freedom of publication. When Gordon invited him to chair the 1936 conference, Baekeland expressed the hope that "the discussions shall be purely scientific and not degenerate into a publicity stunt for various manufacturers." The more serious concern of his firm, however, was that its researchers might divulge confidential knowledge to rivals. At the 1939 symposium, for example, Bakelite's two paper presenters—Bender and Ivey Allen Jr.—needed the prior permission of their superiors. In his request for approval, Bender clarified that his paper on "The Physical Viewpoint of Resinous Particles as to Size and Linkage" was "much like the one" given at a 1937 meeting of the ACS, "plus the one on Molecular Measurements you now have before you"

Figure 9.3
Bakelite's delegation at the 1936 Gibson Island symposium. Bender (third from the right) stands between Baekeland and Turkington.
Courtesy of Leo Baekeland Papers, Archives Center, National Museum of American History, Smithsonian Institution, AC-0005-0000002.

(i.e., the article that would appear in the *Journal of the American Chemical Society* later that year).[68] Together with other evidence, this reference to earlier conference papers and publications suggests that all public disclosures by Bakelite's R&D staff were subjected to control. Baekeland's diaries, moreover, reveal that this review process regularly led to "corrections."[69]

The chronology of Bender's work on urea plastic resin throws more light on the factors determining whether and when research findings could be shared with company outsiders. From 1932 until 1934, Baekeland was involved in negotiations aimed at acquiring US patents of Kurt Ripper, the inventor of a urea-formaldehyde molding mixture promoted under the brand name of "Resopal," to strengthen the Bakelite Corporation's position in this field. Yet, at the same time, he urged Bender to carry on his earlier research on urea resin, dating back to the 1920s, "with renewed zeal and increased efforts." When Ripper visited Bakelite's Bloomfield laboratories on 12 October 1933, "Bender was not called in so as to let him continue his work on urea products according to his own invention and patents.... His results and recent progress are very encouraging and from present indications was an improvement of decided importance."[70] By 20 October, Bender had been able to make further advances. This led Baekeland to conclude that, while Ripper's Resopal seemed technically superior to all the urea molding mixtures then commercially available, Bender's product might become even better. Baekeland personally went through the patent specifications drawn up by Bender to secure proprietary control over the improvement. Bender's patent application of 7 November 1933, however, was not granted until June 1937,[71] and he did not make himself known as an expert in this commercially important area of research by publishing his findings.

Strategic Silence

A similar delay in publication can be observed in the case of Victor Turkington, the colleague of Bender who was also reprimanded for his "pedantic" use of the term "resinoid." Turkington was in charge of a highly productive R&D program on synthetic resin varnishes, one of the projects through which his employers sought to compensate for the reduced earnings in core product lines after the basic Bakelite patents had begun to expire. Turkington also published seminal research in this field. In 1941, for example, he coauthored "the first comprehensive review of the properties and general preparation of oil-soluble phenolic resin, including a rather general evaluation of the varnishes made from these resins."[72] By this time, however, Turkington had been working on this type of resin varnish for over a decade. Moreover, as late as 1951, it was indicated that "very little has been published on methods of manufacture and physical properties of the resins. What has been published is mainly patent literature, and many preparative techniques and results claimed are questionable."[73]

In this climate of secrecy, and given the high stakes involved, Baekeland even considered the release of technical information via international patents intolerable. In September 1931, he became enraged by the "untimely publication" of patents by Turkington and William Butler, also of the Bloomfield laboratories, which "allows all our competitors here and abroad to copy our methods and thus kill all our long and expensive preparations for the varnish trade." To Baekeland's anger, Turkington and Butler's inventions in the area of oil-soluble resins and varnishes had been patented in various European countries, "including such... as Belgium, France, Italy," whose patent offices did not conduct novelty and nonobviousness examinations and had therefore needed comparatively little time to process these applications, "thus divulging everything to competition and infringers long before the

U.S. patents require publication." Given that the violation of these American patents would be hard to prove, Baekeland claimed that "this is murdering our best hope in the future of our enterprise where our other products submitted to fierce competition are ceasing to pay us profits."[74]

Baekeland recalled two earlier instances in which such an "absurdity" had been committed. Importantly, he blamed these on researchers who had seemed to prioritize the pursuit of scientific and inventive credit over the use of patents as a tool for securing proprietary control of the commercial use of an invention. In his own words, Baekeland reasoned that "the desire of some of our men to see their name in print at the head of published patents, regardless [of] its killing effect on our program, has something to do with this." "Deeply disgusted and discouraged," he concluded that his associates could not be trusted and he decided to take matters into his own hands. As a result, in the following year, he was showered with letters from his company's patent department asking whether international patents should be applied for, which European patents should be extended, and which ones should be dropped.[75]

On the whole, Bender's and Turkington's experiences suggest that the freedom of Bakelite researchers to publish was negatively correlated with the commercial importance of the subjects they were working on. It is clear that patent protection typically had to be secured first. Because of the considerable time lag between the initial filing of a patent application and its eventual granting—no fewer than eight years in the case of Turkington's coating composition patent (see table 9.3)—this could require scientific employees to forgo potential reputational rewards resulting from exposing their discoveries or inventions to professional colleagues.

Baekeland's wariness was certainly not unusual. For instance, as historians David Hounshell and John Smith have described, Du Pont withheld information about nylon and neoprene from Imperial Chemical Industries although it had concluded a technology transfer agreement with this British chemical conglomerate in 1929.[76] Even at General Electric, whose publication policy was more liberal than that of, among others, Westinghouse and American Telephone & Telegraph, "researchers always had to wait until all related patent applications had been filed before submitting their articles, and complete disclosure was not always possible."[77] Moreover, three leading members of Bakelite's research division—Redman, Weith, and Ralph C. Shuey—early in their careers had participated in an industrial research

Table 9.3
Research papers and US patents of Bender or Turkington during employment with Bakelite Corporation.

Year	Author(s) and title
1925	Baekeland and Bender, "Phenol Resins and Resinoids"
1930	Turkington, R. C. Shuey, and W. H. Butler, "Phenol Resinoids in Oil Varnishes"
1931	Turkington, Shuey, and Butler, "Viscosity Increase and Gelation in Phenolic Resin Varnish Cooking"
1932	Bender, "Synthetic Resins," in *Colloid Chemistry: Theoretical and Applied*, vol. 4, edited by Jerome Alexander, New York: Chemical Catalog Company
1933	Bender, US Patent 1,922,272 ("Air drying composition") (filed 1932)
1934	Bender, H. F. Wakefield, and H. A. Hoffman, "Colloid Developments in Synthetic Resins," *Chemical Reviews*
	Bender, US Patent 1,944,143 ("Moldable composition and its product") (filed 1929)
	Bender, US Patent 1,953,892 ("Nonfibrous compositions") (filed 1929)
	Turkington, US Patent 1,958,452 ("Shock resistant material") (filed 1928)
1935	Turkington et al., "Phenolic Resin Varnishes: Influence of Linseed Oil on Film Properties"
	Turkington and William H. Butler, US Patent 1,988,615 ("Drying oil composition) (filed 1930)
	Turkington, US Patent 1,988,616 ("Floor covering") (filed 1930)
1936	Bender, US Patent 2,041,453 ("Resinous composition and process of making same") (filed 1931)
	Bender, US Patent 2,064,876 ("Air drying composition of the urea resin type") (filed 1933)
1937	Bender, "The Structure of Heated Resinous Films"
	Turkington, US Patent 2,070,148 ("Phenolic resinous products and compositions containing them") (filed 1932)
	Turkington, US Patent 2,075,839 ("Varnish compositions") (filed 1934)
	Turkington, US Patent 2,076,507 ("Coating composition") (filed 1929)
	Bender and Harry A. Hoffman, US Patent 2,084,314 ("Urea resins") (filed 1933)
	Bender, US Patent 2,101,635 ("Resin dehydration") (filed 1931)
1938	Turkington, US Patent 2,125,827 ("Coating articles") (filed 1936)
	Bender, US Patent 2,136,329 ("Flexible compositions") (filed 1933)
1939	Bender, "Molecular and Cybotactic Weight Measurements of Resinous and Crystalline Materials," *Journal of the American Chemical Society*.

Note: All papers appeared in the *Journal of Industrial and Engineering Chemistry* unless otherwise noted.

fellowship program at the University of Kansas that was characterized by relatively extensive publication restrictions. Therefore, it should be no surprise that, although the Bakelite Corporation did "do some molecule chasing and let a few men have their heads in work along lines in which they might feel inclined to do something," it did not allow for levels of openness that could not even be taken for granted in an academic context.[78]

It is perhaps more relevant to evaluate the position of Bakelite's corporate researchers in light of Baekeland's own experiences in the photographic and electrochemical industries and as an independent research chemist and consultant. Such a comparison on the one hand reveals that the Bakelite Corporation in some respects offered a relatively unrestrictive work environment. For example, despite the general trend toward stronger protection of corporate IP in late nineteenth- and early twentieth-century America, Baekeland and the other members of the firm's Executive Committee agreed in 1922 to shorten the time span of the noncompete clauses in their technical personnel's employment contracts from five years to one or two. While this requirement still limited the ability of these workers to change jobs, it was a far cry from the 20 years requested by Eastman Kodak in the years around 1900. But, on the other hand, it is clear that Baekeland wanted Bakelite researchers such as Turkington to be more secretive about their R&D work than he himself had been around 1909. His anger about Turkington and Butler's international synthetic varnish patents suggests that this change in attitude partly followed from his own disillusionment with—and improved understanding of—various patent regimes. In addition, however, his point of view had primarily become that of the industrial executive who not only found it "quite natural" that his research personnel assigned the ownership rights on their inventions to their employer but also expected them to withhold all confidential technical information.[79] Indeed, by the late 1920s, Baekeland evidently considered an eagerness to rapidly publish inventions or discoveries—a core feature of academic reward systems based on priority of publication—to be childish and disloyal in a corporate context.

At the same time, it is clear that theoretical reflections could be published even when they concerned fields in which the Bakelite Corporation had been investing heavily. To give a final example, in a 1934 article, three Bakelite researchers offered a broad survey of the "Chemistry of Synthetic Varnish Resins." In this paper, Ivey Allen Jr., Virgil Meharg, and John Schmidt reviewed theoretical work with regard to the chemical structures of nine sorts of synthetic resins, including the type invented by their company's president. They also included the chemical formula that Baekeland had "tentatively proposed" in 1909 and briefly discussed his later publication with Bender.[80]

Allen, Meharg, and Schmidt were probably not yet aware that Baekeland had prepared a final research paper related to the chemistry of Bakelite earlier the same year. Kurt Meyer, a chemistry professor in Geneva and one of the most learned critics of Staudinger's macromolecular theory, had invited Baekeland to participate in a meeting of the International Union of Pure and Applied Chemistry in Madrid in April 1934. Although he had entered his seventies, Baekeland remained as apt as ever to cross the boundaries between academia and industry. For example, while his business associates arranged for the distribution of Bakelite Corporation pens at the conference, Baekeland made it clear that he wanted to be listed on the program as an honorary professor of Columbia. Moreover, as he did not touch on any confidential issues, he was anxious to get his paper published at the earliest possible date.[81]

Meyer's invitation also motivated Baekeland to familiarize himself with recent research on the structure and formation of Bakelite. At Baekeland's request, Archie Weith, associate director of the Bloomfield laboratories, put together a list of publications that had come out in previous years. Several of these works offered new insights into the nature of the intermediate reaction products on which Columbia's Bakelite fellows had also focused their attention. In his own conference paper, however, Baekeland did not discuss whether the findings of experts such as British government researcher N. J. L. Megson and German industrial chemist Max Koebner had changed his views on the formulas of the intermediate and, consequently, final varieties of Bakelite. Although he had not published any speculations about the chemical identity of C-stage Bakelite since the 1925 paper with Bender, he now simply reiterated that "notwithstanding strenuous activity in trying to unravel the theoretical side, the latter has not shown as much progress as the industrial developments. The investigation of these complex substances is incomparably more difficult than for ordinary well defined organic compounds. We have to deal here with complex mixtures of amorphous substances of which the chemical constitution is easily influenced by physical as well as by chemical agents."[82] While undeniably true, by then this statement had almost become a platitude. It was therefore unlikely to satisfy some of the more academically oriented people on the extensive list of contacts to whom Baekeland mailed copies of his paper. This category included not just Meyer but also Staudinger, whose latest publications were also discussed at the 1936 Gibson Island symposium.[83]

It would take approximately one more decade before further theoretical insights into the final stages of the Bakelite process created a closer nexus between scientific theory and technological improvement, a development to which Bender and his team of collaborators in the Bakelite Division of Union Carbide continued to make significant contributions. However, as this chapter has shown, this did not mean that the "rather fantastic" early speculations had not given satisfying returns.[84] As we have seen, rewards could take the form of legal IP protection, academic prestige, or the ability to attract talented employees. Moreover, as chapter 10 argues, there were various other ways in which science shaped the records and organizational cultures of the Bakelite companies, starting with the creation of Bakelite Gesellschaft and the General Bakelite Company before World War I.

10 A Company of Scientists and Engineers

During the 1909 trip to Germany in which Baekeland negotiated the formation of a European Bakelite company, he also visited several chemical, electrical, and engineering firms as part of a consulting assignment for the Development & Funding Company. On these visits, however, his attention more than once diverted from the topic he and Elmer Sperry had been asked to study—the liquefaction and storage of chlorine gas—to the overall business methods and philosophy of these knowledge-intensive enterprises. As Baekeland admitted to Elon Hooker, the president of Development & Funding, he had "involuntarily…dilated on matters which have no relation to our own business, but I could not help doing so, impressed as I am with what constitutes German superiority in business enterprises based on scientific principles." In a previous paragraph in the same letter, he had remarked,

> I am surely sorry that you are not with me here in Germany, as I have no doubt that the direct and thorough business methods displayed here by big chemical and engineering concerns would open your eyes wide and would show you the secret of the tremendous success these people are achieving in these particular lines. Instead of having merely bankers or so called business men at the head of their enterprises they have the most thoroughly technically and scientifically developed staff. In fact, the majority of the board of directors is made up of people who have achieved success through practical scientific men. This explains their tremendous daring in business enterprises based on scientific facts.[1]

Given the ubiquity of high-technology firms in today's economy, the question of whether scientists or engineers make superior corporate executives remains as pertinent as ever. But how can whether their priorities differed from those of "bankers or hard boiled businessmen," as Baekeland repeatedly maintained, be tested?[2] And does Baekeland's differentiation of German from American chemical and engineering firms on account of the former's "science-based" nature accord with the available historical evidence?

The approach of this chapter is to concentrate on two areas in which, as is clear from earlier research, the preferences and actions of scientists and engineers may well have been shaped by their educational and professional backgrounds.[3] I focus first on their strategies for technological innovation and second on the remuneration schemes and career trajectories of their fellow scientists and engineers in the enterprises for which they were in charge. In this latter respect, I pay special attention to marketing and distribution, as Baekeland was proud that "even our [the Bakelite Corporation's] sales department is entirely manned by engineers or chemists."[4] Selling is another area of business that underwent dramatic transformations at the time of the second industrial revolution, when sizable manufacturing corporations grew even larger by investing in the creation of their own marketing and distribution units and facilities.[5] Particularly when it came to reaching mass consumer markets, US firms are widely seen as having been the prime movers. Historian Walter Friedman, for example, has gone so far as to contend that "the development of modern sales management is a uniquely American story."[6] However, when it came to selling science-related goods to other industrial manufacturers rather than to final consumers, German firms may have been at least as innovative. In particular, the provision of knowledge-intensive customer service was one of the factors behind the late nineteenth- and early twentieth-century rise to global dominance of the German synthetic organic chemicals industry, which employed a substantial number of scientists and engineers as traveling technicians.[7]

In the United States, chemical companies, together with electrical and early information technology firms, were also among the first to offer customers technical advice on the use of their products.[8] As emphasized in an in-house company history of the German American Roessler & Hasslacher Chemical Company, this policy required the creation of relationships of personal trust and brought about mutual dependencies between sellers and buyers exchanging confidential knowledge and know-how with one another. According to this source, "in the [eighteen]eighties the users of chemicals [in the United States] were very secretive and became suspicious when any seller made a suggestion; the plan of servicing was a direct guarantee that inside information given to the service man would never be used to the disadvantage of the customer. The partners [R&H] supported this guarantee so well by their honest endeavor that they established a reputation for integrity and bound their customers to them by ties of common interest."[9]

Given the perceived innovativeness of this policy, as well as Baekeland's claims for the superiority of German science-related firms quoted earlier, this chapter investigates whether Baekeland's association with Roessler & Hasslacher influenced the business strategies and organizational cultures of the American Bakelite companies. In the view of Rupert Lowe, an engineer who started working for Baekeland in 1923, General Bakelite was run as a de facto subsidiary of R&H up to World War I, when R&H came under scrutiny on account of its connections to the Deutsche Gold- und Silber-Scheideanstalt vormals Roessler AG—the chemical and metallurgical firm better known as Degussa.[10] Still, from its creation in 1922 until the end of 1926, the Bakelite Corporation's highest decision-making organ—the Executive Committee—was chaired by R&H's Philipp Schleussner, a second-generation immigrant from Germany. Another member of the Executive Committee, Louis M. Rossi, was a former employee of R&H. Moreover, not only Schleussner but also William Hamann, R&H's president from 1926 until 1928, sat on the Bakelite Corporation's board of directors from 1917 until Baekeland bought out the lion's share of R&H's interest in Bakelite in 1928.

Research and Development

In 1930, relatively shortly after the close affiliation between Bakelite and R&H ended, R&H was taken over by chemical giant E. I. du Pont de Nemours and Company. Among other things, Du Pont had been attracted to R&H's "strong technical organization and ... aggressive research policy," through which the firm had sought to exploit its "large research facilities."[11] Without a doubt, the business culture of R&H had been technically oriented since the firm's establishment in 1885 as a partnership between Jacob Hasslacher and Franz Roessler, a younger brother of two of Degussa's founders.[12] R&H's commitment to in-house research and development, in comparison, had strongly increased after World War I, when its linkages with Degussa and other European sources of invention had been curtailed. As a partial reflection of that trend, the number of chemists and engineers engaged by R&H and its subsidiaries rose from about 20 around 1924 to almost 200 in 1930. The latter number represented approximately 10 percent of the firm's total personnel.[13]

Partly to anticipate the expiration of its basic patents on the Bakelite process, the Bakelite Corporation also seriously invested in R&D during the 1920s. In fact, according to Baekeland's diary, the firm's annual expenditures on research and development work in the first decade of its existence, from 1922 until 1931, averaged almost $500,000.[14] A budget cut of about $150,000 was then implemented, but in view of the Bakelite Corporation's relatively small size—by August 1934 it employed roughly 450 workers—the remaining amount still was remarkably large.[15] For instance, an internal report by longtime Bakelite chemist Lawrence Byck suggests that it represented more than 5 percent of the net value of the firm's sales. In 1933, moreover, Baekeland calculated "that if we compare our net assets with these of Dupont de Nemours at the rate of money we spent on research in [Bakelite's laboratories in] Bloomfield then Dupont ought to spend at least 26 million dollars annually."[16] In reality, however, Du Pont's R&D budget totaled about six million dollars around this time, representing approximately 3–4 percent of its annual sales.[17]

What did Baekeland think of these high research expenditures? In an improvised speech in April 1927 on "Hard Boiled Business or Technologists at the Head of Chemical Enterprises?," he justified them by means of an analogy that has since become associated with Eastman Kodak's research director Kenneth Mees. Baekeland claimed that "Research work is my gamble," before clarifying "why I never have been interested in other gambling [such as] stock speculation [or] horse races. It is the noblest gambling any body can undertake and it is our duty to teach our nation that if they want a real national gambling we should make money spent on research our national lottery system."[18] During the decade up to 1931, this motto was reflected in Bakelite's financial policies. For example, on the initiative of Baekeland, and in defiance of requests for more profit sharing, extensive cash reserves were accumulated in the first half of this period.[19] In the years that followed, moreover, investments in R&D were largely financed out of these reserves.[20] Hence, not least because of Baekeland's financial conservatism, Bakelite initially weathered the Great Depression relatively well. As the chemist explained in a November 1930 letter to a former colleague at the State University of Ghent and the Teachers Training College of Bruges,

> Business is undergoing the results of the depression of some of the industries which are customers of Bakelite (electrical, automotive, steel and other metals, etc., etc.) [W]e were prepared for what happened.... We even increased the personnel of

> our research laboratory, which now counts a personnel of over 150, of which more than one third are chemists, physicists, or engineers, who have graduated from our Universities and Engineering Schools, and we are spending $200,000 on an additional three-story laboratory building which is being erected along the former research laboratory buildings [in Bloomfield]. ... I consider that the present business depression is a blessing in disguise and will correct the extravagance and general absurd over-confidence which reached its climax in 1929.[21]

Other top executives of Bakelite had argued that their firm's independent research laboratories should be run more efficiently and at a lower cost. In March 1928, for example, four of those executives—Philip Schleussner, Sandford Brown, Lansing Hays, and George Baekeland, who had joined his father's company less than six years earlier—had alleged that the head of these laboratories, Leo Baekeland's former rival Lawrence Redman, was a spendthrift. However, Leo had countered that "none of them knew anything about research nor how to conduct it and I would not consider their complaints until Redman had his say nor would I advocate any reduction until I had consulted Redman."[22] In a follow-up meeting the next day, Leo had repeated that he would not take into consideration the opinions of "unqualified" collaborators. On this occasion, Wall Street lawyer Lansing Hays, who, according to Baekeland, "is the least competent in research matters," was the one who proposed the most drastic measures. Otherwise Baekeland generally appreciated Hays's counsel and entrusted him with consequential private affairs, such as his purchase of a luxurious holiday property in Coconut Grove, Miami, besides various other personal investments.[23] But he now complained that Hays failed to appreciate the peculiarities of the synthetic plastics industry. More specifically, at a time when a growing number of US chemical firms were contemplating the introduction of methods of mass manufacturing, Baekeland criticized the fact that Hays "would want at the head of B.C. a man who is a composite of a head-bouncer and traffic policeman who would handle our staff as a foreman handles white and black labor in a brick yard. ... Following his advice or that of Schleussner on this subject would kill our business in a short time. They overlook [that] our business is not a Ford factory or a Brickyard or coal mine."[24]

In spring 1928, Baekeland felt "very proud of Redman and the men in our staff in Bloomfield,"[25] but by early 1931, when the Bakelite Corporation had ceased being profitable owing to the Great Depression, he agreed that the expenses of Redman's division were "disproportionate ... compared to Dupont and other large corporations and how if we continue we shall get short of money in our mfg [manufacturing] business."[26] Therefore, he became more reluctant to invest in projects that were unlikely to yield short-term commercial benefits, including the Bakelite fellowships at Columbia. For instance, while he had previously highlighted the challenges involved in turning initial inventions into commercially viable innovations, he now complained that "all what he [Redman] has to bring forward is hope and expectation. 'Give us two years more and you will see' etc." Baekeland even became "much disgusted" with "the general extravagance and lack of sense of proportions of the Bloomfield Laboratories," adding that "Redman never was a manager and never will be, and [Archie] Weith had no power, nor did he dare to do anything contrary to the wishes of R. [Redman]."[27] He strongly urged Redman and Weith, the department's associate director, to make significant budget cuts and, on the basis of a series of critical inspections, contributed several suggestions of his own. Among other things, Baekeland advised "that much work of the trained chemists in Bloomfield might well be performed by laboratory helps under supervision and direction of the chemists instead of the latter losing time and attention to minor manipulations."[28]

As part of the reorganization of Bakelite's research department, some of the firm's most highly qualified scientists were dismissed. These included physicist T. S. Taylor, who had previously led research on electrical insulation at Westinghouse in East Pittsburgh and had been on the teaching faculties of the University of Illinois and Yale. In 1925, Baekeland had been favorably impressed by Taylor and the way in which he had "built up his [physical research] laboratory showing resourcefulness." Nonetheless, Taylor was among the research managers who were asked to lay off staff in 1931.[29] In summer 1932, moreover, Bakelite's Operating Committee—a managerial body that was subordinate only to the Executive Committee—decided also to fire Taylor himself. Among other things, the physicist was reproached for having pursued overly academic R&D programs: in September, Baekeland informed him "that we wanted to reorganize the physical department into what it always was meant to be a *testing* department. That at present it was organized out of proportion to our needs and income." This comment was in line with the advertising of Bakelite's laboratory work as a form of customer service (see figure 10.1). It was also consistent with Baekeland's broader criticism that the construction of the Bloomfield laboratories had involved "much unnecessary expenses and unnecessary equipment put as if it were a University funded lab" and anticipated his son's

Figure 10.1

Promotional brochure for Bakelite Corporation (1928). In the booklet for which this image was the front page, it was emphasized that Bakelite's research laboratories were "more than just the private laboratories of the Bakelite Corporation. They are in effect the common laboratory for every user of Bakelite materials—a part of your plant though not in your plant."

Courtesy of Trade Literature Collection, National Museum of American History Library, Smithsonian Institution.

previously quoted comment that only a few of Bakelite's researchers were allowed to engage in "molecule chasing."[30]

In a context of prolonged economic distress, the case of Austin V. H. Mory, Bakelite's assistant director of research in charge of personnel and public relations, provided another "bad example of extravagance."[31] Mory was a leading member of the ACS in Chicago, where he had directed the testing laboratories of Sears, Roebuck, and Company from their establishment in 1911 until 1920.[32] Baekeland and Redman had recruited their professional colleague for Bakelite in late 1923. They had set his annual salary at $5,500, with an additional, performance-based bonus of at least $500. In Baekeland's opinion, "If he is not worth more [than $6,000 a year] he is not the man we want but of course [he] will be of limited use before he gets acquainted with his job." According to a letter by Baekeland's son, however, Mory had not yet demonstrated his value to the company by 1929. Moreover, in a 1932 diary entry, Baekeland wrote that Mory had personally indicated to him that he "had never earned his salary.—Altho he tried to find work to warrant it.—That he really was a 'glorified secretary to Dr Redman' attending to odds and ends."[33] To Baekeland's surprise, these assignments even included the ghostwriting of speeches and publications that Redman had agreed to author. Mory also contributed promotional writings under his own name, including a 1929 article discussing the creation of synthetic plastic resins in a grandiloquent style.[34] In addition, he claimed to have indirectly furthered the interests of his employer through his active participation in the ACS. Yet, this did not suffice to convince top managers of Bakelite such as George Baekeland and the brothers Gordon and Sandford Brown—sons of the president of Condensite, with whom General Bakelite and Redmanol had merged in 1922–1923—that Mory's remuneration accurately reflected his job responsibilities.

The reliance on respectable scientists for public relations and marketing purposes has received much less scholarly attention than the emergence of industrial research. It was, however, a notable aspect of the late nineteenth- and early twentieth-century science-industry nexus, which contributed to a proliferation of formal and informal advertising restrictions.[35] The practice occurred on both sides of the Atlantic but may have been especially common in Germany, where academic scientists and engineers enjoyed an "aura of incorruptible neutrality and objectivity" that, ironically, could be exploited for commercial goals.[36] Accordingly, it may be no coincidence that an immigrant from Germany, Jacob Hasslacher, had encouraged Baekeland to become president of General Bakelite for reasons of scientific prestige. Baekeland strongly valued the advice of Hasslacher, who had become R&H's president in 1885 and had consistently reached out to the New York chemical community in the following decades. In this respect, Hasslacher may have been inspired by Degussa's Heinrich Roessler, an older brother of his partner Franz Roessler. Heinrich Roessler supported the pursuit of science in Frankfurt am Main, the city in which Degussa was headquartered, and maintained friendly relations with prominent chemists such as Wilhelm Ostwald. In recognition of his scientific philanthropy and his own research accomplishments, Roessler was granted the title of professor in 1909, the year before Hasslacher and Baekeland negotiated the formation of General Bakelite.[37]

However, like the sales strategies discussed in this chapter's next section, other "ornamental" uses of science in the histories of the American Bakelite companies cannot be traced back to any obvious German connections. These instances involved Bakelite's own directors and employees as well as company outsiders such as Michael Pupin, the Columbia professor and personal friend of Baekeland. In a 1927 folder entitled "Read what one of America's most noted Radio Scientists says about BAKELITE," for example, the Bakelite Corporation had reprinted complimentary remarks by Pupin, which had previously appeared in the *Boston Evening Transcript*. Baekeland also tried to use his influence to have academic honors conferred upon Redman, his company's director of research. In June 1924, he attempted to accelerate the administrative procedures of the University of Toronto for granting Redman a PhD degree. In addition, in November 1924—that is, toward the end of his presidency of the ACS—Baekeland subtly proposed Redman's name for an honorary degree from the Mellon Institute of Industrial Research at the University of Pittsburgh.[38] As the prestige associated with such awards could indirectly benefit the Bakelite Corporation, Baekeland's lobbying efforts can be seen as further examples pointing to the industrial value of scientific honors and credentials. At the same time, however, this type of advantage was not enough to explain why the number of academically trained scientists and engineers hired by Bakelite and other chemical firms was remarkably high by the standards of the time.

Technical Selling

In the decade from 1919 until 1929, the US Census Bureau stressed almost annually that the synthetic organic chemicals and dyestuffs industry employed "probably a larger proportion of technically trained men than will be found in any other important manufacturing industry in the United States."[39] Over this period, the share of such workers in this sector gradually augmented from 10.5 percent to 15.3 percent. No separate data on the Bakelite companies are available. Yet, according to Baekeland's diary, the segment of contract employees at the Bakelite Corporation and its subsidiaries in the United States also constituted approximately 15.3 percent around the beginning of the Great Depression.[40] It is clear that formally schooled scientists and engineers and other white-collar workers were most likely to be offered salaried, contractual positions to gain their goodwill as well as to prevent them from divulging confidential knowledge and know-how. Therefore, Baekeland's contention that his company's sales department was "*entirely* manned by engineers or chemists" (emphasis mine) must probably be interpreted as referring solely to contract employees. If thus understood, this claim is certainly plausible.

In many ways, providing technical customer service had been a natural corollary to Baekeland's decision to concentrate on the early phases of the production process, namely the chemical preparation of phenolic resins up to the A stage. As a case in point, the further processing of Bakelite by customers such as Boonton Rubber and the Weston Electrical Instrument Company had occurred in close consultation with Baekeland and his technical staff. This is not to say that the American Bakelite companies completely refrained from manufacturing ready-to-use products themselves. In 1920, for example, General Bakelite briefly made molded applications of C-stage Bakelite, with the aim of "inducing others to do their own molding and to nurse them thru [the] initial period before they have their plant ready."[41] In 1923, Baekeland and his associates did not dare initiate additional work of this kind in the United States out of fear of customer resentment, but they tried to circumvent these national-level restrictions by offshoring production to their newly acquired molding facilities in Canada.[42] Likewise, around 1930, the Bakelite Corporation temporarily manufactured the new synthetic resin varnishes invented by Victor Turkington and his team in the Bloomfield laboratories on a commercial scale. In this case, too, the goal was not so much to gain direct financial benefits as to demonstrate the merits of Bakelite's products and, in the process, establish a market for them. Accordingly, as soon as what Baekeland described as the "stubbornness, ignorance or lack of enterprise of average varnish or paint manufacturers" had been overcome, Bakelite again restricted itself to providing unprocessed varnish resins along with technical advice.[43] In this way, it aimed at establishing collaborative—instead of competitive—relations with industry incumbents. Bakelite's approach was similar to how the Pittsburgh Reduction Company had created early markets for aluminum in the 1890s. As Baekeland's son George explained in 1939, "There were about 1,100 companies in the paint and varnish business and we wanted them as customers and not competitors.… But… we did go into the varnish business

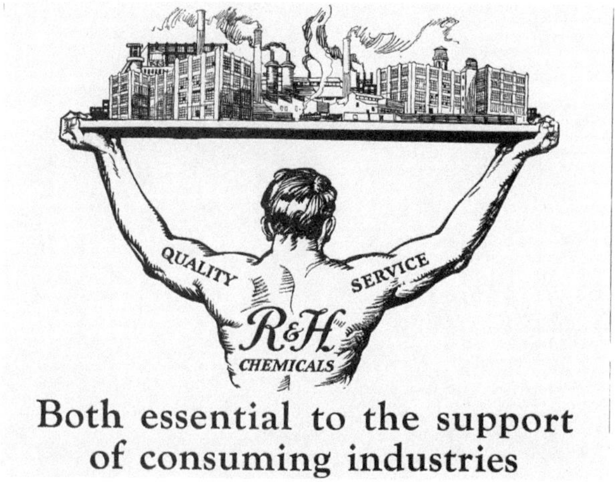

Figure 10.2
Advertisement of R&H, 1925. Another ad in the same brochure indicated that "the unusual sales and distributing facilities of The Roessler & Hasslacher Chemical Company are now offered manufacturers of non-competing products who do not wish to undertake the tedious and expensive process of building up a competent sales staff." As R&H's evocation of male strength implies, this staff does not seem to have included any women.
Courtesy of Hagley Museum and Library.

temporarily, solely for this purpose: to take business away from existing varnish companies, and having taken a man's business away from him, he then takes you seriously. So we went at it."[44]

As these episodes demonstrate, the Bakelite companies' reliance on technical salesmen must be understood in connection with these firms' position in the production chain. Perhaps most important, it corresponded to their catering to the needs of industrial clients in knowledge-intensive sectors. In this respect, too, Bakelite's approach resembled that of Roessler & Hasslacher. R&H had been able to establish a reputation for quality and reliability among its industrial clients in its first years of existence and, by and large, had been able to maintain it up to 1930, when the firm was taken over by Du Pont. Selling had been one of R&H's core competencies (see figure 10.2). Besides manufacturing an increasingly large number of products of its own, it had acted as the US sales agency for Degussa, as well as several additional German and other European companies.[45] By choosing to associate himself with R&H, Baekeland gained access to a market-proven sales and distribution

Figure 10.3
Bakelite Corporation advertisement, 1927.
Courtesy of Science History Institute.

network in addition to other important advantages, such as R&H's formaldehyde supplies.

As mentioned before, Baekeland liked to call his salespeople "service engineers" (figure 10.3). He expected them to set themselves apart from low-status "drummers" and was convinced that the health of his enterprise largely depended on their technical competencies.[46] Some leading salesmen were not able to live up to his demanding standards. Herbert May, for example—an engineering graduate from MIT engaged by General Bakelite in 1912—was informed at the beginning of 1922 that none of Bakelite's directors were satisfied with the performance of the sales department. May was therefore replaced as head of this unit by Hylton Swan, a British American electrical engineer whose career will be discussed in some detail. By the same token, Baekeland considered a "lack of technical knowledge or experience on which all our selling is based" a reason for dismissing a sales executive who had previously worked for Redmanol in late 1929.[47] Overall, however, Baekeland believed that the Bakelite Corporation's service and sales capabilities were a key strength that could and should be exported to some of the firm's affiliates, including Bakelite Gesellschaft mbH in Germany.

If technically oriented business strategies and organizational cultures like those of Bakelite and R&H were more common in Germany than in the United States, the phenolic resin industry seems to have been atypical.[48] Baekeland considered an (alleged) lack of technological knowledge in general and sales competencies in particular to be a major liability of Bakelite Gesellschaft, the Berlin-based enterprise whose post–World War I reconnection efforts he finally acceded to in late 1926. Accordingly, he clarified the Bakelite Corporation's "educational work thru our service sales dept" in a letter to Max Weger, a longtime executive of Bakelite Gesellschaft, in July 1929. In addition, from September until November 1929, Ernst Elbel, a PhD chemist and Weger's son-in-law, stayed in the United States to examine the technical sales strategies and tactics of the Bakelite Corporation.[49] Still, to Baekeland's mind, Bakelite Gesellschaft continued to "stubbornly resist our advice" in the following months and years.[50] Consistent with his belief that the performance of a chemical firm was strongly shaped by the profile of its leader, he primarily blamed Edgar Sachs, the president of Bakelite Gesellschaft. In Baekeland's opinion, Sachs was "hopelessly ignorant of technical matters and hence does not have the sense of proportion between big and small matters."[51] In June 1933, during the second of two visits to Germany in a relatively short period of time (figure 10.4), Baekeland judged that Sachs and, to a lesser degree, Weger "had too much of a preponderant power in the organization, and practically did what they wanted and got used to it." At the same time, he was aware that, from a business point of view, "the fact is that under the present Hitler regime, Sachs as a Jew at the head of B.G. is a very serious drawback. Even were he not such a bossy excitable pugnacious man, making many enemies by his ways, he would be a detriment to Bakelite."[52] (To his credit, Baekeland later helped another German Jew, his former rival Hans Lebach, emigrate from his home

(a)
(b)
OPTIK
BAKELITE
UNTERGRUND · MESSHALLE MARKT

country.)[53] According to Baekeland's diary, Sachs personally informed him that an expert of Deutsche Bank, "which has much to say at Rütgerswerke," the parent company of Bakelite Gesellschaft, had criticized the sales methods and organization of Bakelite as part of a broader evaluation of its performance.[54] After this inspection, the contracts of Sachs and Weger had been changed so as to authorize the supervisory board of Bakelite Gesellschaft to terminate their employment after 1934 if deemed desirable. By September 1934, it had become clear that Sachs, unlike Weger, would effectively be laid off. Baekeland primarily attributed Sachs's professional downfall to his (alleged) incompetence. Yet, Sachs's Jewishness had doubtlessly been taken into consideration, too. Shortly after the Nazi takeover of power in 1933, for example, a competitor of Bakelite Gesellschaft with ties to the Nazi party had already apparently alleged that the leadership of Bakelite Gesellschaft was composed of "enemies of the Vaterland-Industrie [domestic industry]."[55]

Earlier in the twentieth and in the late nineteenth centuries, the US branch of Bayer had repeatedly been reprimanded for filling job vacancies for technical salesmen with Germans and German Americans rather than locally trained people.[56] Owing to Baekeland's collaboration with R&H, first- and second-generation immigrants from Germany had also been well represented on General Bakelite's payroll. Yet, unlike with Bayer's subsidiary, they certainly had not dominated the sales departments of either General Bakelite or the Bakelite Corporation. The backgrounds of several of the about 30 sales technicians that the latter firm employed in 1930 are documented in sources such as Bakelite's in-house company magazine *Trefoil*. These records provide no evidence that a familiarity with German sales models had been among the criteria influencing their recruitment.[57] Instead, they strongly suggest that Bakelite preferred to hire workers acquainted with key markets

Figure 10.4

(a) Baekeland in Berlin Zoo, May 1932, and (b) the modernist Bakelite tower at the Leipzig trade fair of 1937. Baekeland did not personally travel to Nazi Germany after June 1933 but continued to correspond with the leadership of Bakelite Gesellschaft and relied on information provided by other Bakelite executives, including his British associates at Bakelite Ltd.

Sources: (a) Courtesy of Leo Baekeland Papers, Archives Center, National Museum of American History, Smithsonian Institution, AC-0005-0000119. (b) Courtesy of Leo Baekeland Papers, Archives Center, National Museum of American History, Smithsonian Institution, AC0005-0000080.

for its products, such as the electrical, radio, and automotive industries. Charles ("Chick") Norris Jr. (b. 1896), for example—who held a master of industrial engineering degree from Cornell University—had worked as an automotive mechanic before joining Bakelite in 1930. Clinton ("Clint") Blount (b. 1901; studied at Newark College of Engineering, Columbia University, New York University, and the United States Naval Academy) and David ("Dave") Buchanan (b. 1898; master of engineering from Brooklyn Polytechnic Institute) had worked for one or several radio companies prior to beginning their employment at Bakelite in 1924 and 1926, respectively. William B. Hoey (b. 1898; bachelor's degree in electrical engineering from the University of Delaware), who joined Bakelite in 1925, previously was on the payroll of the Electric Service Supplies Company in Philadelphia.[58]

The rewards and limitations of a technical sales and marketing approach targeting other industrial manufacturers are particularly clear in the career of Hylton Swan, the first general manager of the Bakelite Corporation's sales department. Swan, a graduate of the Cooper Institute of New York—which Baekeland valued as a US equivalent to the Industrial School of Ghent—also had a background in the electrical industry.[59] Shortly after the public announcement of the invention of Bakelite in 1909, he had visited Baekeland to receive more information about the material. Around late February 1910, he was hired by the Protal Company, an early user of Bakelite, on the recommendation of Baekeland, who knew and respected Hylton Swan's uncle, photographic and electrical innovator Sir Joseph Swan. Hylton Swan then entered the direct employment of General Bakelite in 1911 at a yearly salary of $2,500.[60] In the decade that followed, he continued to impress Baekeland, who successfully nominated Swan for the position of head of a reorganized sales department in the context of the merger with Condensite and Redmanol.

Swan's increases over his initial earnings can serve as an indicator of his performance at Bakelite, as well as of the career opportunities for technically trained workers within the company. In 1929, Redman and Louis M. Rossi—the head of manufacturing and another longtime Bakelite employee—were the only Bakelite managers to receive a bonus higher than Swan's allotment of $27,000, the equivalent of about $396,000 in 2018 US dollars. In 1931, moreover, the salaries of Swan—as well as those of Sandford Brown and George Baekeland—were fixed at the same level as those of Rossi and Redman. There can therefore be no doubt that Swan's superiors appreciated his work.

Still, in the second half of 1932 and the first half of 1933, Rossi and George Baekeland advised their company's president to replace Swan with either Allan or Gordon Brown—the brothers of Sandford—as head of sales. Swan was apparently aware of these plans, for he expressed concern about being crowded out. Even the trade press reported on Swan's uncertain employment situation and speculated that he might offer his resignation. Leo Baekeland, for his part, assured Swan that his contract would not be terminated, while at the same time observing that "something ought to be done to vivify the sales Dept at a time when there were increasing indications of a betterment in business."[61] Accordingly, it was agreed that Swan would concentrate on the commercial development of a printing process patented by a British inventor, in which the Bakelite Corporation had been investing since 1930, so as to create a "new enormous outlet for laminated Bakelite."[62] Gordon Brown took over the management of Bakelite's sales group, while Allan Brown continued to lead the advertising department.

This reshuffling of personnel did not mean that Swan had fallen out of favor, but it did reflect broader challenges to the technically oriented corporate identity that Baekeland had tried to cultivate. As historian Jeffrey Meikle has described, in the early 1930s Allan Brown led a series of ambitious promotional programs targeted at end consumers, including a pioneering collaboration with industrial designers.[63] In line with his earlier comment that "we chemists do not want advertising as film stars or actors or writers or authors and that any such thing does us much harm just among our professional friends," it is clear that Baekeland was not entirely comfortable with these initiatives. As late as September 1935, for example, he insisted that

> Allan Brown is agitating all the time the idea of celebrating our 25th year by a big jamboree. I am against it.—We are not in the same category of company who want to take every opportunity to advertise themselves to the public, because they sell their products to the ultimate consumer, and resort to boundless advertising. We on the contrary have to sell our materials on the merit of best quality and price.
>
> Others can humbug their customers.—We cannot even if we tried it.—We have to deal with engineers and chemists or technically competent people who can easily determine the technical merits of our products as compared to all the humbugging of tooth powders, patent medicines, beauty parlors etc.[64]

Brown's success in promoting the Bakelite brand to a mass audience was, however, undeniable, and Baekeland must also have been aware that his executive was highly regarded in the plastics industry. As early as 1930, for instance, the journal *Plastics and Molded Products* had praised Brown for

Figure 10.5
Technical and commercial Bakelite advertisements from (a) 1918 and (b) 1926.
(a) Courtesy of Reindert Groot (amsterdambakelitecollection.com). (b) Courtesy of National Museum of J. Harry DuBois Collection on the History of Plastics, Archives Center, National Museum of American History, Smithsonian Institution, AC0008-0000015.

his wit and energy and listed his professional distinctions, which included a vice presidency of the National Industrial Advertisers Association and a directorship of the Association of National Advertisers. In March 1936, *Modern Plastics*, the successor to *Plastics and Molded Products*, even hailed Brown—"the best golf player of the Brown brothers family"—as a "premier idea man in the plastics field."[65] Still, Allan Brown's remuneration for 1936 was substantially lower than Swan's: the former's salary and bonus amounted to $19,231, whereas the latter's was $25,875. As the next section of this chapter demonstrates, the determination of bonuses was among the areas of policy on which Baekeland left his mark. Accordingly, it seems fair to suggest that the chemist's preference for "technical" over "commercial" business activities (figure 10.5) contributed to Swan's higher earnings.[66]

Bonuses: "What a Company to Plunder!"

According to Baekeland's philosophy of business, it was better not to "attempt much if you have as…associates only men whom you cannot trust, or who do not trust you, or who irritate or depress you, or who are dangerous by their greed, their recklessness, their superficiality, their weak character, or their general unreliability."[67] Early leaders of R&H, such as Jacob Hasslacher and Hugo DuBois, certainly did not fall into this category. Even more than a decade after his collaboration with them had ended, Baekeland recollected with fondness how gratifying it had been. With later representatives of R&H, however, he did not get along as well. He was particularly annoyed by Philip Schleussner's recurring annual demands for higher top executive remuneration.[68] In parallel to Schleussner's requests, Baekeland came to identify a "plundering" of corporate earnings by a clique of top managers as a key policy of R&H, harmful to the interests of outside shareholders. In March 1928, for example, he expressed concern that R&H and Adolph Karpen of S. Karpen & Brothers, the main investor in Redmanol, could "run the company for the benefit of a coterie of directors instead of for the benefit of the stockholders just as R&H is doing now by its own company. 'What a marvelous City to plunder' exclaimed the Prussian General [Gebhard Leberecht] Blücher when he visited London."[69] More specifically, Baekeland worried that "Karpen, Schleussner and the others might make a voting trust and swallow up most of the profits before they declared dividends [of which Baekeland was the largest recipient]. Just as R&H is doing now where Schleussner, [Hector R.] Carveth and Hamann each receive a slice of over $100,000 of the profits before stockholders get anything.… What a company to plunder! if they had access to our 6 million $ invested fund!"[70]

By the standards of the time, these sums were just enormous. To put things in perspective, two-thirds of the technically trained employees in the US coal tar products industry—and over 99.5 percent of the nontechnically trained workers—received a weekly salary of under $75 in the year when Baekeland expressed his concern.[71]

But was Baekeland right about the size of the bonuses allocated to R&H's chief managers? From the establishment of R&H in 1885 until 1929, about one-third of the firm's net profits were typically paid out as special compensation to the firm's executives. These executive bonuses—known as *Tantièmes* at Degussa and R&H—peaked around the time when Baekeland made the quoted comments. In 1928, for example, the astronomical sum of $595,050 was divided among 14 officials. Baekeland was not the only person to express reservations against this habit. For instance, in early 1930—as part of the negotiations that led to the takeover of R&H by Du Pont—a committee was appointed to study R&H's remuneration practices. Significantly, it reported that "in no instance have we been able to find a bonus policy as liberal as the one pursued for a number of years by our Company." In line with this committee's recommendations, the share of R&H's earnings put aside for bonus purposes was reduced to 10 percent in 1929. Nonetheless, Carveth, who had become president of R&H in 1928, received personal compensation of $60,000 for his performance during 1930, the year of the takeover.[72] Hence, it seems fair to say that although Baekeland's reference to sums of more than $100,000 was probably an overestimation, he was right in associating Schleussner's insistence on higher *Tantièmes* with well-established financial practices at the German American enterprise.

Not coincidentally, Baekeland possessed a sufficiently large share of stock in the Bakelite Corporation to be able to veto any major statutory or policy change, including increases in executive remuneration such as those proposed by Schleussner and Karpen. To acquire this power of veto, a minimum of three out of ten director votes was needed, and R&H had only two delegates on Bakelite's board. Accordingly, Baekeland had sound reasons for judging that it would not be worth the expense to buy out R&H.[73]

In the course of the years 1927–1928, however, it became Baekeland's goal to make the Bakelite Corporation entirely independent from directors whose primary allegiance was elsewhere. This change was partly inspired by Baekeland's perception of a growing conflict of interest between Bakelite and R&H—a view fueled by a series of incidents and misunderstandings. For example, according to Baekeland's diary, the (erroneous) impression that Bakelite was controlled by R&H had hindered his enterprise in purchasing raw materials at competitive prices from firms other than the German American enterprise. But Baekeland did not undertake action until March 1928, when R&H repeated its preference to turn Bakelite, a closely held corporation, into a publicly listed company.[74] The firm's first and only president was not willing to loosen his grip and found the prospect of having to make concessions to any "pirates" who might purchase Bakelite stock terrifying. Therefore, he personally made a "money sacrifice" of almost two million dollars to acquire 5,000 of R&H's 6,000 shares in

Bakelite.[75] In principle, R&H could still have laid claim on a single director if it accumulated enough votes from other smaller shareholders. However, unlike Baekeland, it does not seem to have campaigned to this end. Consequently, at the annual stockholders' meeting in December 1928, Baekeland succeeded in having both Schleussner and Hamann voted off Bakelite's board of directors.

For obvious reasons, the purchase of R&H's stock temporarily weakened Baekeland's financial position. On the day when he presented his short list for the reshuffled directorate to Schleussner, for instance, the chemist also explained to George Pegram of Columbia "the tight position I was in at present after paying two millions to R&H for their stock and had to be economical unless I wanted to be in the hands of money lenders who might at any time assail my strong position in Bakelite and thus spoil all what I had aimed at in showing a successful chemical enterprise run by Chemists and Engineers without hard boiled business men and bankers."[76]

Nor did Baekeland trust all the newly elected board members. In fact, as early as March 1929—when Sandford Brown's name was put forward as a vice president—he once again feared a conspiracy aimed at undermining his position of control. (Accordingly, in return for supporting Brown's promotion, he requested a change in bylaws reducing the clout of the vice presidents.)[77] Such doubts even persisted after Baekeland and other US-based owners of the Bakelite Corporation unanimously agreed to exchange their stock in Bakelite for shares of the Union Carbide and Carbon Corporation, thus turning Bakelite into a division of Union Carbide. In particular, Baekeland feared that his son, whom he envisioned as head of Union Carbide's Bakelite branch, would be crowded out by Sandford Brown and other executives. Still, he was largely satisfied with the sellout, which concluded his career as a scientific entrepreneur. In November 1939, for example, Baekeland proudly noted that the number of workers employed by the Bakelite Corporation had risen to 1,350. In addition, he felt that Union Carbide's reputation for being a science-driven organization was justified. Like other observers of the US chemical industry, he had noticed that Carbide's personnel included many chemists and engineers, including in upper managerial positions and in areas such as selling. Accordingly, he personally vouched for the "high quality" of Union Carbide, which was not to be confused with similarly named companies that had fallen into "bondage" to financial speculators and unscrupulous businessmen.[78]

Revering Science

Why was it so important for Baekeland that scientists and engineers occupy dominant positions in the Bakelite companies and the firms interested in acquiring them? This preference in part reflected his previously discussed technocratic views. That is, like economist Thorstein Veblen and many other early twentieth-century intellectuals, Baekeland believed that society would benefit if scientists and engineers applied their supposedly objective expertise in the realms of industry and government. They defended this idea in a critical period in the development of capitalism in the United States, characterized by an increase in "conspicuous consumption" and by what scholar Lawrence E. Mitchell has described as the gradual triumph of stock market speculation over technically oriented corporate strategies.[79]

As with many other themes discussed in this book, however, Baekeland's actions and preferences were also shaped by the dual allegiances resulting from his roles as scientist and industrialist. As a member and patron of Columbia's Chemical Engineering Department and as (former) president of professional societies such as the ACS and the American Electrochemical Society, he was in a privileged position to attract science graduates to his company. Moreover, he had the opportunity to take concrete steps to make academic curricula in chemical engineering better match the needs of firms such as his own. Conversely, however, scientific societies and establishments such as the National Research Council tried to pressure industrial manufacturers into hiring an ever-growing number of scientists and engineers and to offer them respectable and relatively well-paid positions. The scrutiny of these associations contributed to a climate in which not just Baekeland but also other chemical industrialists were eager to demonstrate that their firms dedicated a relatively large share of their resources to R&D work.[80] Yet, in light of Baekeland's extensive involvement in the chemical community, it would have been especially embarrassing if the Bakelite companies were perceived as being unsupportive of science and industrial research.

The banker-led National Aniline & Chemical Company was the favorite target of this criticism in the US chemical industry, but Baekeland also publicly faced it on at least one occasion. Moreover, his reaction illustrates the significance that he attached to this issue. At a 1926 meeting of the ACS's Division of Chemical Education, Harold Hibbert, a professor in industrial and cellulose chemistry, made what Baekeland privately described as "a very light headed remark about manufacturers not giving enough attention to research work and gave as an example the Bakelite Corporation which sometimes might find itself superseded by another and better product." Characteristically, Baekeland promptly responded that his firm funded research at not less than four universities—a group that included Columbia and the Mellon Institute at the University of Pittsburgh, with which the Redmanol Chemical Products Company had previously maintained close ties. In addition, he was able to say off the top of his head that there were "45 chemists, physicists, and engineers daily at work at our laboratory, in research and development aside of our other chemists and engineers engaged in manufacture, administration and service etc." According to Baekeland's diary notes, his vigorous rebuttal to Hibbert was greeted with applause. Moreover, Hibbert himself apologized and promised not to mischaracterize Bakelite's commitment to R&D in the future.[81]

On this and most other occasions where Baekeland highlighted the science-oriented nature of the Bakelite companies, he was addressing academically trained scientists and engineers. For obvious reasons, this was an audience that, if not literally at least figuratively, was likely to applaud his message. If he had spoken to a less credentialed public, Baekeland might have added that Bakelite's research and sales departments were not alone in their stress on scientific and technological literacy. For example, in the Bakelite Corporation's central manufacturing plant in Bound Brook, New Jersey, classes were given to instruct foremen in the "basic fundamentals of science," attempting to "instill an interest in the sciences and their relations to one's everyday work."[82] A reading of Bakelite's in-house magazine *Trefoil*, moreover, suggests that a science-oriented spirit permeated the culture of the entire company. As a case in point, in a 1938 issue of this publication, a man named W. H. Rugen began a poem about the "boys who man the stills" with the following verse:

When our Scientists decide,
To take us for a ride,
They pick a resin steed for us to tame;
The subjections if the cuss,
Is put squarely up to us,
Tho' we boast no B.Sc's behind our name.
And yet Research, we're advised,
Has their products civilized,

All primed to take us safely, anywhere,
Still they turn the critters loose,
With no more of an excuse,
Than a theoretic process—and a prayer.[83]

As this chapter has demonstrated, it would be wrong to interpret the reverence for science captured in this poem solely in terms of public relations and scientific veneer. Until 1931, for instance, Baekeland shielded the Bakelite Corporation's research laboratories from the scrutiny of associates whom he deemed insufficiently qualified, insisting that R&D should not be held accountable to standards of cost-effectiveness applied to other industrial activities. Likewise, the statement that Bakelite sales engineers "can talk to any manufacturer in his own [technical] language," made in a promotional brochure from 1927, plainly captured a criterion by which these employees were actually evaluated.[84]

The central place of science and technology in the organizational cultures and innovation strategies of General Bakelite and the Bakelite Corporation accorded with Baekeland's strongly worded praise for the German chemical and electrical industries, exemplified at the beginning of this chapter as well as in earlier parts of this book. There were, moreover, many similarities between Bakelite and the German American Roessler & Hasslacher Chemical Company. Among other things, both enterprises provided technical service to industrial customers, advertised their products in scientific journals, and employed scientists and engineers in middle as well as upper management positions. Still, a few exceptions aside, it would be a stretch to directly attribute Bakelite's business model and philosophy to German models and examples. The policy that Baekeland most unequivocally identified with R&H—the firm's generous profit sharing in the form of *Tantièmes*—was indeed one from which he dissociated himself.

How unusual was Bakelite's approach to science in early twentieth-century corporate America? As late as 1929, Baekeland urged the editors of *Who's Who* to include a profile of Hector Carveth—the president of R&H and formerly a professor at Cornell—in the next edition of their biographical dictionary on the grounds that "unfortunately it does not happen often enough in the United States that large business enterprises here should have able scientists as their executive head."[85] By the end of his career, however, it had become hard to justify this type of claim. For example, not just Union Carbide but also several other firms that expressed an interest in merging with the Bakelite Corporation met that criterion. Accordingly, Baekeland ended up negotiating the possible sale of his firm with professional colleagues like Milton C. Whitaker, who had been on Columbia's faculty until 1917 and had also edited the *Journal of Industrial and Engineering Chemistry*. Whitaker had risen to the position of vice president at American Cyanamid, the most serious contender to Carbide's bid.[86]

Baekeland's argument that corporate executives with a technical background were more appreciative of the value of science and industrial research had also become harder to sustain by the late 1930s. Partly thanks to the promotional efforts of scientists in both academia and industry, investing in R&D had simply become fashionable in the previous years and decades. Another sign of the expansion of the science-industry nexus was that the number of scientists and engineers employed in industry had reached record levels, notwithstanding the devastating impact of the Great Depression. The stock market valuation of industrial corporations, moreover, had started to take into account the perceived potential of these firms to come up with new and promising knowledge-intensive technologies.[87] In short, as Baekeland and millions of other visitors to the 1939–1940 New York world's fair could observe, it had become common to gamble on science.

Afterword

In a 1944 obituary, Céline Swarts emphasized that her husband, who had died on 23 February of that year, had never lost the remarkable degree of independence that he had gained early in his life. Swarts had written the short biographical profile at the suggestion of Baekeland's former associates at Bakelite Limited, who were planning a commemorative event to honor the "father" of their industry. In it, she mainly referred to her spouse's ability to think autonomously, which she related to his scientific and inventive accomplishments. A desire for independence, however, had also been a major motivation guiding Baekeland's decisions both as an entrepreneur and as a father and husband.[1] Self-help was, moreover, among the values that he had tried to pass on to his children and grandchildren. Consistent with this outlook, an additional argument Baekeland made in favor of merging the Bakelite Corporation into Union Carbide and Carbon had been the prospect that Bakelite would continue to be run largely autonomously in this new constellation, under the direction of his own son. His concern that George Baekeland might be crowded out, however, rapidly proved justified. As Leo indicated in one of the last diary entries that he wrote in his life, chemist James Rafferty, a longtime Carbide executive, had been appointed as president of Bakelite. About a decade after Leo's death, moreover, George voluntarily retired from business, and no other member of the Baekeland family would join the rapidly expanding plastics industry afterward.[2]

Meanwhile, Leo Baekeland's own renown had begun to decline somewhat in the context of World War II and the early Cold War, when the science-industry nexus experienced another dramatic expansion and transformation. This was not so much because of the integration of collaborative R&D activities into large industrial corporations and the challenges that this much-discussed development posed to "heroic" inventors and scientists who preferred to work more independently.[3] After all, the rise of large-scale R&D had started several decades earlier. Moreover, the fact that the Bakelite Corporation partly counted on scientific teamwork to remain competitive after the expiration of Baekeland's patents did not prevent professional colleagues from continuing to characterize the chemist as a "genius."[4] Even knowledge-intensive corporations that were much larger than Bakelite and Union Carbide continued to highlight the role of specific individuals for identity-building and public relations purposes, among other reasons. These firms had good reasons for crafting narratives that presented the discoveries of industrial researchers such as Irving Langmuir, the first US industrial scientist to receive a Nobel Prize, as the foundation of new technologies and industries. For example, such accounts could reinforce legal narratives about the originality and novelty of these scientists' work. Then as now, to many audiences they also proved more captivating than more sober analyses of how technological innovation and entrepreneurship have always proceeded in a collective manner.[5]

Given the collective nature of innovation, it is more relevant to ask *which* of the many individual contributors to the creation of science-related technologies and industries were and were not getting intellectual credit for their work. As this book has argued, Baekeland was celebrated as the "father of plastics" not just because of the invention of Bakelite. Among other things, he also benefited from his background in both science and business and his gradually acquired skill in crossing boundaries between academia and industry, with the aim of securing the highest possible level of intellectual property. Baekeland had continuously sought the recognition of other chemists and chemical engineers in the United States, who had been eager to show that their accomplishments at least matched the better-known achievements of German chemists. By the second half of the twentieth century, however, such concerns about the international competitiveness of the US chemical community had largely disappeared. In addition, as a more than century-old trend of growing specialization in science gained further speed, the work of professional chemists was increasingly being evaluated by the members of relatively narrowly defined disciplinary and interdisciplinary communities, including the newly emerging field of polymer science.

In the United States and internationally, polymer science was firmly established as an academic discipline in the years and decades following

World War II—well after the invention of not just Bakelite but also later blockbuster plastics such as PVC and polystyrene.[6] The builders and promoters of this new and still male-dominated field came up with their own set of "fathers," including Hermann Staudinger, Wallace Carothers, Kurt Meyer, and Herman Mark. This change affected the standing of plastic inventors such as Baekeland, who had been only marginally involved in the development of macromolecular theory. This was partly because at least some outside observers did not differentiate between the creation of novel materials, theories, and disciplines in an era where the belief in the technological potential of scientific research was reaching new heights. When Staudinger received the 1953 Nobel Prize in chemistry "for his discoveries in the field of macromolecular chemistry," for example, the *Washington Post* and *Christian Science Monitor* simply referred to him as the "father of plastics."[7] In short, some of the same reasons that had put Baekeland at an advantage over earlier inventors such as John Wesley Hyatt, who had been largely self-trained in science, worked against him as soon as the study of plastics was perceived as the territory of professional polymer scientists.

Most historians who have studied the relationships between academia, industry, and government in the twentieth-century United States agree that World War II and the beginning of the Cold War constituted a major turning point.[8] The inflow of unprecedentedly large amounts of government funding, mainly distributed by military and semimilitary agencies, had both obvious and more subtle impacts on the development of the sciences. In this novel environment, a type of academic entrepreneur that had already been common in European countries such as France and Germany, where the state had long been a crucial patron of the sciences, now also rose to the forefront in the United States. Individuals such as Stanford's Frederick Terman and Boulder's Walter Orr Roberts, for example, were not just skilled in crossing boundaries between industry and the higher education sectors. They also managed to channel extensive federal resources to their institutions and research programs.[9]

Before the end of the Cold War, however, another highly consequential set of changes to the science-industry nexus had already started to occur, first in the United States and then elsewhere. From the late 1970s onward, for example, it became a popular—but controversial—policy prescription to encourage universities to support the economy and reduce the financial pressures they were facing by patenting academic inventions and selling patent licenses. At the same time, firms in high-technology sectors ranging from the pharmaceutical to the microelectronics industries began to more critically evaluate the returns on their in-house R&D expenditures and to outsource part of this work.[10] Some scholars have argued that the novel science-industry relationships emerging from such changes reflected a fundamentally new orientation to the market and the start of a "pro-patent" era. Others, however, have interpreted them as a return to practices that were common in the decades before World War II. Accordingly, these latter authors consider the period of the early Cold War, rather than the "neoliberal" late twentieth- and early twenty-first centuries, as exceptional from a historical point of view.[11]

Does the evidence discussed in this book support the argument that we have recently witnessed a return to the pre–World War II past? A central claim behind this view is that, in both the decades around 1900 and in our own times, firms heavily relied on market transactions to acquire and develop new technologies, including by purchasing patents and patent licenses. In the early Cold War era, by comparison, the process of innovation was, to an unprecedented degree, internalized within large industrial corporations with their own in-house R&D facilities. One problem with this analysis is its US-centricity: there were stronger elements of continuity in countries such as Belgium and France, where the impact of the unfolding Cold War on the funding and organization of research and innovation systems was less dramatic than in America.[12] Even when it comes to the United States, moreover, it remains relatively poorly understood how firms' use of academic and other outside sources of knowledge and intellectual property changed in the 1950s and 1960s.[13] Still, this book's analysis also points to striking similarities between the early and late twentieth century, ranging from the participation of academic scientists in various types of commercialization activities to the openness of knowledge-intensive firms such as Bakelite to collaborating with their customers and with external scientists and patentees.

I have also shown that Baekeland, like a substantial share of today's scientists, used patents and publications in a complementary way. Then as now, this often involved strategic delays of publication besides other forms of secrecy. A focus on securing patent protection could also interfere with scientists' desire to get credit from their professional colleagues. Researchers in both industry and academia, however, have also used patents as a tool for getting intellectual recognition. Moreover, even in research areas where the historically unstable legal distinction between patentable "inventions" and unpatentable scientific "discoveries" has been particularly blurry, the use of patents did not necessarily imply that a lesser amount of scientific information was entering the public domain.[14]

On the other hand, it is notable that Baekeland and his industrially oriented fellow scientists, particularly those based in academia, generally kept their business activities more low profile than their late twentieth- and early twenty-first-century counterparts. This difference partly reflected changes in the cultural, economic, and moral environments in which these individuals operated. Not coincidentally, it was from the 1970s and 1980s onward, when high-tech innovation and new firm creation were presented as keys to economic competitiveness in a rapidly globalizing world, that the entrepreneur emerged as a new scientific persona. That is to say, performing entrepreneurial work became part of the culturally visible role and identity of a specific category of scientists.[15] This greater awareness of the existence of scientific entrepreneurs did not necessarily imply a positive evaluation of commercialization activities. Over the past few decades, for example, science policymakers, journal editors, and the public at large seem to have become more sensitive to conflicts of interest associated with the industrial funding of scientific research. On the other hand, earlier reservations about the legitimacy of wealth creation as an academic "mission" of not only lower-tier schools but also leading research universities appear to have disappeared.

A related development is that several of the boundaries that Baekeland crossed so frequently seem to have increasingly become blurred. Before World War II as well as in the early Cold War era, keeping commercial activities at arm's length could be a strategy for gaining academic status and attracting funding from philanthropic or government sources, as the histories of the Massachusetts Institute of Technology and the Mellon Institute can illustrate.[16] Since then, however, the direct involvement of various types of higher education institutions in such work has assumed ever-larger proportions. As this book has shown, Baekeland repeatedly maintained that the economic use of his own university training had been very limited. Ironically, however, if he were still alive, it is not unlikely that he would be critical of some of the current intermingling of the academic and business worlds and oppose the reduction of science to "technoscience."[17] In his view, which took into account the perspectives of scientifically "purer" colleagues inside and outside the US chemical community, universities should avoid drifting away from their core educational and research duties. In late 1927, for example, he repeated a comment that he had previously made to Columbia's George Pegram in a cautionary note to a colleague at the National Research Council:

> Research is a broad and indefinite word, which covers a multitude of sins. In some instances, some of our American Universities, under the pretext of industrial fellowships, have merely lent their name, staff, and equipment for the benefit of promoters or commercial enterprises. In some cases merely compilation or tests which a foreman or competent workman could have attended to at any factory, has been called industrial research. Columbia should aim in such cases to pick out the purely scientific part of any industrial research problem. If this cannot be done, it should not lend its equipment or organization to anything which is not consistent with the aims of a true University.[18]

When Baekeland made these remarks, the industrial research fellowship was still a relatively new feature of the American academic landscape, which he had personally helped introduce at Columbia in his dual roles as honorary professor and president of General Bakelite. Today, such research fellowship programs are overshadowed by collaborative university-industry centers of a much larger scale. Such centers, moreover, are supplemented by various other institutional innovations, ranging from science and technology parks to business incubators and entrepreneurship courses specifically targeting science and engineering students. The question of where to draw the line between proper and improper academic activities has therefore only become more pressing.

Notes

List of Abbreviations

ABP	Association belge de photographie
ACS	American Chemical Society
AEG	Allgemeine Elektrizitätsgesellschaft
CCP	Charles F. Chandler Papers
D&F	Development and Funding Co.
DR	Du Pont & Co. Records
ESP	Elmer A. Sperry Papers
GEASC	George Eastman Archives and Study Center
HDC	J. Harry DuBois Collection on the History of Plastics
Indus	École industrielle de Gand
IP	intellectual property
LBP	Leo H. Baekeland Papers
NAA	National Academies Archives
NACP	The National Archives at College Park, MD
NAS	National Academy of Sciences
NCB	Naval Consulting Board
NMAH	National Museum of American History
NRC	National Research Council
R&D	Research and Development
R&H	Roessler & Hasslacher Chemical Company
TEP	Thomas A. Edison Papers

1 Introduction

1. Duranti, "Utopia, Nostalgia and World War."

2. Baekeland, diary 59 (7 July 1939), in Leo H. Baekeland Papers (hereafter LBP), Archives Center, National Museum of American History (hereafter NMAH), Smithsonian Institution, Washington, DC, series 4, box 22, folder 2. All ensuing citations of "diary" refer to diary entries by Baekeland in LBP, series 4, boxes 18–22.

3. Diary 59 (7 July 1939).

4. Nasby, *Rolph Scarlett*, 47; Anon., "Father of Plastics."

5. Monaghan, *Official Guide Book of the New York World's Fair*, 186.

6. Monaghan, *Official Guide Book of the New York World's Fair*, 182; Baekeland, diary 59 (28 July 1939).

7. Baekeland, diary 59 (22 June 1939).

8. Anon., "Father of Plastics," 56.

9. Anon., "Society of Chemical Industry," 49.

10. E.g., Brown, "Bakelite—What It Is," 17; Nasby, *Rolph Scarlett*, 47.

11. Rhees, "The Chemists' Crusade"; Steen, *American Synthetic Organic Chemicals Industry*, chap. 8.

12. Baekeland, diary 59 (25 July 1939).

13. Bijker, *Of Bicycles, Bakelites, and Bulbs*, chap. 3.

14. Kuznick, "Losing the World of Tomorrow"; Rydell, "The Fan Dance of Science"; Kargon et al., *World's Fairs on the Eve of War*, chap. 4.

15. Marsch, *Zwischen Wissenschaft und Wirtschaft*; Hounshell, "Evolution of Industrial Research."

16. As recently emphasized in Bonneuil and Pestre, "Introduction," 18–19.

17. On the industrial and technological value of eighteenth- and early nineteenth-century science, especially chemistry, see, e.g., Klein, *Nützliches Wissen*; Miller, *James Watt, Chemist*; Tomory, *Progressive Enlightenment*; Ó Gráda, "Did Science Cause the Industrial Revolution?" Historians do not agree on the extent to which the late nineteenth and early twentieth centuries represented a turning point in the history of science-industry relations in the field of chemistry. For instance, there is a clear contrast between the emphasis on gradual change in Homburg, "Chemistry and Industry," 569–574, and the characterization of the 1870s as a "watershed" in Rocke, "Theory versus Practice in German Chemistry." My own interpretation mostly agrees with that of Ernst Homburg but covers the period up to the 1930s and attaches more importance to institutional and organizational changes other than the emergence of the industrial research laboratory.

18. E.g., Marsch, *Zwischen Wissenschaft und Wirtschaft*; Swinfin, "Chemists Employed in the Manchester Area." On the disproportionate amount of attention to laboratory research, see Fox and Guagnini, *Laboratories, Workshops, and Sites*; Choi, "The Boundaries of Industrial Research"; Bouvier et al., *De l'atelier au laboratoire*.

19. Jarausch, "Higher Education and Social Change"; Fox and Guagnini, *Education, Technology and Industrial Performance*; Trischler and Kohlrausch, *Building Europe on Expertise*, chaps. 1, 2.

20. Cahan, "Institutions and Communities"; Knight and Kragh, *The Making of the Chemist.*

21. As Szöllösi-Janze, "Science and Social Space," emphasizes with respect to Germany.

22. Marcovich and Shinn, "Regimes of Science Production and Diffusion"; Pestre, "Regimes of Knowledge Production in Society."

23. On these new engineering specialties, see, e.g., Fox and Guagnini, *Laboratories, Workshops, and Sites*; Kargon and Knowles, "Knowledge for Use," 8–15; Rosenberg and Steinmueller, "Engineering Knowledge." None of this, however, is to say that formal training in science became a prerequisite for a successful career in business in this period. For relevant discussions, see Khan, *The Democratization of Invention*, 207–219; Israel, *From Machine Shop to Industrial Laboratory*, chap. 6.

24. Carlson, "Academic Entrepreneurship and Engineering Education"; Johnson, "Academic-Industrial Symbiosis"; Harwood, "Institutional Innovation"; Hounshell, "Evolution of Industrial Research."

25. Harwood, "Institutional Innovation."

26. E.g., Galvez-Behar, "Chapitre 9"; Mowery et al., *Ivory Tower and Industrial Innovation*. For additional references, see chapter 7, this volume.

27. On hybrid experts and careers, see, e.g., Klein, "Artisanal-Scientific Experts"; Kranakis, "Hybrid Careers"; Roberts, Schaffer, and Dear, *The Mindful Hand*. On the commercial activities of academic chemists in the eighteenth century and first half of the nineteenth century, see Anderson, "Relations between Industry and Academe"; Gustin, "Emergence of the German Chemical Profession," 142–146; Klein, "Technoscience avant la lettre."

28. On geology, see Lucier, *Scientists & Swindlers*; Tweedale, "Geology and Industrial Consultancy." Pre–World War I Hungary is an example of a country whose electrical firms and inventors were internationally respected for their electrical engineering accomplishments but whose chemical industry was considered to be relatively insignificant. See Berend, "Hungary," 285; Hidvégi, *Anschluss an den Weltmarkt*, chap. 2; Haber, *The Chemical Industry*, 164–166.

29. Classic contributions to this historiography are Beer, "Coal Tar Dye Manufacture"; Meyer-Thurow, "The Industrialization of Invention"; Homburg, "Emergence of Research Laboratories"; Travis, *The Rainbow Makers*; Reinhardt, *Forschung in der chemischen Industrie*; Steen, *American Synthetic Organic Chemicals Industry*; Murmann, *Knowledge and Competitive Advantage*.

30. On the development of industrial job markets for chemists in Germany and Britain, see, respectively, Homburg, *Van beroep "Chemiker"*; Donnelly, "Defining the Industrial Chemist." On the situation in Belgium and the United States, see chapters 2 and 3, this volume.

31. Bijker, *Of Bicycles, Bakelites, and Bulbs*, chap. 3; Meikle, *American Plastic*, chap. 2; Kaufmann, "Grand Duke." An exception is Morris, "Resource Networks," chap. 5, which, however, is largely limited to a specific episode in Baekeland's career (see chapter 4, this volume).

32. With the exception of the chapter by Henry Etzkowitz, "The Triple-Helix and the Rise of the Entrepreneurial University," these themes are also mostly absent from an important 2004 edited volume on the twentieth-century science-industry nexus. See Grandin, Wormbs, and Widmalm, *The Science-Industry Nexus*. See, however, the references in the remainder of this section, as well as in several later chapters of this book, for various exceptions to this trend.

33. On the cultural authority of science, as compared with technology, see Forman, "Primacy of Science."

34. Etzkowitz, "Entrepreneurial Scientists and Entrepreneurial Universities," 198. For a more comprehensive discussion of this question, see chapter 3, this volume.

35. Guagnini, "Ivory Towers?"; Arapostathis and Gooday, *Patently Contestable*.

36. Lucier, *Scientists & Swindlers*; Morris, "Resource Networks," 21, 96–97; Lucier, "The Professional and the Scientist," 712; Johnson, "History of Science from the Perspective of Applied Science."

37. Shapin, *Never Pure*, chap. 1.

38. On Thomson, see Smith and Wise, *Energy and Empire*; Guagnini, "Ivory Towers?" On Pasteur, see, e.g., Cassier, "Producing, Controlling, and Stabilizing Pasteur's Anthrax Vaccine"; Galvez-Behar, "Louis Pasteur, entrepreneur." On Rowland and Noyes, see, respectively, Morris, "Resource Networks," chap. 3; Servos, *Physical Chemistry*, chap. 3. On Curie, see Boudia, *Marie Curie et son laboratoire*; Roqué, "Marie Curie and the Radium Industry."

39. Lucier, *Scientists & Swindlers*. See also Morris, "Resource Networks."

40. Weiner, "Patenting and Academic Research"; Kaataja, "University Researchers Contributing to Technology Markets"; Mercelis, Galvez-Behar, and Guagnini, "Commercializing Science." Weiner specifically referred to the first two decades of the twentieth century, but various later scholars have interpreted his claim more broadly. On academic patenting, see also chapters 3 and 4, this volume.

41. Guagnini, "Ivory Towers?" For additional examples of late nineteenth- and early twentieth-century academic engineers and physicists who participated in the

creation of firms, see Guagnini, "A Bold Leap into Electric Light"; Katzir, "Technological Entrepreneurship"; König, "Engineering Professors as Entrepreneurs"; Morris, "Commerce and Academe."

42. For recent calls to study entrepreneurship in relation to historical change, see Wadhwani and Jones, "Schumpeter's Plea"; Wadhwani and Lubinski, "Reinventing Entrepreneurial History." MacLeod and Radick, "Claiming Ownership in the Technosciences," have recently proposed a broadened agenda for investigating intellectual property in science.

43. Baekeland, diary 36 (12 July 1923). See also Meikle, *American Plastic*, 60–62.

44. Baekeland, diary 42 (7 September 1926).

45. The term "ornamental" is adopted from David, "The Historical Origins of 'Open Science.'" Jonathan Liebenau has similarly referred to the "scientific veneer" used by late nineteenth-century pharmaceutical firms; see Liebenau, *Medical Science and Medical Industry*, 7, 41–47. On the reputational risks associated with such strategies, see Mercelis, "The Scientist and the Advertisement," and the references given there.

46. Merton, "The Normative Structure of Science"; Merton, "Priorities in Scientific Discovery."

47. Lavoisier, *Oeuvres de Lavoisier*, vol. 4, 623, as translated in Van den Belt, "Robert Merton, Intellectual Property, and Open Science," 187–188. Merton did not mention Lavoisier as an example but, as Van den Belt points out, the French chemist's letter quoted earlier supports several of his points.

48. Merton, "The Normative Structure of Science," 275; Dasgupta and David, "Toward a New Economics of Science"; David, "The Historical Origins of 'Open Science.'"

49. E.g., Metlay, "Reconsidering Renormalization"; Miller, "Intellectual Property and Narratives of Discovery/Invention"; Shapin, *The Scientific Life*, chaps. 4 and 5; Vermeir and Margócsy, "States of Secrecy." For a recent study dealing with the relationships between formal IP and scientific credit before and especially after the period covered in this book, see Biagioli and Galison, *Scientific Authorship*.

50. Baekeland, "Protection of Intellectual Property," 52. See also, e.g., Baekeland, "United States Patent System," 204; Baekeland, "Abuses of Our Patent System," 333; Baekeland, "Incongruities of American Patent Litigation," 789.

51. Picard and Olin, *Traité des brevets d'invention*, 10.

52. Redman, "Research as a Fixed Charge," 114. That economic and legal professionals in the United States (and in Germany) were still discussing the appropriateness of the "intellectual property" notion by the outbreak of World War I is evident in Ely, *Property and Contract*, 152. On the earlier history of this discussion, see Bracha,

Owning Ideas. As Catherine Fisk has pointed out, trade secret law descended from unfair competition law; see Fisk, *Working Knowledge*, 15–16.

53. Messamore, "Review of 'Colonial Lives across the British Empire,'" 116; Deacon, Russell, and Woollacott, *Transnational Lives*; Saunier, *Transnational History*.

54. Veysey, *Emergence of the American University*, 121–133; Donovan, "Education, Industry, and the American University," 265–267; Hannaway, "German Model of Chemical Education."

55. Dolby, "Transmission of Two New Scientific Disciplines"; Servos, *Physical Chemistry*; Wise, *Whitney*, 93.

56. On Eastman Kodak, see Jenkins, *Images and Enterprise*, chap. 14. On Du Pont, see Hounshell and Smith, *Science and Corporate Strategy*.

57. Smith, "The Scientific Tradition in American Industrial Research," 124; Hounshell, "Evolution of Industrial Research."

58. Marsch, "Transferring Strategy and Structure," 217.

59. For an exploratory analysis, see Mercelis, "Corporate Secrecy and Intellectual Property in the Chemical Industry," 41–45.

60. For other examples, see Diogo and Van Laak, *Europeans Globalizing*, 94–95.

61. [Thomas Edison], handwritten note on "Lawrence Redman, Archie Weith, and Frank Brock, 'Synthetic Phenol Resins,' *Journal of Industrial and Engineering Chemistry (JIEC)* 8: 1077," in Thomas A. Edison Papers, Rutgers University (TEP), reel 266, frames 240–244 (at 240) (emphasis in original).

62. Baekeland to Edouard Remouchamps, 20 January 1929. Baekeland's original letters to Remouchamps are held by the Ghent University Museum. For published transcripts, see Gillis, *Leo Hendrik Baekeland*.

63. On the liberal ideology of self-made manhood, see Hilkey, *Character Is Capital*; Sandage, *Born Losers*.

64. On the late nineteenth- and early twentieth-century "robber barons," whom Baekeland repeatedly condemned, see, e.g., John, "Robber Barons Redux."

65. Baekeland to Florence Boonroy, 5 February 1933, published in Gillis, *Leo Hendrik Baekeland*.

66. Baekeland to Bob Holland c/o *The Miami Herald*, 31 December 1931, LBP, series 3, box 10, folder 4.

67. Morris, "Resource Networks," 218.

68. Baekeland, diary 50 (15 March 1932). For earlier interpretations of the sale, see Bijker, *Of Bicycles, Bakelites, and Bulbs*, 135; Kaufmann, "Grand Duke," 47–48; Meikle, *American Plastic*, 35; Morris, "Resource Networks," 201. See also chapter 4, this volume.

69. Nathaniel Thurlow to Baekeland, 30 November 1904, LBP, series 8, box 15, folder 12; Baekeland, diary 8 (9, 11, and 18 October 1910).

70. In the mid-1990s—when historians Wiebe Bijker and Jeffrey Meikle had completed their acclaimed studies of the invention and cultural significance of Bakelite—these diaries had not yet been made publicly accessible. See Meikle, *American Plastic*, 319n5).

71. On these social groups, see Bijker, *Of Bicycles, Bakelites, and Bulbs*, chap. 3.

2 Social Mobility through Education

1. Baekeland, diary 19 (25 December 1915 and 10 January 1916).

2. Baekeland, "Address of Acceptance of Perkin Medal," 152.

3. See Haynes, *This Chemical Age*, 336; Wyllie, "The Businessman Looks at the Higher Learning"; Bijker, *Of Bicycles, Bakelites, and Bulbs*, 128–129; Kaufmann, "Grand Duke," 22–23; Morris, "Resource Networks," 186. Early in his career, Baekeland typically gallicized his name as "Léon Baekelandt" or "Backelandt."

4. Mercelis, "Commercializing Academic Knowledge."

5. Baekeland, "Edward Weston's Inventions," 485; Tocqueville, *Democracy in America*, vol. 2, chap. 10.

6. Clarke, *Report on the Teaching of Chemistry and Physics*, 35.

7. Frerichs, "Report of Committee on Chemical Engineering Education," 123.

8. Frerichs, "Report of Committee on Chemical Engineering Education," 122.

9. Parlette, *The University of Hard Knocks*.

10. Dhondt, "Een tweevoudig compromis," 171–184; Vanpaemel, "Jean-Servais Stas," 49; Van Tiggelen, "De scheikundige theorieën," 194.

11. Sarachek, "American Entrepreneurs," 443, 445.

12. Baekeland to Edouard Remouchamps, 18 July 1899.

13. Baekeland to Remouchamps, 29 July 1899.

14. Céline Swarts to Baekeland, 3 October 1890, in LBP, series 3, box 13, folder 5. Baekeland had five siblings, two of whom died shortly after their births; see Wautier and Segers, "Nothing Succeeds like Success," 126. Regarding the remaining three, I have only found references to Leonie (b.1869) and Rachel Baekeland in Leo's personal papers, therefore suggesting that Elodie (b.1866) had also died prematurely, perhaps as a result of the cholera epidemic that struck Ghent in the year of her birth.

15. Céline Swarts to Baekeland, 16 and 19 March 1889, Nina Swarts to Baekeland, 16 March 1889, and Baekeland to Céline Swarts, 21 March 1889 (all in LBP, series 3,

box 13, folder 2); Baekeland to Remouchamps, 18 June 1899, 18 July 1899, 20 January 1937, and 5 January 1938, and Baekeland to Florence Boonroy, 5 February 1933 (all published in Gillis, *Leo Hendrik Baekeland*); Baekeland, diaries 15 (3 June 1916) and 59 (6 August 1939).

16. Rosalie Merchie to Baekeland, 10 November 1910, in LBP, series 3, box 14, folder 10.

17. Baekeland to Remouchamps, 20 August 1908.

18. Kaufmann, "Grand Duke," 9; Weber, *Die protestantische Ethik*, 78.

19. Baekeland, "Career of a Research Chemist," 159. Baekeland also referred to Franklin in Baekeland, "Science and Industry," 49, Baekeland, "The Engineer," 424, and Baekeland, diary 50 (15 March 1932).

20. Pupin, *From Immigrant to Inventor*, 14.

21. Baekeland, diary 52 (10 March 1933); Baekeland, "Natuurkunde: Nijverheidschool van Gent (schooljaar 1878–79)" (unnumbered lecture notes), in LBP, series 5, box 23, folder 2; Baekeland, diary 9 (29 March 1911). On Edison and Bell's generation of "heroic" inventor-entrepreneurs, see Hughes, *American Genesis*.

22. Laird, *Pull*, 11.

23. Baekeland, diaries 4 (7 May 1909), 24 (3 February 1918), 47 (31 October 1929), 57 (16 June 1936); Baekeland to Boonroy, 23 January 1938.

24. Baekeland to Arthur P. O'Toole, 1 July 1931, in LBP, series 3, box 11, folder 4; Baekeland to Remouchamps, 2 August 1898; diary 12 (23 March 1912).

25. Baekeland, diary 6 (8 February 1910); see also diary 17 (3 February and 12 June 1917).

26. Baekeland to Remouchamps, 20 August 1908.

27. Lamoreaux, "Entrepreneurship in the United States," 368. On nineteenth-century artisan values, see Crossick, *An Artisan Elite*.

28. Buyse, *Méthodes américaines*; Manny, "American Schools."

29. Baekeland to Remouchamps, 20 August 1908 (emphasis in original).

30. Masquelez, "L'institut industriel, agronomique et commercial," 63; D'hoker, "Nijverheids- en beroepsonderwijs." I am indebted to Stéphane Lembré for bringing Masquelez's report to my attention.

31. Baekeland, diary 59 (9 December 1939).

32. École industrielle de Gand, "Certificat de Capacité d'Élève en Industries Chimiques," in LBP, box 49.

33. Bergmans, *Notice historique*, 42–44, 46–47 (quotation at 42).

34. D'hoker, "Nijverheids- en beroepsonderwijs"; Bertrams, Coupain, and Homburg, *Solvay*, 55–64.

35. Baekeland, diary 12 (20 and 22 March 1913).

36. École industrielle de Gand, "Certificat de Capacité"; Baekeland, "Chimie: 1879–80," in LBP, series 5, box 23, folder 7.

37. Baekeland, "Chimie: 1879–80"; Anon., *Dubbele wegwyzer*. See also Verbruggen, *De stank bederft onze eetwaren*, 93–94.

38. Baekeland, "Natuurkunde"; Whitaker, "New Chemical Engineering Course and Laboratories," 167. On the importance of technical drawing, see Frerichs, "Remarks"; Olsen, "A Course in Chemical Engineering Education," 273–274.

39. Delaere, *Notice sur l'École industrielle et commerciale*, 7.

40. Baekeland, "Chimie: 1879–80"; Bergmans, *Notice historique*, 152; École industrielle de Gand, "Certificat de Capacité."

41. Baekeland, diary 59 (9 December 1939); Sacré, *A la mémoire de Gustave De Vylder*, 6, 27; Warnerke, "Photographic Technical Instruction," 35.

42. Baekeland, diary 59 (9 December 1939); Gillis, *Leo Hendrik Baekeland*, 22–24 (quotation at 24).

43. Bertrams, *Universités & entreprises*, 30–31; Depaepe, "De toelatingsvoorwaarden tot de universiteit in België"; Dhondt, "Een tweevoudig compromis," 155–170.

44. Kaufmann, "Grand Duke," 8, as well as further references given there; Antheunis, *Leo Baekeland*, 14.

45. Dambre, *August Wagener*; Anon., *Rapport triennal*, ix–x; De Vroede, "Onderwijs 1840–1878"; De Vroede, "Onderwijs 1878–1914."

46. Baekeland to Camiel De Bruyne, 5 November 1930, LBP, series 3, box 8, folder 16.

47. All of Baekeland's scores are published in Gillis, *Leo Hendrik Baekeland*, 26.

48. Baekeland, diaries 27 (23 July 1919) and 54 (23 July 1934); De Bruyne to Baekeland, 3 October 1934, in LBP, series 3, box 8, folder 15.

49. Deelstra, *Het scheikunde-onderwijs*, 125; Van Tiggelen and Deelstra, "Belgium," 24.

50. Deelstra, "De scheikunde," 167–168; Deelstra, "Achtergronden"; Deelstra, *Het scheikunde-onderwijs*, 123–124; Van Tiggelen, "De la chaire aux laboratoires."

51. Baekeland to Charles Chandler, 28 December 1915, in LBP, series 3, box 17, folder 21; De Bruyne to Baekeland, 12 October 1930, in LBP, series 3, box 8, folder 16. See also Baekeland's letter to Chandler of 30 December 1915 in LBP, series

3, box 17, folder 21; Deelstra, "De invloed van J. S. Stas"; Despy-Meyer, "Stas et l'enseignement supérieur."

52. Klinge, "Die Universitätslehrer," 119.

53. Baekeland, diary 19 (21 January 1916); Céline Swarts to Baekeland, 12 December 1890, in LBP, series 3, box 13, folder 6.

54. MacLeod, "Théodore Swarts"; Gustave Wolters to Auguste Wagener, 24 July 1889, in Ghent University Archives, 4 A2/4, box 46.

55. Olsen, "A Course in Chemical Engineering Education," 269.

56. Dhondt, "Een tweevoudig compromis," 180–181; Van Tiggelen, "De la chaire aux laboratoires," 87.

57. Baekeland, "Etudes sur les phénomènes de dissociation" (unpaginated manuscript), in LBP, series 2, box 5, folder 2.

58. Snelders, "Dissociation, Darwinism and Entropy"; Paquot, *Henri Sainte Claire Deville*, 99–110; Wisniak, "Henri Etienne Saint-Claire Deville," 124; Young, "The Sainte-Claire Deville Hot-Cold Tube."

59. Baekeland, "Etudes sur dissociation."

60. Baekeland, "Etudes sur dissociation." On Pfaundler's work, see Berger, "Chemische Mechanik und Kinetik; Jensen and Kuhlmann, "Leopold Pfaundler."

61. The second publication appeared almost 15 years later; see Baekeland, "Dissociation of Lead Nitrate."

62. Addison and Logan, "Anhydrous Metal Nitrates," 120; Baekeland, Sur la dissociation *du nitrate de plomb*, 10n1, 19, 34.

63. Morgan, "Dissociation of Lead Nitrate," 416. By contrast, a summary of Baekeland's paper in *Science* indicated that the chemist "discussed the principles of chemical dynamics involved in the interpretation of his results" as part of his presentation to the New York Section of the American Chemical Society in December 1903. See Anon., "Societies and Academies," 68.

64. Baekeland, "Dreams and Realities," 1005; Snelders, "Jean-Servais Stas," 19; Gillis, *Leo Hendrik Baekeland*, 27–28; Anon., *Situation de l'enseignement supérieur*, cciii.

65. Olsen, "A Course in Chemical Engineering Education," 268; Clarke, *Report on the Teaching of Chemistry and Physics*.

66. Olsen, "A Course in Chemical Engineering Education," 268–269.

67. On the industrial orientation of the American Institute of Chemical Engineers, see Reynolds, "Defining Professional Boundaries."

68. Bertrams, *Universités & entreprises*, 30–32; Vanpaemel, *Wetenschap als roeping*, 76.

69. Mercelis, "Commercializing Academic Knowledge"; Baekeland, "Sur l'oxydation de l'acide chlorhydrique sous l'influence de la lumière," in LBP, series 2, box 5, folder 5 (reprint from *Bulletin de l'Académie royale des sciences, lettres et beaux-arts de Belgique* 55, 3e série, t. 11 [1886]); Mantels, *Gent*, 172–176; Eder, *Des actions chimiques.*

70. Stas, "Recherches de statique chimique," 292; Vogel, "Ueber die verschiedenen Modifikationen," 1170–1171.

71. On this neglect, see Jenkins, *Images and Enterprise*, 32; Bertrams, "Converting Academic Expertise into Industrial Innovation," 820; Edgerton, "Industrial Research in the British Photographic Industry," 109–112; Hens and Roosens, *De Belgische fotochemische industrie*, 66–68; Mercelis, "Commercializing Academic Knowledge."

72. Baekeland, "Address of Acceptance of Perkin Medal," 152.

73. On Van Monckhoven, see Antheunis, Deseyn, and Van Gijsegem, *Focus op fotografie*, 165–188; G. D. V. [Gustave De Vylder], "Dr. D. V. Monckhoven"; Joseph, Faelens, and Roosens, *D. van Monckhoven*; Hens and Roosens, *De Belgische fotochemische industrie*, 99–126.

74. See Van Monckhoven, *De photographie met broomzilvergelatine*, 19. Van Monckhoven's influence may also have been passed on to Baekeland indirectly through De Vylder's photography lectures at the Indus, for De Vylder and Van Monckhoven are known to have been friends (see, e.g., Hens and Roosens, *De Belgische fotochemische industrie*, 102).

75. See, e.g., Céline Swarts, undated letters to Baekeland (ca. 12 and 22 January 1889), in LBP, series 3, box 13, folder 2; Céline Swarts, "Dr. L. H. Baekeland," in LBP, series 1, box 1, folder 7.

76. Baekeland, "Edward Weston's Inventions," 485. Weston moved to New York in 1870; see Woodbury, *A Measure for Greatness.*

77. Wentzel, *Memoirs of a Photochemist*, ix (see also 6–7; British Photographic Research Association, "Report on Experimental Emulsion Manufacture," 1922, in Kodak Historical Collection, British Library, London, file A3022 (2/2).

78. Lundgren, "Transfer of Chemical Knowledge," 775; Rosenberg, "Chemical Engineering."

79. Baekeland to Remouchamps, 27 September 1899; Baekeland, diary 59 (9 December 1939).

80. Several of the relatively low number of research papers that Swarts authored over the course of his career dealt with camphor. See MacLeod, "Théodore Swarts."

81. Kargon and Knowles, "Knowledge for Use"; Fox and Guagnini, *Laboratories, Workshops, and Sites*; Wadhwani et al., "Academic Entrepreneurship."

82. See also Deneckere, *Uit de Ivoren Toren*; Mantels, *Gent.*

3 Crossing Boundaries

1. See Mercelis, Galvez-Behar, and Guagnini, "Commercializing Science," and the references given there.

2. See Brock, *Justus von Liebig*; König, "Engineering Professors as Entrepreneurs"; and the additional references given in chapter 1, this volume.

3. Guagnini, "Ivory Towers?"; Mercelis, "Commercializing Academic Knowledge."

4. For examples of academic scientists who left their employing institutions because of complaints about their commercial work, see Misa, "The Changing Market for Chemical Knowledge"; Morrell, "WH Perkin, Jr."; Mercelis, "Commercializing Academic Knowledge," 35.

5. Diser, *Wetenschap Op de Proef*, 241; Onghena, "Altruïstisch ambtenaar of heroïsch genie?" For different perspectives, see Bertrams, *Universités & entreprises*; Deneckere, *Uit de Ivoren Toren*, chap. 3.

6. E.g., Haynes, "Leo Hendrik Baekeland," 1184–1185; Gillis, *Leo Hendrik Baekeland*, 44–45; Matthis, *Léo-H. Baekeland*, 7–8; Bijker, *Of Bicycles, Bakelites, and Bulbs*, 129–130.

7. Belgian State Archives in Ghent, collection 285, folders 648 and 670 (Archives of Teacher Training College of Bruges, 1879–1888).

8. Antheunis, *Leo Baekeland*, 28; Antheunis, Deseyn, and Van Gijsegem, *Focus op fotografie*, 192–193; Bijker, *Of Bicycles, Bakelites, and Bulbs*, 129–130; Gillis, *Leo Hendrik Baekeland*, 10–11, 32–33; Hens and Roosens, *De Belgische fotochemische industrie*, 152–153; Kaufmann, "Grand Duke," 22–26; Meikle, *American Plastic*, 35; Morris, "Resource Networks," 185–188; Wautier and Segers, "Nothing Succeeds like Success," 146.

9. Morris, "Resource Networks," 187.

10. See Baekeland, Des plaques photographiques développables dans l'eau, Belgian Patent 78,957, filed 22 September 1887; Baekeland, Improved photographic plate to be developed in water, British Patent 1201–1888, filed 26 January 1888; Baekeland, Photographische Trockenplatten, welche durch Eintauchen in Wasser entwickelt werden, German Patent 43,521, filed 18 October 1887, and issued 9 May 1888; Baekeland, Verfahren zur Herstellung von in Wasser entwickelbaren photographischen Trockenplatten, Austro-Hungarian Patent, filed 11 October 1887, and issued 4 May 1889; Baekeland, Nouvelles plaques photographiques développables dans l'eau, French Patent 188,336, issued 25 January 1888. See also Baekelandt et Cie, Nouvelle disposition pour emballer les plaques photographiques, Belgian Patent 80,893, filed 6 March 1888; Baekeland, Préparation de plaques photographiques sèches, qui se développent par l'immersion dans l'eau, Belgian Patent 81,796, filed 14 May 1888. In addition, in the context of the partnership's dissolution, Baekeland's

lawyer, Octave Bruneel, referred to an Italian patent; see Bruneel, letter to unnamed recipient, 31 August [1890], in LBP, series 8, box 34, folder 1.

11. Richard Anthony to Frederick Anthony, 10 September 1889, George Eastman House, Rochester, NY. Illustration reproduced from Baekeland, "Laboratory Notes: 1889–94," 19, in LBP, series 6, box 25, folder 2.

12. Baekeland, Gleesener, and Guequier, "Convention d'exploitation entre les associés," in LBP, series 8, box 34, folder 1.

13. Cities that Baekeland visited included Frankfurt, Leipzig, Berlin, Hamburg, Bremen, and London; see his correspondence with Céline Swarts in LBP, series 3, box 13. There are also indications that he met with August Kekulé at the University of Bonn; see Baekeland to Charles Chandler, 30 December 1915, in LBP, series 3, box 17, folder 21. Unusually, the destinations of Baekeland's scientific missions were not specified in the responsible minister's triennial report. See Anon., *Situation de l'enseignement supérieur*, cii–ciii.

14. Swarts to Baekeland, 8 and 10 October 1888, in LBP, series 3, box 13, folder 1; Swarts to Baekeland, undated (ca. 20 September 1888).

15. Richard Anthony to Frederick Anthony, 10 September 1889; Club der Amateur-Photographen in Wien, "Protokoll der aussenordentlichen Generalversammlung"; Hens and Roosens, *De Belgische fotochemische industrie*, 155; Kaufmann, "Grand Duke," 20; ABP: Section de Bruxelles, "Séance du 12 décembre 1888"; ABP: Section de Bruxelles, "Séance du 9 janvier 1889"; ABP: Section de Gand, "Séance du 4 octobre 1888"; ABP: Section de Gand, "Séance du 4 janvier 1889"; ABP: Section de Gand, "Séance du 7 février 1889"; ABP: Section de Liège, "Séance du 16 janvier 1889"; Anon., "Self-Developing Plates"; Anon., "Water-Developing Plates"; Anon., "Récompenses," 698; Swarts to Baekeland, 7 October 1890, in LBP, series 3, box 13, folder 5. On the whole, the invention seems to have been better received in Belgium than abroad.

16. Diary 3 (24 February 1909).

17. See ABP: Section de Gand, "Séance du 4 avril 1889."

18. Baekeland to Guequier, 8 September 1889, in LBP, series 3, box 9, folder 2. See also Baekeland to Wagener, undated (ca. June 1889), published in Gillis, *Leo Hendrik Baekeland*, 36.

19. Baekeland to Guequier, 8 September 1889.

20. "Constitution de la société en commandite simple Dr. Baekelandt et Compagnie" (31 December 1887), in LBP, series 8, box 34, folder 1.

21. On this position, which was partly modeled after the German *Privatdozent*, see Dhondt, *Un double compromis*, 307–323.

22. Appendix in De Weerdt, *Gentse textielbewerkers en arbeidersbeweging.*

23. Anon., *Dubbele wegwyzer*; Oscar Defer to Octave Bruneel, 6 March 1890, in LBP, series 8, box 34, folder 1.

24. Swarts to Baekeland, 29 August 1890, in LBP, series 3, box 13, folder 4.

25. Hens and Roosens, *De Belgische fotochemische industrie*, 161; Swarts to Baekeland, 4 February 1891, in LBP, series 3, box 13, folder 7; Adrienne Drapier to Baekeland, 13 February 1931, in LBP, series 3, box 9, folder 2. See also Gillis, *Leo Hendrik Baekeland*, 33.

26. See "Reconnaissance de la liquidation de fait" and "Liquidation et dissolution" (both in LBP, series 8, box 34, folder 1); Swarts to Baekeland, 29 August 1890, in LBP, series 3, box 13, folder 4.

27. Swarts to Baekeland, 31 March 1891 (two letters), in LBP, series 3, box 13, folder 8.

28. Bruneel to Baekeland, 6 August 1890; Anon. [Gleesener's lawyer] to Bruneel, 28 July 1890 (both in LBP, series 8, box 34, folder 1).

29. Anon. to Bruneel, 28 July 1890.

30. Antheunis, *Leo Baekeland*, 48; Gillis, *Leo Hendrik Baekeland*, 33; Kaufmann, "Grand Duke," 57; Matthis, *Léo-H. Baekeland*, 11. On the divisiveness of the affair, see also Swarts to Baekeland, 1, 14, and 21 October 1890, in LBP, series 3, box 13, folder 5.

31. See Swarts to Baekeland, 25 and 29 August 1890 (both in LBP, series 3, box 13, folder 4) and 4 February 1891 (in folder 7 in the same box).

32. Swarts to Baekeland, 1 January 1891, in LBP, series 3, box 13, folder 7.

33. Swarts to Baekeland, 29 November 1890, in LBP, series 3, box 13, folder 8.

34. Casson and Godley, "Entrepreneurship and Historical Explanation," 58.

35. Hens and Roosens, *De Belgische fotochemische industrie*, 43; Casson, *The Entrepreneur*, 30; Casson and Godley, "Entrepreneurship and Historical Explanation."

36. For different perspectives on nineteenth-century America's culture of entrepreneurial risk taking, see Galambos, "The Entrepreneurial Culture"; Lamoreaux, "Entrepreneurship in the United States"; Sandage, *Born Losers*.

37. E.g., Swarts to Baekeland, 14 May and 7 June 1889, and Baekeland to Swarts, 3 March and 16 June 1889 (all in LBP, series 3, box 13, folders 2 or 3).

38. See, e.g., Swarts to Baekeland, 5 September and 31 October 1889, in LBP, series 3, box 13, folders 3 and 5.

39. Swarts to Baekeland, 7 March 1891, in LBP, series 3, box 13, folder 8.

40. See Baekeland to Swarts, 16 and 17 June 1889, in LBP, series 3, box 13, folder 2. Baekeland's contact may have been French American Edward Lufbery, the father of Raoul Gervais Victor Lufbery, who would gain fame as an aviator during World War I.

41. Swarts to Baekeland, 14 and 25 February 1891, in LBP, series 3, box 13, folders 7 and 4; diary 12 (20 and 25 March 1913); Swarts to Baekeland, 31 January 1889 and 15 April 1891, in LBP, series 3, box 13, folders 2 and 8; Jane Voortman Lechat to Baekeland, 19 August 1935, in LBP, series 3, box 10, folder 12.

42. Swarts to Baekeland, 14 February 1891, in LBP, series 3, box 13, folder 7.

43. See introduction to this chapter.

44. Anthony, "The Discovery of Dr. Baekeland."

45. Richard Anthony to Frederick Anthony, 10 September 1889.

46. Richard Anthony to Frederick Anthony, 10 September 1889. On Baekeland's work ethic, see also Swarts to Baekeland, undated (probably late 1890 or early 1891), in LBP, series 3, box 13, folder 4. On American "go-aheadism," see Sandage, *Born Losers*.

47. Jenkins, *Images and Enterprise*, 86; Marder, Marder, and Duncan, *Anthony*, 277.

48. Richard Anthony to Frederick Anthony, 10 September 1889; Swarts to Baekeland, 7 March 1891, in LBP, series 3, box 13, folder 8; Swarts to Baekeland, 27 January 1891 (erroneously dated 1890) and 25 September 1890, in LBP, series 3, box 13, folders 4 and 5.

49. Swarts to Baekeland, 19 and 25 September 1889, in LBP, series 3, box 13, folder 5. See also Swarts's letters of 29 August 1890, 5 September 1890 (erroneously dated 1889), 29 November 1890, 3 January 1891 (erroneously dated 1890), and 21 January, 5 February, and 7 March 1891 (all in LBP, series 3, box 13, folders 3, 4, 5, 7, or 8).

50. Swarts to Baekeland, 5 and 25 September and 1 October 1890 (all in LBP, series 3, box 13, folder 5). According to Swarts's calculations, Baekeland needed $264 per year for his rent, $240 to support his parents, $420 for his insurance, and $10 a week to cover his other living expenses.

51. Swarts to Baekeland, 25 February 1891; Baekeland to Swarts, undated (ca. early 1891); Baekeland to Guequier, 8 September 1889.

52. Diary 58 (20 June 1937). An acquaintance of Baekeland, geologist Jules Cornet, discovered rich mineral deposits in Congo in the early 1890s. See Baekeland to Jules Van Aubel, 11 February 1936, in LBP, series 3, box 12, folder 10; Delvaux, "Emile Francqui," 20.

53. Diary 12 (22 March 1913); Swarts to Baekeland, 25 February 1891. See also Swarts's letter of 19 November 1890 in LBP, series 3, box 13, folder 6.

54. See Swarts to Baekeland, 2 September 1890, in LBP, series 3, box 13, folder 5; Baekeland to Remouchamps, 30 August 1898.

55. Diary 12 (24 March 1913). On the "efficiency craze" in the Progressive-era United States, see, e.g., Alexander, *Mantra of Efficiency*, chap. 4.

56. In the following decade, it was more closely monitored. See the section "Negotiating Academia-Industry Boundaries" at the end of this chapter.

57. Wagener to Devolder, 31 May 1889 (published in Gillis, *Leo Hendrik Baekeland*).

58. Wagener to Devolder, 21 November 1889 (published in Gillis, *Leo Hendrik Baekeland*, 41–42).

59. Swarts to Baekeland, 10 December 1890, in LBP, series 3, box 13, folder 6; Swarts to Baekeland, 3 January 1891.

60. Baekeland to Remouchamps, 2 August 1898.

61. E.g., Bijker, *Of Bicycles, Bakelites, and Bulbs*, 129–130; Meikle, *American Plastic*, 35.

62. See Dhondt, "Een tweevoudig compromis," 279–294; Swarts to Baekeland, 24 January 1891, in LBP, series 3, box 13, folder 7; Vanpaemel and Van Tiggelen, "The Profession of Chemistry," 203.

63. Swarts to Baekeland, 18 February 1891, in LBP, series 3, box 13, folder 7.

64. On Effront, see Deelstra, "De scheikunde," 178; Deelstra, "De fermentatiescheikunde in België."

65. Swarts to Baekeland, 12 December 1890, in LBP, series 3, box 13, folder 6.

66. Swarts to Baekeland, 7 January 1891 (but written over several days), in LBP, series 3, box 13, folder 7.

67. Swarts to Baekeland, 14 February 1891.

68. Swarts to Baekeland, 29 October 1890, in LBP, series 3, box 13, folder 5; Swarts to Baekeland, 22 August 1890.

69. Swarts to Baekeland, 5 September 1890 (erroneously dated 1889).

70. Anon., "Frédéric Swarts (1891)."

71. Swarts to Baekeland, 9 January 1891, in LBP, series 3, box 13, folder 7.

72. Swarts to Baekeland, 10 December 1890, in LBP, series 3, box 13, folder 6.

73. Swarts to Baekeland, 19 November 1890. See also Swarts's letter of 25 September 1890.

74. Baekeland to Swarts, undated (ca. early 1891). On Baekeland's dislike of "society," see also, e.g., diary 1 (27 April 1907).

75. E.g., Baekeland to Swarts, 3 March 1889, in LBP, series 3, box 13, folder 2.

76. Baekeland to Swarts, undated (ca. early 1891).

77. Swarts to Baekeland, 20 October 1890, in LBP, series 3, box 13, folder 6. See also, e.g., Swarts to Baekeland, 27 January 1891 (erroneously dated 1890).

78. Swarts to Baekeland, 5 September 1890, in LBP, series 3, box 13, folder 5. See also Swarts's letter of 19 November 1890 in folder 6 in the same box.

79. Baekeland to Swarts, undated (ca. early 1891).

80. See Kaufmann, "Grand Duke," 32; pp. 137–138, this volume.

81. Baekeland, "Address of Acceptance of Perkin Medal," 152; Baekeland, Antiseptic compound, US Patent 500,549, filed 27 June 1892, and issued 4 July 1893.

82. Baekeland, "Use of Fluorides in the Manufacture of Alcohol."

83. See Baekeland, "The Carbon Process"; Swarts to Baekeland, 29 November 1890 (erroneously dated 1891) and 15 April 1891.

84. Diary 54 (19 June 1934). See also Baekeland, "Laboratory Notes: 1889–94," 17, 38.

85. Anon., "Leonard Jacobi," 495; Dinkelspiel, *Towers of Gold.*

86. Whitten, "The Depression of 1893."

87. Baekeland, "Address of Acceptance of Perkin Medal," 152.

88. E.g., diary 41 (30 June 1925). Statistics often cited in entrepreneurship research suggest that "over half of all new ventures will fail within their first four years of existence"; see Nagy and Lohrke, "Only the Good Die Young?," 185.

89. Frans Schollaert to Gustave Wolters, 22 May 1896, in Ghent University Archives, 4 A2/4, box 66, folder 20. I am grateful to Ruben Mantels for bringing to my attention the archival sources on which most of this section is based.

90. Wolters to Schollaert, 23 July 1896, in Ghent University Archives, 4 A2/4, box 66, folder 20.

91. Wolters to the members of the university's teaching and administrative personnel, 6 September 1898 (two letters), in Ghent University Archives, 4 A2/4, box 72, folder 20; Wolters to Schollaert, 23 July 1896.

92. See Wadhwani et al., "Academic Entrepreneurship," and the references given there.

4 Velox

1. Hooker, "Discussion," 449; Walker, "The University and Industry," 65. On Walker, see, e.g., Landau and Rosenberg, "Successful Commercialization," 87–89. Parts of this chapter were previously published in *Knowledge Management and Intellectual Property: Concepts, Actors and Practices from the Past to the Present,* Stathis Arapostathis and Graham Dutfield, eds., 62–84, Northampton, MA: Edward Elgar, 2013, and are reproduced here with permission of Edward Elgar through PLSclear.

2. E.g., Robbins, "Solving the Patent Problem"; Galvez-Behar, "Chapitre 9"; Rasch, "Auf dem Weg zum Diensterfinder."

3. Most recently, this claim has been made in Sá and Kretz, "Technology Commercialization as University Mission," 123, based on Charles Weiner's article "Patenting and Academic Research." On Weiner's argument, see p. 13, this volume.

4. On such patents, see, e.g., Dutfield, *Intellectual Property Rights and the Life Science Industries*; Gabriel, *Medical Monopoly*; Gaudillière, "How Pharmaceuticals Became Patentable." According to US chemist James L. Howe, by 1890 the medical profession was exceptional, if not unique, in treating discoveries as "the property, not of its discoverer, but of all humanity." See Howe, "Aim and Future of Natural Science," 239.

5. Other employees of the University of Ghent who obtained patents included chemist François Donny and electrical engineer Oscar Steels. At the State University of Liège, metallurgy professor Adolphe Lesoinne was a notable example of an academic patentee (and entrepreneur). For cases from several other countries, particularly Britain and Germany, see the articles introduced in Mercelis, Galvez-Behar, and Guagnini, "Commercializing Science," and the references given there. At least two of the chemists who presided over the American Chemical Society in the 1880s, Frederick Genth and John Mallet, were also academic patentees.

6. On patent controversies in other industries, see, e.g., Beauchamp, "The First Patent Litigation Explosion"; Usselman, *Regulating Railroad Innovation*, chaps. 3 and 4. In nineteenth-century Europe, the legitimacy of patent systems was questioned at an even more fundamental level. See, e.g., Galvez-Behar, "Controverses et paradoxes"; MacLeod, *Heroes of Invention*, chap. 9.

7. For a more detailed discussion of these different developments, see Berger, "Photography Distinguishes Itself"; Mercelis, "Public Knowledge and Private Enterprise"; and the references given in these works.

8. Beauchamp, "The First Patent Litigation Explosion."

9. For early historical analyses of strategic, unproductive patenting, see Reich, "Research, Patents, and the Struggle to Control Radio"; Liebenau, "Patents and the Chemical Industry."

10. Wall, *Photographic Emulsions*, 43–44.

11. Fisk, *Working Knowledge*. See also Bracha, *Owning Ideas*, chap. 5; Lamoreaux and Sokoloff, "Inventors, Firms, and the Market for Technology," 44–49; Wise, *Whitney*, 70–72.

12. In Germany, corporate control of innovation led to a backlash after the turn of the twentieth century and especially after the Nazi takeover of power in 1933. See, e.g., Gispen, *Poems in Steel*. For a useful perspective on the legal context in Belgium, see Moreau, *Traité de la concurrence illicite*, 164–171.

13. E.g., Kaufmann, "Grand Duke," 48. Of the other accounts, Susan Morris's stands out for its critical stance toward Baekeland's own explanations. However, I have

found no evidence supporting her hypothesis that Baekeland discussed his strategy with Richard Anthony and, possibly, Columbia's Charles F. Chandler. See Morris, "Resource Networks," 194.

14. See pp. 20–21, this volume.

15. Baekeland, testimony in *Oldfield Revision and Codification of the Patent Statutes*, 29.

16. Baekeland, "Address of Acceptance of Perkin Medal," 154.

17. Baekeland, "Address of Acceptance of Perkin Medal," 154. See also Baekeland, "Statement in re Velox," 492.

18. Baekeland, "Address of Acceptance of Perkin Medal," 154.

19. See Baekeland, "Notebook XVI" (8 April 1899), in LBP, series 6, box 26, folder 2.

20. Eder, "Invention of Chemically Developable Silver Chloride-Gelatin Emulsions," 392. See also Eder's "Geschichte der Erfindung" and his *Geschichte der Photographie*, 607n1, 616–622.

21. In addition to these publications, several articles on Velox had appeared in American and international photographic journals, including the *Jahrbuch für Photographie und Reproductionstechnik*, edited by Eder.

22. Eder, "Invention of Chemically Developable Silver Chloride-Gelatin Emulsions," 392.

23. Eder, "Invention of Chemically Developable Silver Chloride-Gelatin Emulsions," 392. See also Anon., "Protokoll der Plenar-Versammlung."

24. Lüppo-Cramer, "Biographie von Josef Maria Eder," 1043–1044. On Eder, see also Faber, "Josef Maria Eder"; Gröning, "Die Photographische Gesellschaft"; and the numerous entries in the "Biobibliography on Photography in Austria" of the Albertina Museum, Vienna.

25. E.g., Eder, *Photographie als Wissenschaft*. From its opening in 1888 until 1897, the K.k. Graphische Lehr- und Versuchsanstalt had been known as the K.k. Lehr- und Versuchsanstalt für Photographie und Reproduktionsverfahren.

26. Anon., "Ecole royale."

27. R. W. Harrison, unpublished manuscript (reaction to Eder's article), 1930, in LBP, series 3, box 12, folder 8.

28. Khan, *The Democratization of Invention*.

29. Cottrell, "The Research Corporation," 865; *Joint Hearings on S.3223 and H.R. 9932*; Apple, "Patenting University Research"; Cassier and Sinding, "Patenting in the Public Interest." On the relationship between patenting and knowledge sharing, see also Bessen and Nuvolari, "Knowledge Sharing among Inventors."

30. Arapostathis and Gooday, *Patently Contestable*; Apple, "Patenting University Research"; Cassier and Sinding, "'Patenting in the Public Interest."

31. Eder, *Geschichte der Photographie*, 955; Lüppo-Cramer, "Biographie von Josef Maria Eder," 1049–1050. There are no indications that the creation of a new Austrian patent system in 1897–1899 or its reform in 1925 affected Eder's attitude toward patents.

32. Mercelis, "Public Knowledge and Private Enterprise." On the history of the Photographische Gesellschaft, the photographic society in which Eder was most closely involved, see Ponstingl, *Die Explosion der Bilderwelt*.

33. Photographische Gesellschaft in Wien, "Prämien-Verleihung," *Photographische Korrespondenz* 35 (1898): 85.

34. Mercelis, "Commercializing Academic Knowledge."

35. See, e.g., Abney, *Photography with Emulsions*, 82, 164–165; Pritchard, *Photographic Studios*, 242–244; Vogel, Wallace, and Wilson, *The Progress of Photography*, 79, 285–289.

36. Baekeland, "Statement in re Velox," 491. Eder and Pizzighelli's pivotal publications had been the article "Beiträge zur Photochemie des Chlorsilbers" and the booklet *Die Photographie mit Chlorsilber-Gelatine mit chemischer Entwicklung*. For De Vylder's translation, see Eder and Pizzighelli, "Photographie par l'émulsion au chlorure d'argent et le développement chimique."

37. Israel, "Flash of Genius," 8. See also Fisher, "Geistiges Eigentum," 282–284; Fisk, *Working Knowledge*, 41, 109–112; Friedel, "Perspiration in Perspective."

38. On such "reverse salients," see Hughes, *Networks of Power*.

39. Baekeland, "Statement in re Velox," 491.

40. Chandler, "Presentation Address."

41. Diary 19 (26 and 29 December 1915 and 6 and 7 January 1916). Baekeland used the same words in his reaction to Eder's article in the *Photographic Journal*.

42. Désiré Van Monckhoven, as quoted in Anon., "Researches of Monckhoven," *Photographic Times* 17 (1887): 360, included in Charles F. Chandler Papers, Rare Book and Manuscript Library, Columbia University, New York, NY (hereafter CCP), box 241, folder 1. See also Hens and Roosens, *De Belgische fotochemische industrie*, 116, 121.

43. Baekeland, "Les bains d'hydroquinone," 502.

44. Rodolfo Namias, "Velox Type of Chloride of Silver Developing Paper, History of the Invention and Constitution of the Emulsion" (translation of article in *Il Progresso Fotografico* 39 [December 1930]), in LBP, series 3, box 12, folder 8; Mees, "Leo Hendrik Baekeland and Photographic Printing Papers"; Mees, *From Dry Plates to Ektachrome Film*, 17–18.

45. This assumption has been particularly common in theoretical studies of innovation. See, e.g., Hall et al., "Choice between Formal and Informal Intellectual Property," 376.

46. E.g., Swanson, "Emergence of the Professional Patent Practitioner"; Beauchamp, *Invented by Law*; Israel and Rosenberg, "Patent Office Records as a Historical Source."

47. See, e.g., Anon., "Au Cercle d'Etudes photographiques et scientifiques d'Anvers," *Hélios* 7 (1897): 830–832 ; Anon., "Modern Photo-chemical Processes," 353. Most of Baekeland's articles originally appeared in *Wilson's Photographic Magazine* and were then taken over by other photographic journals.

48. Brand, "Letter to the Editor: 'Shall We Make Our Own Plates?,'" 190.

49. See, e.g., the testimony of George Eastman and Frank M. Cossitt in *United States v. Eastman Kodak Co.*, in George Eastman Archives and Study Center, George Eastman House, Rochester, NY (hereafter GEASC), 1915, 842–859, 900–901.

50. Anon., "Regular Meeting, Tuesday Evening, September 10th," 468; Anon., "Black Tones on Velox Paper," 141–142; Drinkwater, "Velox," 80–84; Foxlee, "Velox for Professional Work," 246; Wenzel, "How to Work Velox Paper," 496–498; Wenzel, "Velox Again," 69–70.

51. See Erichsen, "Society News," 426; [Editors], "Editorial Table," 427; Anon., "Printing Methods and Formulae," 59. According to US Trademark 31,058, Nepera began using the word "Velox" on 6 August 1895. The Detroit convention at which Nepera presented "their entirely new product called 'Velox' paper" had begun three days earlier.

52. Mees, *The Fundamentals of Photography*, 84 (emphasis mine); Mees, *From Dry Plates to Ektachrome Film*, 17; McFarland, "Convenient and Rapid Printing Process"; Walmsley, "Admirable Printing Process for Photo-Micrographs"; Foxlee, "Velox for Professional Work," 246; Anon., "Regular Meeting, Tuesday Evening, September 10th," 467; George Eastman to Henry A. Strong, 8 July 1899, in GEASC; Aquinas, "Printing Room Particulars," 69.

53. Baekeland, *Determining the Relative Permanency of Photographic Prints*, 1.

54. Baekeland, *Photochemical Industry of the United States*, 4.

55. See the partial transcripts of various documents related to Nepera's sale in the "Baekeland" folder in GEASC.

56. Anthony, "The Discovery of Dr. Baekeland," 182.

57. Baekeland, testimony in *General Bakelite Co. v. George J. Nikolas*, patent infringement suit at US District Court for Eastern District of New York, 1915, 19, in The National Archives at New York City (equity docket 1–99); Baekeland to C. E. Munroe,

15 June 1926, and Baekeland to Robert Taft, 30 October 1939 (in, respectively, LBP, series 3, box 16, folder 1, and LBP, series 3, box 12, folder 8).

58. Anon., "Biography of Leo Hendrik Baekeland as It Appears in the Book of Enduring Names of the Town Hall, Inc. in New York City," s.d., in LBP, series 1, box 49, folder 2A. See also Chandler, "Presentation Address," 3; Swarts, "Dr. L. H. Baekeland," 1944, in LBP, series 1, box 1, folder 7.

59. See Fisk, *Working Knowledge*, 188–190; Jenkins, *Images and Enterprise*, 152–153.

60. Fisk, *Working Knowledge*, 188, 190.

61. Swarts to Baekeland, 21 January 1891, in LBP, series 3, box 13, folder 7.

62. Eastman Kodak included similar restrictions in its agreements with other competitors; see Judge John R. Hazel, opinion in *United States v. Eastman Kodak*, US Circuit Court, Western District of New York, 1915, in The National Archives at New York City, equity docket A-51, 22, 29.

63. See ABP: Section de Gand, "Séance du 4 janvier 1889," 5 ; ABP: Section de Gand, "Séance du 7 février 1889," 73; ABP: Section de Bruxelles, "Séance du 9 janvier 1889," 66; ABP: Section de Liège, "Séance du 16 janvier 1889," 68.

64. Baekeland, "Notes Concerning Velox and other Photo Papers," s.d., in LBP, series 6, box 26, folder 2, 42; Anthony, "The Discovery of Dr. Baekeland," 182; Jenkins, *Images and Enterprise*, 165.

65. Diary 59 (10 December 1938); Swarts, "Dr. L. H. Baekeland"; Swarts to Baekeland, 21 August 1890, in LBP, series 3, box 13, folder 4.

66. Baekeland, "Laboratory Notes: 1889–94."

67. Swarts to Baekeland, 21 January 1891, in LBP, series 3, box 13, folder 7. See also Swarts's letters of 24 January, 18 and 28 February, 14 and 31 March, and 29 November 1891; Baekeland, "Printing on Glass by the Albumen Bichromate Process," 292–295; Baekeland, *Photochemical Industry of the United States*, 3; Baekeland, "Notes Concerning Velox," 5, 23; Baekeland, "Laboratory Notes: 1889–94"; Baekeland, "The Carbon Process"; Ware, "Positives," 1157.

68. Swarts to Baekeland, 21 January 1891.

69. See the following 1903 papers by Baekeland, which were reprinted in 1904: *Photochemical Industry of the United States*; *Determining the Relative Permanency of Photographic Prints*; *Quantitative Determination of Silver in Photographic Paper*; *Hygrometric Conditions of the Atmosphere in the Manufacture of Photographic Paper*; *Electrolytic Action of Metallic Particles in Sensitized Papers*; *Toning Action of a Mixture of Thiosulphate of Sodium and Alum*; *Photoretrogression*. See also Witt and Pulvermacher, *Bericht*, vol. 4.

70. Testimony of Baekeland in *United States v. Eastman Kodak Co.*, in GEASC, 1915, 273–285; see also diary 50 (15 March 1932).

71. E.g., Testimony of Clifford Alexander in *United States v. Eastman Kodak Co.*, in GEASC, 1915, 432–435.

72. Chas. J. Rodd to Eastman Co., 10 December 1894, and J. W. Millington to George Eastman, 19 May 1892 (both in George Eastman correspondence, GEASC). See also Beauchamp, *Invented by Law*, 73. For an earlier example of patent-related espionage, see Guise-Richardson, "Redefining Vulcanization," 73.

73. E.g., Kingsley, "Gelatine Silver Print," 573.

74. Bancroft, "The Photographic Plate, V," 314–323; Anon., "Latentes Bild," 364; Baekeland, "La photorétrogression"; Baekeland, "Photoretrogression."

75. Witt and Pulvermacher, *Bericht*, vol. 4, 393.

76. Diary 50 (15 March 1932); Baekeland, "Notebook X," in LBP, series 6, box 25, folder 7 (formula of Hahn's metol-quinol Velox developer in entry of 31 October [probably 1895]); Jenkins, *Images and Enterprise*, 90.

77. Baekeland to Remouchamps, 17 March, 30 May, and 6 July 1899.

78. Baekeland to Remouchamps, 30 May 1899; [Editors], "Editorial Notes," 238; [Editors], "Our Table," 313; issue of *Photographic Life* included in LBP, series 2, box 5, folder 6.

5 New Opportunities and Risks

1. Diary 50 (15 March 1932). See also Hooker, "An Appreciation of Dr. Baekeland," 183.

2. See George Eastman's letters to Henry A. Strong, 20 June 1899, and to George Davison, 26 October 1899 (both in George Eastman Papers, George Eastman Archives and Study Center, George Eastman House, Rochester, NY); diary 50 (15 March 1932).

3. E.g., letterhead of Baekeland to Clinton P. Townsend, 26 June 1904, in Elmer A. Sperry Papers, Soda House, Hagley Museum and Library, Wilmington, DE (hereafter ESP), accession 1893, box 16 (electrochemistry files).

4. Gooday, "Liars, Experts, and Authorities," 433.

5. Evans, *They Made America*, 219; Kaufmann, "Grand Duke," 110–111, 140–142.

6. See the German Historical Institute's recently completed project on German American immigrant entrepreneurs, as introduced in Berghoff and Spiekermann, "Immigrant Entrepreneurship."

7. Diary 57 (8 July 1936); Baekeland to Camiel De Bruyne, 23 May 1927, in LBP, series 3, box 8, folder 16.

8. Georges Loppens and Honoré De Swarte, Process of retting or steeping flax or hemp, US Patent 576,867, filed 26 November 1895, and issued 9 February 1897.

For biographical information on Loppens, see H. S., "In Memoriam: Ir. Georges LOPPENS," 125–127. Loppens had been born in the same year as Baekeland and had been studying at the civil engineering school of Ghent University while Baekeland was enrolled in the same institution's natural science faculty.

9. The only additional incorporator of American Retting was notary John Manheimer. I am grateful to Hugh Karraker for providing me with copies of the incorporation records of both American Retting and Continental Flax, part of the Westchester County court records.

10. Besides Baekeland, the founding directors of Continental Flax were Frederick Jacobi, J. Doull Miller, Albert A. Levi, and Eugene E. Spiegelberg.

11. Baekeland to Swarts, 18 February 1902, in LBP, series 3, box 13, folder 9. See also Baekeland's letter of 12 February 1902 in the same folder.

12. H.S., "In Memoriam: Ir. Georges LOPPENS."

13. Diary 14 (10 August 1913).

14. Baekeland to Swarts, 12 February 1902, in LBP, series 3, box 13, folder 9.

15. Swarts to Baekeland, 4, 8, and 10 October 1888, in LBP, series 3, box 13, folder 1.

16. The American Fireproofing and Cement Construction Company, *The De Man System of Fireproof Construction*, 7.

17. Baekeland to Swarts, 20 February 1902, in LBP, series 3, box 13, folder 9. For a more critical evaluation of the technical merits of De Man's system, see Condit, "The First Reinforced-Concrete Skyscraper," 9–10.

18. Slaton, *Reinforced Concrete*, 141–146; Cody, *Exporting American Architecture*, 38.

19. Diary 1 (15 and 25 August 1907).

20. For a critical analysis of the role of science in this industry, see Slaton, *Reinforced Concrete*.

21. For a review of the literature on trust and entrepreneurship, see Welter, "All You Need Is Trust?"

22. Golan, *Laws of Men and Laws of Nature*; Gooday, "Liars, Experts and Authorities"; Hamlin, "Scientific Method and Expert Witnessing"; Essig, "Poison Murder and Expert Testimony"; Arapostathis and Gooday, *Patently Contestable*; Lucier, *Scientists & Swindlers*, chaps. 2 and 6.

23. Golan, *Laws of Men and Laws of Nature*, 140.

24. Fourcade, "Cents and Sensibility"; Jasanoff, *Science at the Bar*.

25. White, "Information, Markets, and Corruption"; Lucier, "The Professional and the Scientist"; Rowland, "A Plea for Pure Science."

26. Baekeland, "Applied Chemistry," 553.

27. Baekeland to Edwin Prindle, 19 December 1928, in LBP, series 3, box 17, folder 14. See also Baekeland, "Applied Chemistry," 553.

28. Baekeland, "Danger of Overspecialization," 849.

29. Trescott, *American Electrochemicals Industry*.

30. Haynes, *American Chemical Industry*, 210–215; Thomas, *Salt & Water, Power & People*. On Sperry, see Hughes, *Elmer Sperry*.

31. Thomas, *Salt & Water, Power & People*, 23.

32. Haber, *The Chemical Industry*, 79–82.

33. Kaufmann, "Grand Duke," 61–64; Thomas, *Salt & Water, Power & People*, 10–11.

34. Sperry to Hooker, 23 February 1904, in ESP, accession 1893, box 17 (electrochemistry files).

35. E.g., Sperry to Hooker, 17 June 1904, in ESP, accession 1893, box 17; Baekeland, "Address of Acceptance of Perkin Medal," 154.

36. Baekeland, "New Electrolytic Alkali-Works," 211–212; Baekeland, Electrolytic diaphragm and method of making same, US Patent 855,221, filed 10 March 1906, and issued 28 May 1907; diary 1 (3, 4, and 28 April, 7, 8, and 10 July, and 8 September 1907).

37. Baekeland, "Address of Acceptance of Perkin Medal," 154; Thomas, *Salt & Water, Power & People*.

38. For a general discussion of university-industry relationships and R&D in the field of electrochemistry, see Trescott, *American Electrochemicals Industry*. On GE's laboratory, see, e.g., Wise, *Whitney*.

39. Knorre, "Über die Bedeutung der Elekrochemie für die Technik," 130; Trescott, *American Electrochemicals Industry*, chap. 8.

40. Diary 4 (24 June 1909).

41. See, e.g., diaries 4 (12 June 1909) and 6 (17, 22 April and 27 May 1910).

42. Trescott, *American Electrochemicals Industry*, discusses the work of some of these former Columbia students, including Anson Betts and Hamilton Castner, but with special attention to Frederick Becket. On Case's ties with Dow, see Kargon and Knowles, "Knowledge for Use," 5–7. Another large electrochemical firm, Alcoa, engaged a leading aluminum expert of Lehigh University as a long-term consultant. See Graham and Pruitt, *R & D for Industry, 58-9.*

43. E.g., Richards, "Niagara as an Electrochemical Centre," 17–18.

44. Diary 2 (20 March 1908); Baekeland to Pellew, 17 November 1907, in ESP, accession 1893, box 16 (electrochemistry files).

45. Elmer Sperry to Elon H. Hooker, 21 March 1904, in ESP, accession 1897, box 17 (electrochemistry files); Wise, *Whitney*, 93.

46. *Transactions of the American Electrochemical Society* 3 (1903).

47. Diary 3 (22 January 1909).

48. For example, there are no indications that he billed D&F separately for his assistance in the suit.

49. Baekeland, testimony in New York State Court of Appeals, *Simmons-Baker v. Development & Funding Company*, 748–784.

50. Diary 1 (8 July 1907).

51. Baekeland, testimony in New York State Court of Appeals, *Simmons-Baker v. Development & Funding Company*, 753–754.

52. E.g., Fressoz, "Beck Back in the 19th Century," 344–346; Wilmot, "Pollution and Public Concern."

53. Baekeland, testimony in New York State Court of Appeals, *Simmons-Baker v. Development & Funding Company*, 751.

54. Baekeland, testimony in New York State Court of Appeals, *Simmons-Baker v. Development & Funding Company*, 752.

55. Diary 2 (7 April 1908).

56. Diary 2 (7 April 1908).

57. Diary 2 (6 April 1908).

58. *Gasse v. Development & Funding Co.*, 135 New York Supplement, 732 (N.Y. Sup. Ct. 1911); *Reports of Cases Decided in the Court of Appeals of the State of New York*, vol. 220, 742–743.

59. *Gasse v. Development & Funding Co.*, 135 New York Supplement, 732 (N.Y. Sup. Ct. 1911).

60. Factory Investigating Commission of the State of New York, *Second Report, vol. 3*. On the Alkali Acts, see Reed, "The Alkali Inspectorate"; Garwood, "Green Crusaders or Captives of Industry?"; Wilmot, "Pollution and Public Concern."

61. Factory Investigating Commission of the State of New York, *Second Report, vol. 3*, 627–643; Thomas, *Salt & Water, Power & People*, 23–24.

62. Diary 1 (22 January 1908).

63. Factory Investigating Commission of the State of New York, *Second Report, vol. 3*, 704, 712; Newman, *Love Canal*, 67.

6 A Technocrat on the Public Stage

1. Israel, *Edison*, chap. 19; Hounshell, "Edison and the Pure Science Ideal."

2. Diary 19 (28 March 1915).

3. Thackray, "University-Industry Connections and Chemical Research," 207–209.

4. Baekeland to Charles Herty, 12 September 1916, as quoted in Rhees, "The Chemists' Crusade," 137. On the notion of rational reform, see Jordan, *Machine-Age Ideology*.

5. Diary 19 (23 November 1925).

6. Diaries 28 (4 October 1919), 45 (11 August 1928), and 46 (4 March 1929).

7. LBP, series 3, box 17, folder 17; diary 42 (7 June 1926).

8. Baekeland, "Applied Chemistry," 547–548.

9. Diary 22 (28 June 1917).

10. Diary 22 (29 June 1917).

11. Diaries 19 (18 April 1916) and 22 (24 May 1917). On Hooker's role in the production of chemical weapons, see also diary 24 (4 and 12 January and 6 and 7 February 1918); Newman, *Love Canal*, 67–71.

12. Diary 22 (29 June 1917); Edgerton, *The Shock of the Old*, chap. 6.

13. Slotten, "Humane Chemistry or Scientific Barbarism?," 485; Brophy, Miles, and Cochrane, *The Chemical Warfare Service*, 18.

14. Diary 23 (2 November 1917). By the end of the war, the number of scientists assigned to this project had risen to about 1200; see Whittemore, "World War I," 150. On Baekeland's role in the production of aviation fuel, see Leslie, *Boss Kettering*, 78–80.

15. Diary 24 (11 December 1917, 4 January 1918, 15 February 1918); see also diary 25 (21 September 1918). On the reuse of chemical weapons as pesticides, see Russell, *War and Nature*.

16. E.g., Adas, *Dominance by Design*, chap. 3.

17. Baekeland, "Introductory Address by the Chairman," 818; Maxim, "Warfare of the Future," 821, 831. In 1898, however, Baekeland had staunchly defended America's decision to enter into war with Spain; see Baekeland to Remouchamp, 30 May, 13 June, and 5 July 1898.

18. Diary 2 (9 October 1908).

19. Baekeland, "Introductory Address by the Chairman," 819–820.

20. Diary 24 (22 December 1917).

21. See Jones, "Role of Chemists in Research on War Gases," chap. 5; Jones, "American Chemists and the Geneva Protocol," 428, 432; Slotten, "Humane Chemistry or Scientific Barbarism?," 492.

22. Slotten, "Humane Chemistry or Scientific Barbarism?," 492.

23. Amos Fries to Baekeland, 18 September 1924, in LBP, series 3, box 12, folder 3; Fries to Baekeland, 30 September 1924, in LBP, series 3, box 16, folder 7. On Fries's advocacy of chemical warfare, see Faith, *Behind the Gas Mask*; Jones, "American Chemists and the Geneva Protocol"; Nader and Marcinko, "'Chlorine Chambers' against Cold and Flu," 301, 309–321.

24. Baekeland, "Prospects and Retrospects," 278–279.

25. Diary 20 (17 June 1916). The growth of the chemical profession is documented in Thackray et al., *Chemistry in America*. Significantly, between 1900 and 1910, ACS membership tripled, and it tripled again between 1910 and 1920.

26. On this campaign, see Reed, *Crusading for Chemistry*; Rhees, "The Chemists' Crusade"; Steen, *American Synthetic Organic Chemicals Industry*. On Baekeland's role, see, e.g., his articles "Renewing the Earth from the Air" and "What Is the Matter with the American Chemist?" See also diaries 17 (6 and 10 February and 6 May 1915), 19 (30 April 1916), and 20 (4, 15, and 17 May and 1–8 June 1916).

27. Baekeland to Charles H. Herty, 2 July 1924, in LBP, series 3, box 12, folder 3; Slotten, "Humane Chemistry or Scientific Barbarism?," 481–482.

28. Baekeland, "Prospects and Retrospects," 278; Baekeland, "Science and Industry," 39.

29. E.g., Bruch, "Der wissenschaftsgläubige Mensch"; Bayertz, "Biology and Beauty," 283–284; Bayertz, "Darwinismus als Weltanschauung," 133, 136–137.

30. Diary 12 (20 March 1913).

31. Baekeland to Remouchamps, 27 September 1899 (my translation from the Dutch).

32. Baekeland, "Danger of Overspecialization," 846. The second part of the quotation was solely Baekeland's.

33. Baekeland, "Science and Industry," 40, 43–44.

34. Diary 3 (28 February 1909).

35. This was even true in Germany, where the ambivalence toward technoindustrial modernity is often said to have been especially pronounced. See Rohkrämer, *Eine andere Moderne?*, pt. 1.

36. Diary 1 (13 April 1907).

37. Diary 1 (5 April 1907). The only reaction in *Science* appears to have been an approving quotation in Woodhull, "What Specialization Has Done for Physics Teaching," 729.

38. In addition, a summary of "The Danger of Overspecialization" had been included in *Electrochemical and Metallurgical Industry* 5 (1907): 391.

39. For important early interpretations of these debates, see Gispen, *New Profession, Old Order*; Herf, *Reactionary Modernism*. For more recent analyses, see Hård, "German Regulation"; König, *Der Gelehrte und der Manager*, 91–119; Rohkrämer, *Eine andere Moderne?*, 56–71, 345; Voskuhl, "Engineering Philosophy."

40. See Baekeland, "Wissenschaft und Industrie," and the reactions by Kissling, the journal editors, and Baekeland on pp. 762 and 965 in the same volume of the *Chemiker-Zeitung*. On the *Chemiker-Zeitung*'s editor-in-chief, Eduard J. von Vietinghoff-Scheel, see pp. 142–145, this volume.

41. Redaktion der *Chemiker-Zeitung*, [Wissenschaft und Industrie]; Kissling, [Wissenschaft und Industrie].

42. Fisch, "Zivilisation, Kultur," 749.

43. Makowka, [Wissenschaft und Technik].

44. Vietinghoff-Scheel, [Wissenschaft und Technik].

45. Schatzberg, *Technology*.

46. On the activism of Whitaker and Herty, see Reed, *Crusading for Chemistry*; Rhees, "The Chemists' Crusade," 21, 77–84, 140, 245; Steen, *American Synthetic Organic Chemicals Industry*, 119–121, 125–126.

47. Baekeland to Arthur D. Little, 14 August 1929, in LBP, series 3, box 10, folder 11.

48. Rhees, "The Chemists' Crusade," 20–22, 46–47.

49. Rhees, "The Chemists' Crusade," 20; Shapin, *The Scientific Life*, 110.

50. Browne and Weeks, *History of the American Chemical Society*.

51. Rhees, "The Chemists' Crusade," 126–132.

52. Layton, *Revolt of the Engineers*; Meiksins, "'Revolt of the Engineers' Reconsidered," 222. See also Knoedler and Mayhew, "Thorstein Veblen and the Engineers"; Noble, *America by Design*.

53. On this "revolutionary" assumption of Cooke, see Layton, *Revolt of the Engineers*, 159.

54. Meiksins, "'Revolt of the Engineers Reconsidered,'" 222–227.

55. Lidbury et al., "Discussion," 477.

56. Diary 19 (23 November 1915); see also entries of 16 and 17 February and 24 March 1916 in the same diary.

57. Diary 19 (22 February 1916).

58. Diary 19 (23 November 1915); Lidbury et al., "Discussion," 1152–1153.

59. Baekeland, "Science and Industry"; Baekeland, "Dreams and Realities." See also diary 17 (30 May and 3 June 1915).

60. Rodgers, "In Search of Progressivism," 120, 126–127; McCormick, "The Discovery That Business Corrupts Politics," 247–249.

61. Van Keuren, "Science, Progressivism, and Military Preparedness," 722. Van Keuren, however, overestimated the differences between Baekeland and Edison.

62. Kevles, "Hale, the First World War, and the Advancement of Science"; Wise, *Whitney*, 186, 197.

63. See the "Baekeland," "Nitrogen," and "Committee on Optical Glass" folders in the Records of the Naval Consulting Board, National Archives, Washington, DC, RG 80, boxes 3, 11, and 48.

64. E.g., Hughes, *American Genesis*, 118–125.

65. Diary 22 (19 April 1917).

66. Diary 19 (24 March 1916). See also Baekeland's letters to Bogert of 7 and 9 April 1917 and to Hale of 10 April 1917 (all in National Academies Archives, Washington, DC [hereafter NAA], Central Policy Files, Committee on Chemistry).

67. Baekeland to Bogert, 1 May 1917, in NAA, Central Policy Files, Committee on Chemistry.

68. Brittain, "Introduction of the Loading Coil," 38.

69. Diary 22 (17 June 1917).

70. Frederick Cottrell to C. D. Walcott, 3 February 1913, as quoted in Cornell, *Establishing Research Corporation*, 132–133. See, e.g., Ralph H. McKee to Baekeland, 3 January 1924, and Wigglesworth to Baekeland, 8 January 1924 (both in LBP, series 3, box 16, folder 4). See also Thomas et al., "Leo Hendrik Baekeland."

71. Diary 20 (17 June 1916). Ironically, it was Johns Hopkins's later president, Isaiah Bowman, who found Baekeland "too loquacious." See Robbins, "Solving the Patent Problem," 269.

72. Diary 22 (21 May 1917).

73. Baekeland to Chandler, 8 November 1917, in CCP, box 1, folder 15.

74. G. S. Bidwell (Baekeland's secretary) to Frank C. Whitmore, 16 December 1927, in LBP, series 3, box 17, folder 14 (quoting a letter that Baekeland had written to Pegram shortly before).

75. Diary 34 (24 November 1924).

76. Arthur W. Hixson, "History of the Department of Chemical Engineering in Columbia University," 1956, 29, in Historical Subject Files, CCP.

77. James B. Conant to Baekeland, 11 September 1932, in LBP, series 2, box 6, folder 7. On Conant, see Hershberg, *James B. Conant*; Hounshell, "Evolution of Industrial Research," 27–28.

78. On the emergence of the "publish or perish" culture in academia, see Clark, *Academic Charisma*. On the expectation to publish in specialized scientific periodicals, see Csiszar, *The Scientific Journal*, chap. 4.

79. Baekeland, "Dreams and Realities," 1003. On this address, see also diary 49 (23 October 1931).

80. In 1930, Baekeland recollected that, "My name had been presented a first time [for membership in the National Academy of Sciences] during the war period but…some German born members had used their influence against my application." See diary 48 (18 December 1930).

81. Kevles, "Hale, the First World War, and the Advancement of Science," 430. See also Wise, *Whitney*, 199.

82. See Israel, *Edison*, 468–469; Hughes, *Elmer Sperry*, 308–309.

83. Diary 18 (6 October 1915).

84. See diary 48 (18 December 1930).

85. Dunn to Baekeland, 13 May 1936, and Jerome Alexander to Baekeland, 13 May 1936 (both in LBP, series 3, box 17, folder 8); Bush, "Gano Sillick Dunn," 38.

86. See Anon., "U.S. National Academy of Sciences"; Anon., "Scientific Notes and News" (1936).

87. By contrast, the academy's own report did mention the Bakelite Corporation; see National Academy of Sciences, *Report for the Fiscal Year 1935–1936*, 28.

7 Bakelite: Transatlantic Patenting

1. Swarts to Baekeland, 11 August 1905, in LBP, series 3, box 13, folder 9. Parts of this chapter were previously published in *Technology and Culture* 53, no. 2 (2012): 366–400, and are reprinted here with the permission of the Johns Hopkins University Press.

2. Baekeland, testimony in *General Bakelite v. Nikolas*, 723–724, and affidavit of 6 July 1911 in file of US Patent 1,038,475, in National Archives at College Park, MD (hereafter NACP), RG 241; Anon., "Break in Camphor Market"; Judge Thomas I. Chatfield, ruling in *General Bakelite Co. v. General Insulate Co.*, patent-infringement suit in US District Court for the Eastern District of New York, 1921, 6, in LBP, series 9, box 36, folder 1. On the importance and dramatic consequences of the global camphor trade, see also Inkster, "Indigenous Resistance and the Technological Imperative."

3. Baekeland, testimony in *General Bakelite v. Nikolas*, 723–724; Nathaniel Thurlow to Baekeland, 30 November 1904, in LBP, series 8, box 15, folder 12; Thurlow, Process of producing camphor, US Patent 698,761, filed 6 August 1901, and issued 29 April 1902. Thurlow's synthetic camphor process, which he appears to have invented during his time at the Ampere Electrochemical Company, had been acquired by the Portchester Chemical Company of New York.

4. Diary 1 (7 April 1907). See also Thurlow, "Assignment," 14 May 1906, and Baekeland and Thurlow, "Memorandum of Agreement Concerning Process of Manufacturing Alcohol Soluble Gums," 12 February 1907 (both in LBP, series 2, box 7, folder 5).

5. Baekeland, testimony in *General Bakelite v. Nikolas*, 724; Kleeberg, "Ueber die Einwirkung des Formaldehyds auf Phenole."

6. See the file of US Patent 949,671 in NACP, RG 241.

7. See the first entries in Baekeland, "BKL 1 (18 June–24 August 1907)," in LBP, series 6, box 26, folder 3. For a detailed analysis of these experiments, see Bijker, *Of Bicycles, Bakelites, and Bulbs*, 146–150. My interpretation is mostly consistent with Bijker's but suggests that Bijker's criticism of Baekeland's retrospective explanation of his interest in wood impregnation was somewhat overblown.

8. Diary 8 (9, 11, and 18 October 1910). On the earlier patent ownership arrangements between Baekeland and Thurlow, see Thurlow to Baekeland, 30 November 1904, in LBP, series 8, box 15, folder 12; and Thurlow, "Assignment," 14 May 1906, and Baekeland and Thurlow, "Memorandum of Agreement Concerning Process of Manufacturing Alcohol Soluble Gums," 12 February 1907 (both in LBP, series 2, box 7, folder 5).

9. Chang, "Intellectual Property Rights and Economic Development," 292.

10. E.g., Herstein, "Patents and Chemical Industry in the United States," 330–331; Liebenau, "Patents and the Chemical Industry," 144; Maestrejuan, "Inventors, Firms, and the Market for Technology," 219; Sáiz, "Why Did Corporations Patent in Spain?" See also Cantwell, *Technological Innovation and Multinational Corporations*, 21.

11. Penrose, *Economics of the International Patent System*, 211.

12. See Beauchamp, *Invented by Law*, chap. 5. On Bell's difficulties in obtaining IP protection in Spain, see Beatty and Sáiz, "Industrial Property Institutions," 20.

13. See Crouch, *The Bishop's Boys*, 417.

14. Anson Betts to Elmer Sperry, 20 June 1907, and Sperry to Betts, 29 June 1907, in ESP, accession 1893, box 19 (electrochemistry files).

15. Besides the studies by Beauchamp and Crouch cited earlier, see also Israel, *Edison*, 125–126, 148–149, 174, 215–216; Hausman, Hertner, and Wilkins, *Global Electrification*, 77, 92; Hughes, *Networks of Power*, 47–78; Guagnini, "Patent Agents," 175; Wilkins, *Emergence of Multinational Enterprise*, 47–68.

16. Diaries 2 (5 October 1908) and 3 (12 and 19 November and 16 December 1908).

17. Diaries 2 (18 March 1908; see also entries of 14 February and 13 June 1908) and 3 (11 December 1908). On the amendments, see Baekeland to Patent Commissioner E. B. Moore, 9, 14, and 17 March 1908 (in the files of US Patents 942,699; 942,700; 942,809; 949,671; 1,213,726), in NACP, RG 241.

18. Baekeland, "Synthesis, Constitution, and Uses of Bakelite," 149–161. On this paper, see also diary 3 (5 and 13 January and 4 and 5 February 1909).

19. Diary 2 (1–3 and 10 June 1908). See also diary 3 (15–17 November 1908).

20. Diaries 3 (19 January 1909), 9 (29 May 1910), and 10 (21 March 1912); Baekeland, "Synthesis, Constitution, and Uses of Bakelite," 154; Baekeland, "Condensation Products of Phenols and Formaldehyde," 518; Ladas, *Patents, Trademarks, and Related Rights*, 344–345, 354–355, 371; Kranakis, "Patents and Power," 703. See also diary 5 (2 and 11 November 1909).

21. Baekeland received the award for two papers: his first public address on Bakelite and its follow-up paper, "Soluble, Fusible, Resinous Condensation Products."

22. Baekeland, diaries 3 (14–16 and 25 February and 17 March 1909), 60 (14 November 1940), and 61 (4 November 1941), Baekeland, "Bakelit, ein neues synthetisches Harz," 317–318, 326–328, 347–348, 358–359; letters from Baekeland to Herman Stadlinger, 29 November 1938, and to H. Potter, 28 November 1938 (both in LBP, series 3, box 8, folder 5). In the diary entry from 1940, Baekeland stated that the paper had also been published in Italian and Russian, among other languages. For the early sales records, see Swarts, "Account Book, 1908–1911," in LBP, series 8, box 34, folder 3.

23. Diaries 2 (5 October 1908) and 3 (12 and 19 November and 16 December 1908).

24. Diaries 2 (18 March 1908; see also entries of 14 February and 13 June 1908) and 3 (11 December 1908). On the amendments, see Baekeland to Patent Commissioner E. B. Moore, 9, 14, and 17 March 1908 (in the files of US Patents 942,699; 942,700; 942,809; 949,671; and 1,213,726) in NACP, RG 241.

25. Diary 1 (4 and 8 November, 18–22 December 1907, and 5–10 January 1908). To better protect his British interests, he later also collaborated with agents in London; see diaries 2 (28 May 1908) and 4 (27 May 1909).

26. See Baekeland, diary 4 (16 June 1909); Osterhammel, *Verwandlung der Welt,* 1029.

27. Diary 4 (23 June 1909).

28. Diary 4 (23, 27, and 29 June and 3 and 7 July 1909).

29. For instance, his diaries suggest that he did not attend any IP-related paper presentations at the Congress of Applied Chemistry in London. The papers are published in Ramsay and MacNab, *Seventh International Congress of Applied Chemistry.*

30. Supplement to diary 4 (31 July 1909).

31. Diary 4 (11 June 1909); see also the entries of 9–10 June in diary 4 and of 3 August 1909 in its supplement.

32. On the content of the dispute, see Bijker, *Of Bicycles, Bakelites, and Bulbs,* 168–169. On Baekeland's and Schweitzer's mutual dislike of each other, see, e.g., Baekeland, diaries 4 (25 May and 6 June 1909), 5 (21 August 1909), 9 (8 July 1911), and 11 (9 September 1912).

33. Diary 4 (7 and 17 June and 1 July 1909). See also the diary 4 supplement (7–9 and 12 July 1909).

34. Diary 4 (14 July 1909) (emphasis in original).

35. Diary 7 (2 July 1910).

36. Diary 7 (12 August 1910) and diary 12 (10 March 1912).

37. Supplement to diary 4 (16–17 July 1909), diary 5 (27 and 31 October and 16 November 1909), and diary 6 (2–3 April and 31 May 1910). In addition to the parties mentioned earlier, Rütgerswerke's subsidiary Chemische Fabrik Lindenhof C. Weyl & Company also invested in Bakelite.

38. Hughes, *American Genesis,* 24; Morris, "Resource Networks," 24; Shapin, *The Scientific Life,* 232, 245, 249–250, 264.

39. Baekeland's correspondence with Bakelite Gesellschaft is included in LBP, series 8, box 15, folders 2 and 3. See also, e.g., diaries 22 (23 January 1923) and 42 (8 March and 30 July 1926); letters from Baekeland to Prof. Pigeon, 3 May 1926, in LBP, series 3, box 11, folder 5, and Baekeland to Jean Gérard, 9 April 1924, in LBP, series 3, box 17, folder 19.

40. Diary 57 (16 June 1936; see also entry of 3 July 1936).

41. On the pioneering nature of this division of labor, see Hemming, *Plastics and Molded Electrical Insulation,* 180, 204–205.

42. Supplement to diary 4 (12 July 1909). On Baekeland's patent-licensing strategy, see diaries 1 (6 December 1907) and 2 (31 January, 18 March, 19 April, and 12 May 1908); Baekeland to F. G. Wiechmann, 30 April 1909, in LBP, series 8, box 15, folder 12.

43. Baekeland to Howard Van Sinderen, 15 September 1910, in LBP, series 8, box 15, folder 6. See also Edward Aiken, report for Frank Dyer, 25 October 1910, in TEP, reel 217, frames 427–433 (at 432).

44. On Baekeland's negotiations with Roessler & Hasslacher, see Hugo DuBois, "Visit of Dr. Baekeland to Our Office," 13 May 1910, and Alexander Schneider, travel notes, 14 and 17 May 1910 (both in Evonik Industries AG, Corporate Archives Hanau, DL 3. Schneider/2); Baekeland, diaries 5 (29 September 1909) and 6 (13 and 15 May 1910). See also diary 2 (27 October 1908), supplement to diary 4 (14 July 1909), and diary 5 (21 August and 3 November 1909). On Baekeland's relationship with Protal, see Meikle, *American Plastic*, 47–48; Bijker, *Of Bicycles, Bakelites, and Bulbs*, 164–165.

45. These were Baekeland, Method of making insoluble products of phenol and formaldehyde, US Patent 942,699, filed 13 July 1907, and issued 7 December 1909; Baekeland, Condensation product of phenol and formaldehyde and method of making the same, US Patent 942,700, filed 4 December 1907, and issued 7 December 1909; Baekeland, Abrasive composition and method of making same, US Patent 942,808, filed 26 October 1907, and issued 7 December 1909; Baekeland, Condensation product and method of making same, US Patent 942,809, filed 15 October 1907, and issued 7 December 1909; Baekeland, Indurated product and method of preparing same, US Patent 942,852, filed 13 July 1907, and issued 7 December 1909.

46. Diary 5 (18 September–2 October 1909).

47. Baekeland (presumed), "Project for the Formation of a Bakelite Company and Possible Sources of Income" (ca. May 1910), in LBP, series 2, box, 7, folder 5.

48. Baekeland (presumed), "Project for the Formation of a Bakelite Company and Possible Sources of Income."

49. Baekeland to Dyer, 9 January 1911, in TEP, reel 217, frames 572–573.

50. Ladas, *Patents, Trademarks, and Related Rights*, 23–24, 352–353; Baekeland, "Abuses of Our Patent System," 336; Baekeland, "Comments on the Report of the Investigation of the United States Patent Office," 419.

51. Diaries 6 (7 April 1910) and 7 (30 June 1910); see also diaries 7 (2 and 11 July 1910) and 9 (8 and 13 July 1911).

52. Baekeland, "Abuses of Our Patent System," 336, and testimony in *Oldfield Revision and Codification of the Patent Statutes*, 32. On the importance of the worldwide novelty requirement, see Mersch, *Die Welt der Patente*.

53. See Meikle, *American Plastic*, 50–57; and Bijker, *Of Bicycles, Bakelites, and Bulbs*, 170–172. On Condensite, see also Israel, *Edison*, 429–431.

54. Diary 7 (2 July 1910).

55. C. E. Leffel to Hylton Swan, 2 October 1916, in Albany Billiard Ball Company Records, Archives Center, NMAH, series 3, box 21, folder 4.

56. Dyer to Smith, 25 March 1910, in TEP, reel 217, frame 325. On the actions taken against Condensite in Europe, see diary 12 (13 and 15 February 1913).

57. Charles Brock to W. H. Wickham, 26 September 1912, in J. Harry DuBois Collection on the History of Plastics (hereafter HDC), Archives Center, NMAH, series 3, box 3. Baekeland's estimate of the cost was in line with patent lawyer Louis Raegener's; see Raegener, "Defects in the Practice of Our Patent System," 201.

58. Diary 11 (24 and 27 September 1912).

59. Diary 11 (30 October 1912).

60. Diary 11 (30 October 1912). On the negotiations in Europe, see diary 12 (27 and 28 February 1913).

61. Baekeland, "United States Patent System," 204; diary 3 (8 January 1909).

62. Moore, "United States Patent System"; Baekeland, diary 12 (28 January 1913); Anon., "Monopoly Victory in Patent Decision"; Benjamin, "Letter to the Editor"; Anon., "Favor Patent Law Revision"; Anon., "Asks Supreme Court to Reverse Itself"; Baekeland, "Abuses of Our Patent System," 337; Baekeland, "Legalized Blackmail: A Sample of Ignorant Patent Legislation" (ca. 1912), in LBP, series 9, box 35, folder 3; Baekeland, "Incongruities of American Patent Litigation," 786–788; Baekeland, "Protection of Intellectual Property," 54–56; diary 12 (28 January 1913); Baekeland, *Report of the Patent Committee of the National Research Council*, 1; Baekeland, diary 24 (20 December 1917 and 17 and 18 May 1918); secretary of the National Research Council to F. E. Stewart, 28 February 1918, in NAA, Central Policy Files, Committee on Patent Office. On the role of the Patent Office Society in the creation of the NRC committee, see Robbins, "Solving the Patent Problem," chap. 3.

63. Diary 3 (8 January 1909); Raegener, "Defects in the Practice of Our Patent System," 202.

64. Diaries 2 (1 May 1908) and 10 (26 January 1912); Baekeland, "United States Patent System," 205; Baekeland, *Report of the Patent Committee of the National Research Council*, 3–5. The quotation is from Baekeland, "Abuses of Our Patent System," 334. Raegener and Baekeland had both participated in a discussion of the US patent system at the Thirteenth General Meeting of the Electrochemical Society on 1 May 1908; see Marvin, "Electrochemical Patents."

65. Baekeland, *Report of the Patent Committee of the National Research Council*, 5–9; Baekeland, "United States Patent System," 205; Baekeland, "Comments on the Report of the Investigation of the United States Patent Office," 418; Baekeland, "Abuses of Our Patent System," 336.

66. See, e.g., Baekeland, "Abuses of Our Patent System," 336. Although he strongly advocated it during the period in which he was actively involved in patent-reform efforts, Baekeland later became more critical of such a system of public opposition; see the letters from Baekeland to H. E. Howe, 7 June 1924, and to Milton Whitaker, 28 April 1932 (both in LBP, series 9, box 35, folder 6).

67. Diary 11 (3 May 1912).

68. Baekeland, "Comments on the Report of the Investigation of the United States Patent Office," 419; Baekeland, "Abuses of Our Patent System," 337; Baekeland, "Protection of Intellectual Property," 53. Open-court testimony had already been introduced by December 1913, when the patent-infringement suit *General Bakelite Co. v. George J. Nikolas* commenced; see [Editors], "Personal Notes," 638.

69. Baekeland, "Protection of Intellectual Property," 53–54.

70. Diary 12 (6, 7, 10, and 12 March 1913); Baekeland, "Comments on the Report of the Investigation of the United States Patent Office," 418.

71. Hesse, "Shall We Have Annual Patent Renewal Fees?," 697; Hesse, "United States Patent System," 203–204.

72. Baekeland, "Dr. Hesse's Suggestion of Renewal Fees," 938, part of Prindle et al., "Symposium on Annual Patent Renewal Fees."

73. Diary 24 (18 May 1918).

74. Noble, *America by Design*, 107–109; Robbins, "Solving the Patent Problem," chap. 3.

75. Baekeland to Whitaker, 28 April 1932.

76. Sources in NAA, Central Policy Files, Committee on Patent Office: Patent Committee National Research Council, *Report on Nolan Patent Office Bill, H.R. 11,984*, 1921; Edwin Prindle, telegram to Frederick G. Cottrell, 24 February 1921; letters from Prindle to C. D. Walcott, 8 February 1921, Clinton P. Townsend to Prindle, 14 March 1921, Baekeland to Cottrell and Prindle, 12 April 1921, H. C. Parmelee to Cottrell, 4 April 1921, Cottrell to Parmelee, 16 April 1921.

77. Prindle to Baekeland, 23 June 1924, in LBP, series 3, box 11, folder 6; Noble, *America by Design*, 107–108; Scherer, "Political Economy of Patent Policy Reform"; Biagioli, Jaszi, and Woodmansee, "Introduction," 6; Lamoreaux and Sokoloff, "Afterword," 472.

78. Grosvenor, "The Relation of the Manufacturer to Our Patent System," 120.

79. Mr. Kemp, Director, Burdett's Research Laboratory, in Stieglitz, "Synthetic Pharmaceuticals," 381–382.

80. On the Temporary National Economic Committee, unofficially known as the "Monopoly Committee," see Hintz, "The 'Monopoly' Hearings"; Lynch, *Concentration*

of Economic Power; Owens, "Patents, the 'Frontiers' of American Invention, and the Monopoly Committee of 1939."

81. George Baekeland, testimony in *Investigation of Concentration of Economic Power.*

82. On the views of the other committee members, see Robbins, "Solving the Patent Problem," chap. 3.

83. Cottrell, testimony in *Joint Hearings on S.3223 and H.R. 9932*; Cottrell, "The Research Corporation."

84. Mowery et al., *Ivory Tower and Industrial Innovation*; LeCain, "The Limits of 'Eco-efficiency.'"

85. Stieglitz, "Synthetic Pharmaceuticals," 378.

86. Stieglitz, "Synthetic Pharmaceuticals," 377.

87. General Bakelite Company, "Minutes of Boards of Directors Meetings, 1910–19," in LBP, series 8, box 34, folder 6, 21. On this anti-Bakelite "propaganda," see also diary 29 (17 March 1920). On the international Bakelite plants, see Chandler, "Presentation Address"; Baekeland to Paul Fredericq, undated (ca. February 1913), in Correspondence of Paul Fredericq, Ghent University Library.

88. Diary 35 (4 April 1923); see also diaries 36 (17–18 and 21 April 1923) and 52 (14 June 1933).

89. Evans, *They Made America*, 219. See also Kaufmann, "Grand Duke," 110–111, 140–142.

90. Baekeland, "Dreams and Realities," 1003.

91. Baekeland to Fredericq, 15 January 1914, in Correspondence of Paul Fredericq.

8 Fancy Consumer Goods in an Industrial Age

1. Baekeland, "Synthesis, Constitution, and Uses of Bakelite," 156, 158.

2. See Friedel, *Pioneer Plastic*, 44–48. Early producers of aluminum faced similar challenges; see, e.g., Smith, *From Monopoly to Competition*, 84.

3. Baekeland, "Synthesis, Constitution, and Uses of Bakelite," 157.

4. Different aspects of this development are discussed in Blaszczyk, *American Consumer Society*, 25, 95–96; Kline, "Construing 'Technology' as 'Applied Science'"; Oldenziel, *Making Technology Masculine*; Schatzberg, "From Art to Applied Science."

5. See Marx, "Technology." On the role of social scientists in filling this void, see Schatzberg, *Technology*.

6. For a similar example, see Israel, *Edison*, 302.

7. Meikle, *American Plastic*, 46, and also 33.

8. Baekeland, *Some Aspects of Industrial Chemistry*, 5.

9. On General Bakelite's access to phenol and formaldehyde during and after World War I, see, e.g., Leo Baekeland to George Baekeland and George Roll, 16 June 1920, in LBP, series 3, box 14, folder 15; Byck, "A Survey of the Bakelite Thermosetting Business," in LBP, series 1, box 2, folder 5, 7; Meikle, *American Plastic*, 326–327 (n. 102). On Dow's production of synthetic phenol and its sales to Bakelite, see, e.g., files 150017, 150018, 150065, 180147, and 250049 in Dow Collection.

10. Anon., "Phenol Resin Industry."

11. E.g., United States Tariff Commission, *Census of Dyes and of Other Synthetic Organic Chemicals, 1927*, 56; United States Tariff Commission, *Census of Dyes and of Other Synthetic Organic Chemicals, 1929*, 59.

12. E.g., Ellis, *Chemistry of Synthetic Resins*, 316; [Editors], "New Era in Field of Phenol Resins"; Bakelite Corporation vs. Brunswick-Balke-Collender Co., 7 F.2d 697 (D. Del. 1925). Another crucial patent was US Patent 942,700, which covered B-stage Bakelite and Baekeland's method for producing it. See Baekeland, Condensation product of phenol and formaldehyde and method of making the same, US Patent 942,700, filed 4 December 1907, and issued 7 December 1909.

13. Diary 4 (7 July 1909).

14. See file of US Patent 942,699, in NACP, RG 241. See also diaries 1 (15 and 25 July 1907 and 1 and 2 October1907) and 2 (5, 6, and 9 March 1908). Baekeland's long laboratory notebook entry of 20 June 1907 leaves no doubt that he was aware of the advantage of applying a countervailing pressure when heating intermediate-stage Bakelite at high temperatures. However, it was not until the following months and years that he started paying more attention to this aspect of the Bakelite process and designed tailor-made autoclaves for this purpose (see figure 8.2).

15. On Bell's telephone patent, see Beauchamp, *Invented by Law*.

16. See file of US Patent 942,699, in NACP, RG 241 (quotation from Edward B. Moore to Baekeland c/o Byrnes & Townsend, 18 May 1909). For more on the relationship between Baekeland's patents and research papers, see chapter 9, this volume.

17. This was Baekeland, Composite cardboard, US Patent 1,019,406, filed 30 November 1910, and issued 5 March 1912.

18. See, e.g., pp. 191–192, this volume

19. Diary 14 (24 October 1913).

20. Baekeland to Chandler, 7 January 1916, in LBP, series 3, box 17, folder 21. For a summary of the case, see Anon., "Bakelite Suit Decision."

21. Baekeland, testimony in *General Bakelite v. Nikolas*, 177–182. See also diary 32 (22 November 1921).

22. See, e.g., file of US Patent 942,700, in NACP, RG 241. See also United States Tariff Commission, *Report to the President of the United States*, 18.

23. For a specific example, see pp. 189–196, this volume.

24. Diary 14 (12 December 1913).

25. [Richard Seabury], "Confidential Report Bakelite Molding Industry," 19 November 1926, 3, in HDC, series 3, box 3.

26. Seabury, "Historical Report of Boonton Rubber Manufacturing Company," 15 April 1916, and letters from Brock to Wickham, 26 September 1912, Seabury to Baekeland, 31 August 1912, Seabury to Charles Skinner, 21 December 1916 (all in HDC, series 3, box 3).

27. General Bakelite Company, "Minutes of Boards of Directors Meetings, 1910–1919," 147; diary 49 (30 July 1931); Jacques LaRoche to Baekeland, 15 July 1932, in LBP, series 3, box 10, folder 12.

28. Byck, "A Survey of the Bakelite Thermosetting Business," 6.

29. Diary 32 (2 August 1921).

30. Crouch, *The Bishop's Boys*, 463. Baekeland sympathized with Orville Wright; see their correspondence in The Wilbur and Orville Wright Papers, Library of Congress, Washington, DC, box 12, folder 3. See also Baekeland, diary 17 (7 April 1917).

31. Crouch, *The Bishop's Boys*, 46–47, 463.

32. Diary 7 (4 August 1910); see also diary 16 (2 December 1914).

33. Baekeland, "Address of Acceptance of Perkin Medal," 155.

34. E.g., Brock to Wickham, 26 September 1912, and Insulating Dept. Westinghouse Electric & Manufacturing Company to Seabury, 22 May 1916; Seabury, handwritten note on letter from Baekeland of 18 April 1912; Anon., "Historical Report Boonton Mfg.," April 1918, 3 (all in HDC, series 3, box 3).

35. Aiken, report for Dyer, 25 October 1910.

36. Dyer to Baekeland, 3 January 1911, in TEP, reel 217, frames 569–571 (at 569).

37. Charles Steinmetz, as quoted in Kline, *Steinmetz*, 155. See also Hemming, *Molded Electrical Insulation*, 85.

38. The tariff commission's annual reports on synthetic organic chemicals are available at www.archive.org.

39. See, e.g., diaries 27 (27 July 1919) and 36 (8 June, 30 July, 22 August, 22 September, and 8 November 1923).

40. Diary 46 (14 September 1928).

41. Foster Sproxton to Baekeland, 8 August 1938, in LBP, series 3, box 17, folder 7.

42. Byck, "A Survey of the Bakelite Thermosetting Business," 16.

43. On the initiatives of one of these brothers, Allan Brown, see Meikle, *American Plastic*; Meikle, "Materia Nova," 38–53; pp. 205–206, 231–232, this volume.

44. George Baekeland, testimony in *Investigation of Concentration of Economic Power*, 1088.

45. Little to Baekeland and Baekeland to Little, 25 and 27 October 1924, in LBP, series 3, box 10, folder 11. See also Baekeland, "Recent Developments in Bakelite," 938; Baekeland, "The Present Status of Phenoplastics in U.S.A.," 1934, 2, in LBP, series 2, box 6, folder 5.

46. Lawrence Byck, as quoted in the anonymous manuscript "Chapter II: The Bakelite Story and Dr. Baekeland," s.d., 26, in LBP, series 1.

47. Diary 2 (25 March 1908).

48. Diary 6 (29 March 1910).

49. See diary 9 (29 February and 22 May 1911).

50. Diary 7 (17 June 1910). In 1919, DeMuth employed about 9,000 workers; see DeMuth, "Industrial Democracy at Work," 175.

51. Diary 10 (15 November 1911; see also entry of 23 November 1911).

52. Diary 11 (31 May 1912). Sum converted to 2018 US dollars on www.measuringworth.com using the Purchasing Power Calculator.

53. Diary 10 (24 and 28 December 1911).

54. See diary 34 (8 April 1922). On the problem of color instability, see also, e.g., diary 9 (22 May, 26 July, 12 and 17 October, and 28 November 1911). For a discussion of General Bakelite's investments in R&D, see pp. 217–223, this volume.

55. Diary 26 (11 February 1919).

56. See *Annual Chemical Directory of the United States, 1918*; 25; *Annual Chemical Directory of the United States, 1919*, 30.

57. Anon., "Chapter II: The Bakelite Story and Dr. Baekeland," 33–34.

58. Ellis, *Chemistry of Synthetic Resins*, 316. See also Ellis, *Synthetic Resins and Their Plastics*, 116.

59. Diary 36 (7 November 1923).

60. [Editors], "'New Era in Field of Phenol Resin.'"

61. See Redman, Weith, and Brock, Phenolic condensation product article and method of manufacture, U.S. Patent 1,424,738, filed 8 September 1921, and issued 1 August 1922; United States Tariff Commission, *Report to the President of the United States*, 17–21.

62. Reiss-Premier Pipe Company et al., "Petition for Change of Classification, Rate of Duty and Basis of Valuation of Bakelite, Redmanol and Condensite Materials and Articles," 1923, in William Burgess Papers, Mudd Library, Princeton University, box 2, folder 5; Anon., "Embargo on Bakelite 'C'"; Anon., "Phenomenal Use of Resinoids."

63. "Summary of Conference in Office Mr. Burgess," 16 July 1923, 4–5, in William Burgess Papers, box 2, folder 5; diaries 10 (4 January 1910), 36 (1 October 1923), 51 (6 June 1932), and 53 (21 June 1933). In December 1929, the Bakelite Corporation's board of directors decided to discontinue the transparent department on account of its subpar performance; see diaries 46 (14 September 1928) and 49 (17 July, 26 September, and 19 December 1929).

64. [Editors], "Unfair Competition in Foreign Trade."

65. The case was summarized in United States Tariff Commission, *Report* to the President of the United States. It also received relatively broad coverage in the journal *Plastics*.

66. Diary 42 (23 May 1926).

67. Lawrence Byck, as quoted in Anon., "Chapter II: The Bakelite Story and Dr. Baekeland," 26.

68. Bijker, *Of Bicycles, Bakelites, and Bulbs*, 175–176; Meikle, *American Plastic*, 57.

69. Pelikan, "The Plastics Industry Abroad," 705.

70. M. L. Hamlin to Baekeland, 1 February 1916, in LBP, series 3, box 17, folder 21.

71. Diary 3 (18 and 24 March 1909).

72. See, e.g., Bijker, *Of Bicycles, Bakelites, and Bulbs*, 109–110; Chandler, "Presentation Address," 156; Friedel, *Pioneer Plastic*, 13; Meikle, *American Plastic*, 10.

73. Friedel, *Pioneer Plastic*, 77–78.

74. A. E. Buttner to Baekeland, 12 August 1932, in LBP, series 2, box 6, folder 7.

75. On the importance of honor and integrity to scientific and other professionals, see Mercelis, "The Scientist and the Advertisement," and the additional references there.

9 A Mysterious Macromolecule

1. Haber, *The Chemical Industry*, 5–6. Haber offered no evidence to support this interpretation.

2. Redman, "Research as a Fixed Charge," 114.

3. Mokyr, *The Gifts of Athena*, 109.

4. Quoted from Hounshell and Smith, *Science and Corporate Strategy*, 225.

5. Megson, "Aldehyde Phenolic Condensations," 336.

6. Megson, "Aldehyde Phenolic Condensations," 336; Carswell, *Phenoplasts*; Martin, *The Chemistry of Phenolic Resins*. On the different meanings of "fundamental research" in the interwar period, see Cerveaux, "Taming the Microworld"; Clarke, "Pure Science with a Practical Aim."

7. Hultzsch, *Chemie der Phenolharze*, 1 (emphasis in original).

8. Baekeland, "Chemical Constitution," 509; International Congress of Pure and Applied Chemistry, *Eighth International Congress of Applied Chemistry*, vol. 27, 73.

9. Hounshell and Smith, *Science and Corporate Strategy*, 223, 290. For a more recent example, see Mody, *Instrumental Community*, 64, 83–84, 127–128.

10. Budinksi and Budinski, *Materiaalkunde voor technici*, 138–140; Carswell, *Phenoplasts*, 12; Wakeman, *The Chemistry of Commercial Plastics*, 117–118; Hultzsch, *Chemie der Phenolharze*, 106–107; interview with Rupert B. Lowe, 18 November 1976, in LBP, series 12, box 46.

11. Hultzsch, *Chemie der Phenolharze*, 5–6. See also Lebach, "Bakelite and Its Applications," 559, 561; Martin, *The Chemistry of Phenolic Resins*, 119; Megson, "Aldehyde Phenolic Condensations."

12. Baekeland, "Synthesis, Constitution and Uses of Bakelite," 160; Baekeland, "Chemical Constitution"; Baekeland and Bender, "Phenol Resins and Resinoids," 225; Harvey and Baekeland, "Further Studies on Phenolic Hexamethylenetetramine Compounds," 135.

13. See p. 19, this volume.

14. Redman, Weith, and Brock, "Synthetic Resins," 12 (see also 7 and 14).

15. In November 1910, Baekeland wrote that the hexamine method "has never appealed to me because it is too expensive but I think we better be on [the] safe side." See diary 8 (29 November 1910 and also 28 November, 3–5 and 17 December 1910). However, his reservations about hexamine's utility later disappeared.

16. See Chatfield, ruling in *General Bakelite v. General Insulate*, in LBP, series 9, box 36, folder 1, 16; Baekeland, Reaction product of hexamethylenetetramin and

phenolic bodies and method of making same, US Patent 1,187,230, filed 13 December 1910, and issued 13 June 1916.

17. Baekeland, "Synthetic Phenol Resins," 569.

18. Baekeland, "Recent Developments in Bakelite," 932–933.

19. Baekeland, "Phenol-Formaldehyde Condensation Products," 737.

20. See diary 10 (21 March 1912); Aylsworth, Phenolic condensation product and method of preparing same, US Patent 1,020,593, filed 11 February 1910, and issued 12 March 1912.

21. Rassow, "Was haben wir Chemiker," 705.

22. Diary 11 (2 September 1912). On 14 October 1912, Baekeland "got news that Raschig controversy before German patent office has been decided in my favor." See entry for that day in diary 11.

23. International Congress of Pure and Applied Chemistry, *Eighth International Congress of Applied Chemistry*, vol. 27, 73. See also Raschig, "Teerphenole."

24. Diary 11 (2 September 1912).

25. Diary 11 (13 October 1912; see also entries of 2 and 9–11 September 1912 in same diary); diary 12 (13–15 February 1913).

26. Baekeland, "Phenol-Formaldehyde Condensation Products," 737. In April 1913, Lebach corroborated Baekeland's position on this point in a paper of his own; see Lebach, "Bakelite and Its Applications."

27. Diary 27 (30 April 1919).

28. Baekeland, testimony in *General Bakelite v. General Insulate*, in LBP, series 9, box 36, folder 2, 82, 93. Lee had asked whether "the methylen radical lets go of the nitrogen and grabs on to the phenol."

29. Baekeland, testimony in *General Bakelite v. General Insulate*, 123.

30. Baekeland, testimony in *General Bakelite v. General Insulate*, 125; Bijker, *Of Bicycles, Bakelites, and Bulbs*, 115–116; Furukawa, *Inventing Polymer Science*, 17–19.

31. Chatfield, ruling in *General Bakelite v. General Insulate*, 2, 7. On Chatfield, see Swayne, *Twenty-Five Year Record, Class of Ninety-Three, Yale College*, 174–175.

32. See Baekeland, Condensation product and method of making same, US Patent 942,809, filed 15 October 1907, and issued 7 December 1909; Baekeland, Process of making insoluble bodies derived from phenol alcohols, US Patent 1,038,475, filed 4 October 1909, and issued 10 September 1912.

33. See Redman, Weith, and Brock, "Synthetic Phenol Resins."

34. Chatfield, ruling in *General Bakelite v. General Insulate*, 20.

35. Chatfield, ruling in *General Bakelite v. General Insulate*, 16–17.

36. Stieglitz, "Chairman's Address," 504 (emphasis in original).

37. E.g. Kline, "Construing 'Technology' as 'Applied Science'"; Gooday, "Vague and Artificial."

38. Diary 13 (16 May 1913); Noyes, "Julius Stieglitz," 289; Stieglitz, "Chairman's Address," 504.

39. Diary 53 (13 September 1933).

40. Baekeland to Fredericq, undated (ca. February 1913).

41. Diary 13 (16 May 1913; see also entries of 8, 10, 12, and 15 May).

42. [Editors], "Willard Gibbs Medal Award," 503.

43. Diary 13 (16 May 1913).

44. Baekeland, "Chemical Constitution," 507.

45. Baekeland, testimony in *General Bakelite v. General Insulate*, 97.

46. See Anon., "Willard Gibbs Medal Award" (1916); Anon., "Willard Gibbs Medal Award" (1920).

47. Hixson, "History of the Department of Chemical Engineering in Columbia University," 29; General Bakelite Company, "Minutes of Boards of Directors Meetings, 1910–1919," 285. In the early 1930s, this amount was reduced to $1,000; see diaries 49 (17 June, 13 July, 17 August, and 16 September 1931) and 50 (6 April 1932).

48. Anon., "Columbia University General Bakelite Company Fellowship." On Du Pont's fellowship program, see Hounshell and Smith, *Science and Corporate Strategy*, 287–293; Steen, *American Synthetic Organic Chemicals Industry*, 261. On the Mellon Institute, see Servos, "Changing Partners," 221–230; Spero, "An Entrepreneurial Opportunity in Process"; Mees, *Organization of Industrial Scientific Research*, 49; Bacon, "Industrial Fellowships of the Mellon Institute." For examples of fellowships that were introduced shortly before General Bakelite's, see Anon., "Research Fellowship State College of Washington"; University of Michigan, "Fellowships in Chemical Engineering."

49. Besides the articles referred to in table 9.3, see Bender, "*Phenol Resins and Resinoids*"; Harvey, "*Further Studies on Phenolic Hexamethylenetetramine Compounds*."

50. Allen, Meharg, and Schmidt, "Chemistry of Synthetic Varnish Resins," 664.

51. Bogert, "Discussion," 446 (emphasis in original); diary 44 (28 October and 21 November 1927).

52. Diaries 28 (19 November 1919) and 34 (28 September and 10 November 1922).

53. For a rare historical study of chemical coauthorship, see Nye, "Mine, Thine, and Ours."

54. Patterson, *Prehistory of Polymer Science*, 17; Megson, "Aldehyde Phenolic Condensations," 336; Weith, "Polymerization and Synthetic Resins," 205.

55. William Hale to Baekeland, 28 July 1925, in LBP, series 3, box 17, folder 14; Barry, Drummond, and Morrell, *Chemistry of the Natural & Synthetic Resins*, 126–132.

56. Diaries 34 (27–28 September, 10 November, and 6 and 10 December 1922), 36 (7 June 1923), and 46 (14 June 1929).

57. Diaries 34 (6 December 1922) and 36 (13 June 1923).

58. Hounshell, "Evolution of Industrial Research," 27–28; Hounshell and Smith, *Science and Corporate Strategy*, 228–229; Furukawa, *Inventing Polymer Science*, 152–161; Kline and Lassman, "Competing Research Traditions," 617–618, 621–623; Wise, *Whitney*, 80–87, 165; Shapin, *The Scientific Life*, chaps. 4 and 5. For earlier complaints regarding the career prospects for industrial research chemists, see, for instance, Rassow, "Was haben wir Chemiker," 706; Duncan, "Problems Connected with the Present-Day Relation between Chemistry and Manufacture."

59. Diary 36 (3 July 1923).

60. E.g., Anon., "Continuing Research Clarifies Polymer Formation and Behavior," 2771–2772. For further indications that Bender's work was much appreciated by his professional peers, see Martin, *The Chemistry of Phenolic Resins* (e.g., ix).

61. Cerveaux, "Taming the Microworld," 273; Ede, *Rise and Decline of Colloid Science*, i. On Baekeland's interest in colloid chemistry and his personal acquaintance with leading colloid chemists such as Wilder Bancroft and Jerome Alexander, see diaries 20 (26 September 1916) and 38 (15 and 18 September and 16 November 1924) and letter to Jean Gérard, 9 August 1933, in LBP, series 3, box 17, folder 19.

62. LBP, series 3, box 10, folder 7; Ulrich, *Macromolecular Science*, 124–125. On the innovative format of these meetings, see also Daemmrich and Shaper, "The Gordon Research Conferences as Scientific Infrastructure."

63. Baekeland and Bender, "Phenol Resins and Resinoids," 225; Mory, "Synthetic Resins," 1106.

64. Diary 47 (18 June 1930).

65. Turkington, Shuey, and Butler, "Phenol Resinoids in Oil Varnishes," 1177.

66. Diary 48 (9 July 1930). On the expansion of Bakelite's trademark, see, for instance, [Editors], "Trade Names"; Tate, "Protective Power of a Good Trade Mark."

67. Diary 57 (22 July 1936).

68. Baekeland to Neil Gordon, 27 January 1936; Bender to Gordon, 23 February 1939 (both in LBP, series 3, box 10, folder 7).

69. E.g., diary 52 (19 and 26 February 1933).

70. Diary 53 (30 August and 12 October 1933); Bender, Moldable composition and its product, US Patent 1,944,143, filed 28 February 1929, and issued 16 January 1934.

71. Diary 53 (5 December 1933); Bender and Hoffman, Urea resins, US Patent 2,084,314, filed 7 November 1933, and issued 22 June 1937.

72. Shreve and Golding, "Oil-Soluble Phenolic Resins," 134; Turkington and Allen, "Oil-Soluble Phenolic Resins."

73. Shreve and Golding, "Oil-Soluble Phenolic Resins," 134.

74. Diary 49 (3 and 16 September 1931).

75. Diaries 49 (3 September 1931), 50 (30 January and 1, 23, 26, and 27 February 1932), and 52 (29 December 1932). See also diary 50 (30 January 1931).

76. Hounshell and Smith, *Science and Corporate Strategy*, 191–205.

77. Reich, *The Making of American Industrial Research*, 110; Kline and Lassman, "Competing Research Traditions," 636.

78. George Baekeland, testimony in *Investigation of Concentration of Economic Power*, 1089; Rhodes, "The American Way in Industrial Research," 551, 553; Thackray, "University-Industry Connections and Chemical Research," 208, 214–215. The Kansas program was relocated to Pittsburgh, where the Mellon Institute was established in 1913.

79. Rogers et al., "Discussion of Professor Duncan's Paper," 185. For further examples, see Baekeland, "Applied Chemistry," 554; diary 50 (5 March 1932); telegram to Bakelite Corporation, 15 March 1935, in LBP, series 3, box 16, folder 3.

80. Allen, Meharg, and Schmidt, "Chemistry of Synthetic Varnish Resins" (quotation at 664).

81. On the Madrid conference, see LBP, series 3, box 17, folder 6. For more information on the publication of a revised version of Baekeland's paper, see his correspondence with Howe in LBP, series 3, box 16, folder 3.

82. Baekeland, "The Present Status of Phenoplastics in U.S.A.," 1–2; correspondence with Weith in LBP, series 3, box 16, folder 3.

83. Baekeland to Staudinger, 4 August 1936, in LBP, series 3, box 16, folder 3, and list of contacts in the same folder.

84. Weith, "Polymerization and Synthetic Resins," 202; Martin, *The Chemistry of Phenolic Resins*.

10 A Company of Scientists and Engineers

1. Baekeland to Hooker, 30 June 1909, in ESP, accession 1897, box 17 (electrochemistry files).

2. Diary 42 (2 June 1926). See also, e.g., diaries 47 (19 June 1930), 48 (5 August and 17 September 1930), 49 (11 December 1931), and 53 (19 September 1933).

3. For a variety of examples, see Dick and Jones, "The Commercialization of Molecular Biology," 139–140; Marsch, *Zwischen Wissenschaft und Wirtschaft*, 183; Thackray et al., *Chemistry in America*, 93; Auletta, *Googled*, xii. Additional references are given later in this chapter.

4. Baekeland to De Waele, 8 October 1928.

5. A classic analysis of this development is Chandler, *Scale and Scope*.

6. Friedman, *Birth of a Salesman*, 4.

7. Engel, "Coloring the World"; Murmann, *Knowledge and Competitive Advantage*, 147; Wetzel, *Naturwissenschaften und chemische Industrie*, 221–226. See also Wengenroth, "Germany," 148.

8. E.g., Cortada, *Before the Computer*, 68; Chandler, *The Visible Hand*, 302–314, 353–355.

9. Anon., "History of The R. & H. Chemicals Department," in Du Pont & Co. Records, Hagley Museum and Library, Wilmington, DE (hereafter DR), accession 500, box 1658, 2.

10. Interview with Lowe, 18 November 1976, in LBP, series 12, box 46.

11. Fin Sparre, "Report on Accomplishment of Objectives," 30 June 1932, 6–7, in DR, accession 500, box 1658; Fin Sparre, "Roessler & Hasslacher Chemical Company" (report of development department to executive committee, 24 February 1930), 8–9, in DR, accession 1813, box 5, folder 15 (Records of Willis F. Harrington).

12. For an overview of the firm's history up to World War I, see Mercelis, "Jacob Hasslacher."

13. Sparre, "Roessler & Hasslacher Chemical Company," exhibit C.

14. Diary 50 (27 March 1932). The same sum is mentioned in diary 49 (30 July 1931). In 1928, on the other hand, Baekeland referred to annual research expenditures of $400,000; see diary 45 (3 August 1928).

15. Diary 51 (7 June and 24 August 1932).

16. Diary 52 (22 July 1933; see also entry of 20 July 1933).

17. Hounshell and Smith, *Science and Corporate Strategy*, 208, 612.

18. Diary 43 (27 April 1927). Mees made the same analogy in a better-known 1935 speech.

19. E.g., diary 41 (16 June 1925).

20. E.g., diaries 45 (27 July and 3 August 1928), 48 (19 October 1930), and 49 (30 July 1931).

21. Baekeland to De Bruyne, 5 November 1930, in LBP, series 3, box 8, folder 16. According to the 1931 *Bulletin of the National Research Council*, the Bakelite laboratories employed 53 technically trained workers and an assisting staff of 117. For earlier data, see also diary 42 (7 September 1926).

22. Diary 45 (10 March 1928).

23. See Baekeland's correspondence with Hays in LBP, series 3, box 14, folders 18 and 19, and series 3, box 9, folder 11.

24. Diary 45 (11 March 1928). On the adoption of mass manufacturing methods, see Steen, *American Synthetic Organic Chemicals Industry*, 267–268, 286.

25. Diary 45 (18 May 1928).

26. Diary 48 (10 February 1931).

27. Diary 49 (17 June and 24 July 1931).

28. Diary 53 (22 July 1933).

29. Diaries 41 (18 November 1925) and 49 (5 August 1931).

30. Diaries 49 (24 July 1931) and 51 (7 September 1932) (emphasis mine); p. 211, this volume.

31. Diary 51 (2 August 1932).

32. See Anon., "A. V. H. Mory Leaves for Cincinnati"; Anon., "Old Friends: A. V. H. Mory"; diary 36 (9 November 1923); Goldstein, *Creating Consumers*, 196–197.

33. Diaries 37 (30 November 1923), 45 (22 January 1928), and 51 (2 August 1932).

34. Mory, "Nature Was Notably Aided." See also, e.g., Redman and Mory, *The Romance of Research*.

35. Mercelis, "The Scientist and the Advertisement."

36. Dienel, *Ingenieure zwischen Hochschule und Industrie*, 546. See also Mercelis, "The Scientist and the Advertisement," and the references given there.

37. Diary 8 (24 September 1910); Wolf, *Im Zeichen von Sonne und Mond*, 80–82.

38. Bakelite Corporation, "Read What One of America's Most Noted Radio Scientists Says about BAKELITE," s.d. [1924], in HDC, series 1, box 1, folder 2; diaries 37 (26 June 1924) and 38 (22 November 1924).

39. E.g., United States Tariff Commission, *Census of Dyes and Coal-Tar Chemicals, 1919*, 51. In 1922, the title of this publication changed to *Census of Dyes and of other Synthetic Organic Chemicals*.

40. To be precise, 113 out of 744 employees (i.e., 15.2 percent) were under contract. See diary 49 (14 October 1931). By 1935, the Bakelite Corporation had six affiliate firms in the United States: Bakelite Dental Products, Incorporated, Bakelite Building Products, Incorporated, Bakelite-Rogers Company, Incorporated, the Autopoint Company, Floor Covering Patents, Incorporated, and the Halowax Corporation; see Anon., "History of the Bakelite Corporation," 18. Baekeland did not specify which Bakelite affiliates were included in his count.

41. Diary 26 (13 February 1919). See also diaries 28 (26 and 28 January 1920), 30 (1 October 1920), 31 (17 and 21–27 November 1920 and 24–28 January 1921), 32 (11 August 1921), and 34 (25 May 1922); Byck, "A Survey of the Bakelite Thermosetting Business," 13.

42. Diary 36 (8 June and 22 August 1923); General Bakelite Company, "Minutes of Boards of Directors Meetings, 1919–1939," in LBP, series 8, box 34, folder 7, 60.

43. Diary 47 (29 August 1929; see also entry of 18 June 1930).

44. George Baekeland, testimony in *Investigation of Concentration of Economic Power*, 1087; Smith, *From Monopoly to Competition*, 83–84, 93.

45. See Mercelis, "Jacob Hasslacher."

46. See, for instance, diaries 41 (1 October 1925) and 46 (4 March 1929). On the social status of salesmen in nineteenth- and twentieth-century America, see, e.g., Friedman, *Birth of a Salesman*; Greene, "Attention to 'Details.'"

47. Diaries 33 (3 January 1922) and 47 (20 December 1929).

48. On this argument, see, e.g., Fear, "Constructing Big Business," 552, 562, 567.

49. Diary 47 (19 July and 10 September 1929).

50. Diary 47 (1 November 1929). See also the entry of 18 June 1930 in the same diary and diary 57 (3 January 1937).

51. Baekeland to Emile de Cartier de Marchienne, 30 June 1935, in LBP, series 2, box 7, folder 9; diary 47 (23 October 1929). See also, e.g., diary 43 (10 November 1926).

52. Diary 53 (18 June 1933). See also diary 50 (24 December 1931).

53. Diaries 58 (6 September 1938) and 59 (23 October 1938); H. V. Potter to Baekeland, 3 October 1938, in LBP, series 8, box 15, folder 4.

54. Diary 52 (16 June 1933). See also the entry of 18 June 1933 in the same diary and diaries 42 (20 October 1926), 43 (10 November 1926), and 53 (30 June 1933).

55. Diaries 53 (20 and 26 June 1933) and 54 (28 June and 18 July 1934).

56. Reimer, "Bayer & Company in the United States," 69–72.

57. Diary 51 (24 August 1932).

58. See Anon., "Molding Material Salesmen," *Trefoil* 3, no. 4 (April 1938): 10–11, 13 (in HDC, series 1, box 1, folder 2); interview with Charles Norris, 1952, in Plastics Pioneers Association Interviews, Archives Center, NMAH.

59. See Baekeland to Wiechmann, 2 September 1909, in LBP, series 8, box 15, folder 14; diary 3 (24 February 1909); Anon., "Hylton Swan."

60. Wiechmann to Baekeland, 15 March 1910, in LBP, series 8, box 15, folder 15; General Bakelite Company, "Minutes of Boards of Directors Meetings, 1910–1919," 75.

61. Anon., "Plastics in the News," 290; diaries 51 (15 September 1932) and 52 (18 May 1933).

62. Diary 48 (22 August 1930). On the disappointing outcome of this project, see Baekeland, "The President's Report to the Directors of Bakelite Corporation at the Directors' Meeting of December 20, 1934," in LBP, series 8, box 15, folder 1.

63. Meikle, *American Plastic*, 104–118; Meikle, "Materia Nova."

64. Diaries 36 (12 July 1923) and 55 (5 September 1935).

65. Anon., "ADVERTISING," 632; [Editors], "If We Were Giving Medals," 27; interview with Allan Brown, 1952, in Plastics Pioneers Association Interviews.

66. Diaries 57 (31 December 1936) and 36 (29 July 1923).

67. Baekeland, "Address of Acceptance of Perkin Medal," 156.

68. Diaries 34 (7–8 December 1922), 37 (28 November 1923), 38 (23 October and 15 December 1924), 41 (17 December 1925), 43 (27 December 1926 and 12 March 1927), and 44 (14 December 1927). See also diary 53 (15–16 November 1933).

69. Diary 45 (5 March 1928).

70. Diary 45 (11 March 1928).

71. See United States Tariff Commission, *Census of Dyes and of other Synthetic Organic Chemicals, 1928*, 77.

72. See Albert Frankel, "Report of Treasurer," 10 March 1930, and Lansing Hayes [Hays], Randolph Paul, and George E. Warren, "The Roessler & Hasslacher Chemical Company: Report of Committee on Bonus Distribution for the Year 1929," 13 January 1930 (both in DR, accession 1813, box 5, folder 13 [Harrington records]). See also Mercelis, "Jacob Hasslacher," and references given there.

73. Diary 43 (20 March 1927).

74. Diaries 43 (6 March 1927), 45 (5–6, 8, 11–13, 29, and 31 March 1928), and 46 (10 December 1928).

75. Diary 45 (3–9, 18, 21–24, and 28–29 August 1928). Quotation from diary 43 (24 March 1927).

76. Diary 46 (6 December 1928).

77. Diaries 46 (19–20 March 1929), 49 (11 and16 December 1930), and 50 (17, 24, and 28 December 1930 and 5 and 14 January 1931).

78. Diaries 59 (20 and 23 September and 2 and 8 November 1939) and 62 (18 September 1942); Cohoe, "Leo Hendrik Baekeland," 231; Steen, *American Synthetic Organic Chemicals Industry*, 281.

79. Mitchell, *The Speculation Economy*.

80. E.g., Steen, *American Synthetic Organic Chemicals Industry*, 272.

81. Diary 42 (7 September 1926). See also diary 36 (24 September 1923); Wolfrom, "Harold Hibbert," 154; Steen, *American Synthetic Organic Chemicals Industry*, 250–255.

82. Jacobsen, "Signs of Progress"; interview with Rupert B. Lowe, 18 November 1976, in LBP, series 12, box 46.

83. Rugen, "The 'X' in Research."

84. Bakelite Corporation, "-BUT we've always done it that way" brochure, 1927, in Trade Literature Collection, NMAH.

85. Baekeland to Messrs. A. N. Marquis Company, 10 September 1929, in LBP, series 3, box 8, folder 14.

86. E.g., diary 59 (2, 3, and 5 October 1938). The idea of a merger with American Cyanamid had come up repeatedly since 1930. Other firms that showed at least a limited interest in acquiring the Bakelite Corporation included Du Pont and Monsanto.

87. E.g., Hounshell, "Evolution of Industrial Research," 34–41; Usselman, "Research and Development."

Afterword

1. Céline Swarts, "Dr. L. H. Baekeland," in LBP, series 1, box 1, folder 7. For a classic claim about independence as an entrepreneurial motive, see Schumpeter, *Theory of Economic Development*, 93. For more recent evidence, see, e.g., Chell, *The Entrepreneurial Personality*, 133–134; Gerwen and de Goey, *Ondernemers in Nederland*, 13.

2. Diary 62 (18 September 1942); Anon., "George Baekeland Is Dead." See also diaries 59 (29 November 1939) and 62 (21 August 1942).

3. For different perspectives on this development, see, e.g., Beauchamp, *Invented by Law*; Giffard, *Making Jet Engines*; Hintz, "Portable Power"; Hughes, *American Genesis*; Lamoreaux and Sokoloff, "Decline of the Independent Inventor."

4. Anon., "Grasselli Medal Award."

5. Miller, "Political Economy of Discovery Stories." On the cultural and other dynamics contributing to changes in the reputations of inventors and scientists, see, e.g., MacLeod, *Heroes of Invention*; Swanson, "'Great Men.'" On the collective nature of technological innovation and entrepreneurship, see, e.g., Brandt, "Envisioning a National Infrastructure for Science"; Caron, "L'innovation collective"; Galambos and Sewell, *Networks of Innovation*; Wadhwani and Lubinski, "Reinventing Entrepreneurial History," 778–779.

6. Furukawa, *Inventing Polymer Science*; Hollestelle, "Academische polymeerwetenschap."

7. Anon., "German, Dutchman Given Nobel Prizes"; Anon., "Staudinger Wins Chemistry Prize."

8. E.g., Galison and Hevly, *Big Science*; Leslie, *The Cold War and American Science*; Mirowski, *Science-Mart*. Other authors have identified the 1930s as an important period of transition. See, e.g., Kargon and Hodes, "Karl Compton, Isaiah Bowman, and the Politics of Science"; Owens, "MIT and the Federal 'Angel.'"

9. On the role of the state in shaping entrepreneurial processes in France and Germany, see, e.g., Galvez-Behar, "Institutional Enterprise as a Compromise"; Lenoir, "Revolution from Above." On Terman's and Roberts's successes in securing government funding, see, respectively, Lowen, *Creating the Cold War University*; Bassi, *A Scientific Peak*.

10. For different perspectives on these and related changes in the United States, see, e.g., Berman, *Creating the Market University*; Mirowski, *Science-Mart*; Mody, *The Long Arm of Moore's Law*; Mowery et al., *Ivory Tower and Industrial Innovation*. For a transatlantic comparison, see, e.g., Teixeira, "Two Continents."

11. E.g., Martin, "Are Universities and University Research under Threat?" and, from a different perspective, Mowery, *Plus ça change*. For recent reviews of parts of the extensive historical literature concerning these themes, see Mercelis, Galvez-Behar, and Guagnini, "Commercializing Science"; Wadhwani et al., "Academic Entrepreneurship."

12. Pestre, "Scientists in Time of War," offers a comparative evaluation of the impact of World War II and the Cold War in France and the United States.

13. This is partly because the literature on academia-industry collaboration in this period has focused on a limited number of institutions, most notably MIT and Stanford.

14. On the contemporary relationships between patents and publications, see Stephan, *How Economics Shapes Science*, 57, and the references given there. On the

history of the distinction between patentable inventions and unpatentable discoveries in US patent law, see Simon, "Reinventing Discovery." In the aftermath of World War I, this distinction received special scrutiny from members of an international movement for the protection of "scientific property"; see, e.g., Miller, "Intellectual Property and Narratives of Discovery/Invention."

15. Daston and Sibum, "Introduction"; Shapin, *The Scientific Life*, chap. 7.

16. The Mellon Institute, however, was much less successful in bringing about such a reorientation than MIT had been in the 1930s. See Lécuyer, "The Making of a Science Based Technological University"; Servos, "Changing Partners"; Fishman, "MIT Patent Policy."

17. On the origins and initial meanings of the term "technoscience," see, e.g., Bensaude-Vincent, "Technoscience and Convergence"; Forman, "Primacy of Science," 5–6. Some historians have traced the origins of "technoscience" to the late nineteenth and early twentieth centuries; see, e.g., Channell, *A History of Technoscience*. In contrast, the approach of this book has been to explore the significance and changing meanings of older concepts such as (pure/applied) "science," "technology," and "research" in relation to an expanding science-industry nexus.

18. Bidwell (Baekeland's secretary) to Whitmore, 16 December 1927, in LBP, series 3, box 17, folder 14.

Bibliography

Archives and Papers

Belgian State Archives in Ghent
Archives of Teacher Training College of Bruges, 1879–1888

British Library, London
Kodak Historical Collection (file A3022)

Evonik Industries AG, Corporate Archives Hanau (Holdings Degussa AG)
DL 3. Schneider/2

George Eastman Archives and Study Center (GEASC), George Eastman House, Rochester, NY
George Eastman Papers

Ghent University Archives, Ghent
4 A2/4, boxes 39, 46, 66, and 72

Ghent University Library
Correspondence of Paul Fredericq

Ghent University Museum
Baekeland, correspondence with Edouard Remouchamps and Florence Boonroy

Hagley Museum and Library, Wilmington, DE
Du Pont & Company Records (DR) (accession 500, box 1658; accession 1676, box 6; accession 1813, box 5)

Elmer A. Sperry Papers (ESP) (electrochemistry files)

Library of Congress, Washington, DC

The Wilbur and Orville Wright Papers (box 12)

Mudd Library, Princeton University, Princeton, NJ

William Burgess Papers (box 2)

National Academies Archives (NAA), Washington, DC

Central Policy Files, Committee on Chemistry

Central Policy Files, Committee on Patent Office

National Museum of American History (NMAH), Smithsonian Institution, Washington, DC

Archives Center

Leo H. Baekeland Papers (LBP)
- Audio materials (series 12)
- Business and legal records (series 8)
- Correspondence (series 2, 3, and 8)
- Diaries (series 4)
- Laboratory notes (series 6)
- Patents (series 9)
- Published and unpublished writings (series 2)
- Reading and lecture notes (series 5)
- Reference materials (series 1–3)

Albany Billiard Ball Company Records

J. Harry DuBois Collection on the History of Plastics (HDC)

Plastics Pioneers Association Interviews

Trade Literature Collection

Bakelite Corporation brochures

Rare Book and Manuscript Library, Columbia University, New York, NY

Charles F. Chandler Papers (CCP) (boxes 1 and 241)

Historical Subject Files

Rutgers, the State University of New Jersey, Piscataway, NJ

Thomas A. Edison Papers (TEP) (microfilm reels 217 and 266)

Science History Institute, Philadelphia, PA

Dow Collection (files 150017, 150018, 150065, 180147, and 250049)

US National Archives

At College Park, MD

RG 241: Records of the Patent and Trademark Office

At New York, NY

General Bakelite Co. v. George J. Nikolas, trial transcripts (equity dockets 1–99).

United States v. Eastman Kodak, US Circuit Court, Western District of New York, 1915 (equity docket A-51).

At Washington, DC

RG 80: Records of Naval Consulting Board

Secondary Literature and Published Sources

Abney, William de W. *Photography with Emulsions*. New York: Scovill Manufacturing, 1886.

Adas, Michael. *Dominance by Design: Technological Imperatives and America's Civilizing Mission*. Cambridge, MA: Belknap Press of Harvard University Press, 2006.

Addison, Cyril C., and Norman Logan. "Anhydrous Metal Nitrates." In *Advances in Inorganic Chemistry and Radiochemistry*, vol. 6, edited by Harry J. Emeléus and Alan G. Sharpe, 71–142. New York: Academic Press, 1964.

Albertina Museum, Vienna. "Biobibliography on Photography in Austria." http://fotobiobibliografie.albertina.at/d/fotobibl/einstieg.html.

Alexander, Jennifer K. *The Mantra of Efficiency: From Waterwheel to Social Control*. Baltimore: Johns Hopkins University Press, 2008.

Allen, Ivey, Jr., Virgil E. Meharg, and John H. Schmidt. "Chemistry of Synthetic Varnish Resins." *Industrial and Engineering Chemistry* 26 (1934): 661–669.

The American Fireproofing and Cement Construction Company. *The De Man System of Fireproof Construction*. New York: American Fireproofing and Cement Construction Company, 1901.

American Institute of Electrical Engineers. *The Niagara Falls Electrical Handbook*. Niagara Falls, NY: American Institute of Electrical Engineers, 1904.

Anderson, Robert G. W. "Relations between Industry and Academe in Scotland, and the Case of Dyeing: 1760 to 1840." In *Compound Histories: Materials, Governance and Production, 1760–1840*, edited by Lissa Roberts and Simon Werrett, 333–353. Leiden: Brill, 2018.

Anon. "ADVERTISING: The Vital Force in the Plastic Molding Industry and some of Its DIRECTORS." *Plastics and Molded Products* 6 (1930): 632–633, 654–655.

Anon. "Asks Supreme Court to Reverse Itself." *New York Times*, 23 December 1912.

Anon. "A. V. H. Mory Leaves for Cincinnati." *Chemical Bulletin* 7 (1920): 17.

Anon. "Bakelite Suit Decision." *Metallurgical and Chemical Engineering* 13 (1915): 411–414.

Anon. "Black Tones on Velox Paper." *Wilson's Photographic Magazine* 33 (1896): 141–142.

Anon. "Break in Camphor Market." *New York Times*, 9 April 1904.

Anon. "Au Cercle d'Etudes photographiques et scientifiques d'Anvers." *Hélios* 7 (1897): 830–832.

Anon. "Columbia University General Bakelite Company Fellowship." *Journal of Industrial and Engineering Chemistry* 11 (1919): 798.

Anon. "Continuing Research Clarifies Polymer Formation and Behavior." *Chemical and Engineering News* 31 (1953): 2771–2772.

Anon. *Dubbele wegwyzer der stad Gent en der provincie Oost-Vlaenderen*. Ghent: Vanderhaeghen, 1878 and 1888.

Anon. "Ecole royale et impériale de Photographie de Vienne." *Bulletin de l'Association belge de photographie* 15 (1888): 205–211.

Anon. "Embargo on Bakelite 'C.'" *Plastics* 2 (1926): 268.

Anon. "Father of Plastics." *Time* 35, no. 21 (20 May 1940): 56.

Anon. "Favor Patent Law Revision: Oldfield Bill to Prevent Monopoly Reported to the House." *New York Times*, 9 August 1912.

Anon. "Frédéric Swarts (1891)." In *Université de Gand: Liber memorialis*. t. 2. Ghent: Vanderpoorten, 1913.

Anon. "George Baekeland Is Dead." *New York Times*, 1 February 1966.

Anon. "German, Dutchman Given Nobel Prizes." *Washington Post*, 5 November 1953.

Anon. "Grasselli Medal Award." *Industrial and Engineering Chemistry* 24 (1932): 112.

Anon. "The History of the Bakelite Corporation." *Bakelite Review* 7 (1935): 5–18.

Anon. "Hylton Swan." *New York Times*, 17 December 1960.

Anon. "Latentes Bild." Jahrbuch für Photographie und Reproduktionstechnik 19 (1905): 363–364.

Anon. "Leonard Jacobi." In *Biographical History of Westchester County, New York*, vol. 2, 494–496. Chicago: Lewis Publishing, 1899.

Anon. "Modern Photo-chemical Processes." *American Druggist and Pharmaceutical Record* 33 (1898): 353–354.

Anon. "Monopoly Victory in Patent Decision." *New York Times*, 12 March 1912.

Anon. "Old Friends: A.V.H. Mory." *Chemical Bulletin* 13 (1926): 48.

Anon. "The Phenol Resin Industry and the Future." *Plastics* 3 (1927): 114.

Anon. "Phenomenal Use of Resinoids." *Plastics* 3 (1927): 566.

Anon., "Plastics in the News," *Plastic Products* 9, no. 7 (September 1933): 284–290.

Anon. "Printing Methods and Formulae." *Photographic Mosaics* 23 (1896): 58–61.

Anon. "Protokoll der Plenar-Versammlung vom 7. December 1880." *Photographische Korrespondenz* 17 (1880): 246.

Anon. *Rapport triennal sur l'état de l'enseignement moyen en Belgique, présenté aux Chambres legislatives*. Brussels: Gobbaerts, 1890.

Anon. "Récompenses." *Bulletin de l'Association belge de photographie* 16 (1889): 694–701.

Anon. "Regular Meeting, Tuesday Evening, September 10th." *American Amateur Photographer* 7 (1895): 467–468.

Anon. "Research Fellowship State College of Washington." *Industrial and Engineering Chemistry* 10 (1918): 753.

Anon. "Scientific Notes and News." *Science* 83, no. 2158 (8 May 1936): 430–433.

Anon. "Self-Developing Plates." *American Amateur Photographer* 1 (1889): 79–80.

Anon. *Situation de l'enseignement supérieur donné aux frais de l'état: rapport triennial, 1886–88*. Brussels: Gobbaerts, 1889.

Anon. "Societies and Academies." *Science* 19, no. 471 (8 January 1904): 67–74.

Anon. "Society of Chemical Industry: The Annual Meeting at Liverpool." *Chemical Age* 35 (1936): 49–50.

Anon. "Staudinger Wins Chemistry Prize; Zernike, Physics." *Christian Science Monitor*, 5 November 1953: 12.

Anon. "U.S. National Academy of Sciences: New Members." *Nature* 137 (1936): 859.

Anon. "Water-Developing Plates." *Photographic Times* 22 (1890): 323.

Anon. "Willard Gibbs Medal Award." *Journal of Industrial and Engineering Chemistry* 8 (1916): 559–564.

Anon. "Willard Gibbs Medal Award." *Journal of Industrial and Engineering Chemistry* 12 (1920): 697–700.

Antheunis, Georges. *Leo Baekeland*. Ghent: Stad Gent, Dienst Leefmilieu, 1988.

Antheunis, Georges, Guido Deseyn, and Marc Van Gijsegem. *Focus op fotografie: Fotografie te Gent van 1839–1940*. Ghent: MIAT, 1987.

Anthony, Richard A. "The Discovery of Dr. Baekeland." *Journal of Industrial and Engineering Chemistry* 8 (1916): 182–183.

Apple, Rima D. "Patenting University Research: Harry Steenbock and the Wisconsin Alumni Research Foundation." *Isis* 80, no. 3 (1989): 374–394.

Aquinas, Thomas. "Printing Room Particulars." *Photographic Mosaics* 23 (1896): 68–69.

Arapostathis, Stathis, and Graham Dutfield, eds. *Knowledge Management and Intellectual Property: Concepts, Actors and Practices from the Past to the Present.* Northampton, MA: Edward Elgar, 2013.

Arapostathis, Stathis, and Graeme Gooday. *Patently Contestable: Electrical Technologies and Inventor Identities on Trial in Britain.* Cambridge, MA: MIT Press, 2013.

L'Association belge de photographie (ABP): Section de Bruxelles. "Séance du 12 décembre 1888." *Bulletin de l'Association belge de photographie* 16 (1889): 2–3.

L'Association belge de photographie (ABP): Section de Bruxelles. "Séance du 9 janvier 1889."*Bulletin de l'Association belge de photographie* 16 (1889): 65–67.

L'Association belge de photographie (ABP): Section de Gand. "Séance du 4 avril 1889." *Bulletin de l'Association belge de photographie* 16 (1889): 215–218.

L'Association belge de photographie (ABP): Section de Gand. "Séance du 4 janvier 1889." *Bulletin de l'Association belge de photographie* 16 (1889): 4–5.

L'Association belge de photographie (ABP): Section de Gand. "Séance du 4 octobre 1888." *Bulletin de l'Association belge de photographie* 15 (1888): 516–519.

L'Association belge de photographie (ABP): Section de Gand. "Séance du 7 février 1889." *Bulletin de l'Association belge de photographie* 16 (1889): 72–74.

L'Association belge de photographie (ABP): Section de Liège. "Séance du 16 janvier 1889." *Bulletin de l'Association belge de photographie* 16 (1889): 67–72.

Aylsworth, Jonas W. Phenolic condensation product and method of preparing same. US Patent 1,020,593, filed 11 February 1910, and issued 12 March 1912.

Auletta, Ken. *Googled: The End of the World As We Know It.* New York: Penguin Press, 2009.

Bacon, Raymond F. "The Industrial Fellowships of the Mellon Institute." *Journal of Industrial and Engineering Chemistry* 11 (1919): 371–374.

Baekeland, Leo H. Abrasive composition and method of making same. US Patent 942,808, filed 26 October 1907, and issued 7 December 1909.

Baekeland, Leo H. "The Abuses of Our Patent System." *Journal of Industrial and Engineering Chemistry* 4 (1912): 333–337.

Baekeland, Leo H. "Address of Acceptance of Perkin Medal." *Metallurgical and Chemical Engineering* 14 (1916): 151–158.

Baekeland, Leo H. Antiseptic compound. US Patent 500,549, filed 27 June 1892, and issued 4 July 1893.

Baekeland, Leo H. "Applied Chemistry." *Science* 42, no. 1086 (22 October 1915): 547–555.

Baekeland, Leo H. "Bakelit, ein neues synthetisches Harz." *Chemiker-Zeitung* 35 (1909): 317–318, 326–328, 347–348, 358–359.

Baekeland, Leo H. "'Bakelite,' a Condensation Product of Phenols and Aldehydes, and Its Uses." *Journal of the Franklin Institute 169*, no. 1 (1910): 55–60.

Baekeland, Leo H. "The Carbon Process." *American Amateur Photographer* 4 (1892): 204–206.

Baekeland, Leo H. "The Career of a Research Chemist." *Journal of the Franklin Institute* 230 (1940): 159–161.

Baekeland, Leo H. "The Chemical Constitution of Resinous Phenolic Condensation Products." *Journal of Industrial and Engineering Chemistry* 5 (1913): 506–511.

Baekeland, Leo H. "Comments on the Report of the Investigation of the United States Patent Office." *Journal of Industrial and Engineering Chemistry* 5 (1913): 417–421.

Baekeland, Leo H. Composite cardboard. US Patent 1,019,406, filed 30 November 1910, and issued 5 March 1912.

Baekeland, Leo H. Condensation product and method of making same. US Patent 942,809, filed 15 October 1907, and issued 7 December 1909.

Baekeland, Leo H. Condensation product of phenol and formaldehyde and method of making the same. US Patent 942,700, filed 4 December 1907, and issued 7 December 1909.

Baekeland, Leo H. "Condensation Products of Phenols and Formaldehyde." *Journal of Industrial and Engineering Chemistry* 3 (1911): 518–520.

Baekeland, Leo H. "The Danger of Overspecialization." *Science* 25, no. 648 (31 May 1907): 845–854.

Baekeland, Leo H. Des plaques photographiques développables dans l'eau. Belgian Patent 78,957, filed 22 September 1887.

Baekeland, Leo H. "The Dissociation of Lead Nitrate." *Journal of the American Chemical Society* 26 (1904): 391–399.

Baekeland, Leo H. "Dreams and Realities." *Journal of Chemical Education* 9 (1932): 1000–1009.

Baekeland, Leo H. "Dr. Hesse's Suggestion of Renewal Fees." *Journal of Industrial and Engineering Chemistry* 11 (1919): 938.

Baekeland, Leo H. "Edward Weston's Inventions." *Science* 41, no. 1057 (2 April 1915): 484–492.

Baekeland, Leo H. *Electrolytic Action of Metallic Particles in Sensitized Papers*. Berlin: Deutscher Verlag, 1904.

Baekeland, Leo H. Electrolytic diaphragm and method of making same. US Patent 855,221, filed 10 March 1906, and issued 28 May 1907.

Baekeland, Leo H. "The Engineer; Human and Superior Direction of Power." *Science* 54, no. 1041 (4 November 1921): 417–424.

Baekeland, Leo H. Improved photographic plate to be developed in water. British Patent 1201–1888, filed 26 January 1888.

Baekeland, Leo H. "The Incongruities of American Patent Litigation." *Journal of Industrial and Engineering Chemistry* 4 (1912): 785–789.

Baekeland, Leo H. Indurated product and method of preparing same. US Patent 942,852, filed 13 July 1907, and issued 7 December 1909.

Baekeland, Leo H. "La photorétrogression ou la disparition de l'image photographique." Moniteur scientifique du Docteur Quesneville 17 (1903): 808–811.

Baekeland, Leo H. "Les bains d'hydroquinone." *Bulletin de l'Association belge de photographie* 16 (1889): 495–504.

Baekeland, Leo H. *A Method for Determining the Relative Permanency of Photographic Prints*. Berlin: Deutscher Verlag, 1904.

Baekeland, Leo H. Method of making insoluble products of phenol and formaldehyde. US Patent 942,699, filed 13 July 1907, and issued 7 December 1909.

Baekeland, Leo H. "The New Electrolytic Alkali-Works at Niagara Falls." *Electrochemical and Metallurgical Industry* 5 (1907): 209–212.

Baekeland, Leo H. "New York Section of the American Chemical Society: Introductory Address by the Chairman," *Science* 28, no. 728 (11 December 1908): 817–820.

Baekeland, Leo H. Nouvelles plaques photographiques développables dans l'eau, French Patent 188,336, issued 25 January 1888.

Baekeland, Leo H. "On Soluble, Fusible, Resinous Condensation Products of Phenols and Formaldehyde." *Journal of Industrial and Engineering Chemistry* 1 (1909): 545–549.

Baekeland, Leo H. *On the Influence of Hygrometric Conditions of the Atmosphere in the Manufacture of Photographic Paper*. Berlin: Deutscher Verlag, 1904.

Baekeland, Leo H. *On the Toning Action of a Mixture of Thiosulphate of Sodium and Alum*. Berlin: Deutscher Verlag, 1904.

Baekeland, Leo H. "Phenol-Formaldehyde Condensation Products." *Journal of Industrial and Engineering Chemistry* 4 (1912): 737–743.

Baekeland, Leo H. *The Photochemical Industry of the United States*. Berlin: Deutscher Verlag, 1904.

Baekeland, Leo H. Photographische Trockenplatten, welche durch Eintauchen in Wasser entwickelt werden. German Patent 43,521, filed 18 October 1887.

Baekeland, Leo H. "Photoretrogression, oder das Verschwinden des latenten photographischen Bildes." *Zeitschrift für wissenschaftliche Photographie Photophysik und Photochemie* 3 (1905): 58–67.

Baekeland, Leo H. *Photoretrogression, or the Disappearance of the Latent Photographic Image. Berlin: Deutscher Verlag, 1904.*

Baekeland, Leo H. *A Practical Method for the Quantitative Determination of Silver in Photographic Paper*. Berlin: Deutscher Verlag, 1904.

Baekeland, Leo H. Préparation de plaques photographiques sèches, qui se développent par l'immersion dans l'eau. Belgian Patent 81,796, filed 14 May 1888.

Baekeland, Leo H. "Printing on Glass by the Albumen Bichromate Process." *International Annual of Anthony's Photographic Bulletin* 5 (1892): 292–295.

Baekeland, Leo H. Process of making insoluble bodies derived from phenol alcohols. US Patent 1,038,475, filed 4 October 1909, and issued 10 September 1912.

Baekeland, Leo H. "Prospects and Retrospects." *Science* 60, no. 1552 (26 September 1924): 275–279.

Baekeland, Leo H. "Protection of Intellectual Property in Relation to Chemical Industry." *Journal of Industrial and Engineering Chemistry* 5 (1913): 51–57.

Baekeland, Leo H. Reaction product of hexamethylenetetramin and phenolic bodies and method of making same. US Patent 1,187,230, filed 13 December 1910, and issued 13 June 1916.

Baekeland, Leo H. "Recent Developments in Bakelite." *Journal of Industrial and Engineering Chemistry* 3 (1911): 932–938.

Baekeland, Leo H. "Renewing the Earth from the Air: The Romance of the Search for Nitrogen." *Scribner's Magazine* 60, no. 5 (November 1916): 547–556.

Baekeland, Leo H. *Report of the Patent Committee of the National Research Council*. Reprint and Circular Series of the National Research Council, no. 1. Washington, DC: National Research Council, 1919.

Baekeland, Leo H. "Science and Industry." *Transactions of the American Electrochemical Society* 17 (1910): 37–53.

Baekeland, Leo H. *Some Aspects of Industrial Chemistry*. New York: Columbia University Press, 1914.

Baekeland, Leo H. "Statement in re Velox." *Photographic Journal* 70 (1930): 491–492.

Baekeland, Leo H. *Sur la dissociation du nitrate de plomb*. Brussels: Hayez, 1890.

Baekeland, Leo H. "The Synthesis, Constitution, and Uses of Bakelite." *Journal of Industrial and Engineering Chemistry* 1 (1909): 149–161.

Baekeland, Leo H. "Synthetic Phenol Resins." *Journal of Industrial and Engineering Chemistry* 8 (1916): 568–570.

Baekeland, Leo H. "The United States Patent System, Its Use and Abuse: The Inventor's Standpoint." *Journal of Industrial and Engineering Chemistry* 1 (1909): 204–205.

Baekeland, Leo H. "The Use of Bakelite for Electrical and Electrochemical Purposes." *Transactions of the American Electrochemical Society* 15 (1909): 593–615.

Baekeland, Leo H. "The Use of Fluorides in the Manufacture of Alcohol." *Journal of the American Chemical Society* 14 (1892): 212–220.

Baekeland, Leo H. Verfahren zur Herstellung von in Wasser entwickelbaren photographischen Trockenplatten, Austro-Hungarian Patent, filed 11 October 1887, issued 4 May 1889.

Baekeland, Leo H. "What Is the Matter with the American Chemist?" *Harper's Monthly Magazine* (April 1917): 707–714.

Baekeland, Leo H. "Wissenschaft und Industrie." *Chemiker-Zeitung* 34 (1910): 725.

Baekeland, Leo H., and Howard L. Bender. "Phenol Resins and Resinoids." *Industrial and Engineering Chemistry* 17 (1925): 225–237.

Baekelandt et Cie. Nouvelle disposition pour emballer les plaques photographiques. Belgian Patent 80,893, filed 6 March 1888.

Bakelite Corporation v. Brunswick-Balke-Collender Co. 7 F.2d 697 (D. Del. 1925).

Bancroft, Wilder D. "The Photographic Plate, V." *Journal of Physical Chemistry* 15 (1910): 313–366.

Barry, T. Hedley, Alan A. Drummond, and Robert S. Morrell. *The Chemistry of the Natural & Synthetic Resins*. London: Ernest Benn, 1926.

Bassi, Joseph P. *A Scientific Peak: How Boulder Became a World Center for Space and Atmospheric Science*. Boston: American Meteorological Society, 2015.

Bayertz, Kurt. "Biology and Beauty: Science and Aesthetics in *fin-de-siècle* Germany." In *Fin de siècle and Its Legacy*, edited by Mikulas Teich and Roy Porter, 278–295. Cambridge: Cambridge University Press, 1990.

Bayertz, Kurt. "Darwinismus als Weltanschauung." *Theoretische Geschiedenis* 10 (1983): 131–146.

Beatty, Edward, and Patricio Sáiz. "Industrial Property Institutions, Patenting, and Technology Investment in Spain and Mexico, c.1820–1914." Economic History Working Paper 02/2007, Universidad Autónoma de Madrid, 2007, http://www.uam.es/departamentos/economicas/analecon/especifica/mimeo2/wp_2007_02.pdf.

Beauchamp, Christopher. "The First Patent Litigation Explosion." *Yale Law Journal* 125 (2016): 848–944.

Beauchamp, Christopher. *Invented by Law: Alexander Graham Bell and the Patent That Changed America*. Cambridge, MA: Harvard University Press, 2015.

Beer, John J. "Coal Tar Dye Manufacture and the Origins of the Modern Industrial Research Laboratory." *Isis* 49, no. 2 (1958): 123–131.

Bender, Howard L. Air drying composition. US Patent 1,922,272, filed 8 October 1932, and issued 15 August 1933.

Bender, Howard L. Air drying composition of the urea resin type. US Patent 2,064,876, filed 11 July 1933, issued 22 December 1936.

Bender, Howard L. Flexible compositions. US Patent 2,136,329, filed 16 February 1933, and issued 8 November 1938.

Bender, Howard L. Moldable composition and its product. US Patent 1,944,143, filed 28 February 1929, and issued 16 January 1934.

Bender, Howard L. "Molecular and Cybotactic Weight Measurements of Resinous and Crystalline Materials." *Journal of the American Chemical Society* 61 (1939): 1812–1816.

Bender, Howard L. Nonfibrous compositions. US Patent 1,953,892, filed 17 May 1929, and issued 3 April 1934.

Bender, Howard L. "Phenol Resins and Resinoids." PhD diss., Columbia University, 1925.

Bender, Howard L. Resin dehydration. US Patent 2,101,635, filed 16 April 1931, and issued 7 December 1937.

Bender, Howard L. Resinous composition and process of making same. US Patent 2,041,453, filed 1 June 1931, and issued 19 May 1936.

Bender, Howard L. "The Structure of Heated Resinous Films." *Industrial and Engineering Chemistry* 29 (1937): 1130–1134.

Bender, Howard L. "Synthetic Resins." In *Colloid Chemistry: Theoretical and Applied*, vol. 4, edited by Jerome Alexander, 351–362. New York: Chemical Catalog Company, 1932.

Bender, Howard L., and Harry A. Hoffman. Urea resins. US Patent 2,084,314, filed 7 November 1933, and issued 22 June 1937.

Bender, Howard L., H. F. Wakefield, and H. A. Hoffman. "Colloid Developments in Synthetic Resins." *Chemical Reviews* 15 (1934): 123–137.

Benjamin, George H. "Letter to the Editor." *New York Times*, 14 March 1912.

Bensaude-Vincent, Bernadette. "Technoscience and Convergence: A Transmutation of Values?" Summerschool on Ethics of Converging Technologies, Dormotel Vogelsberg, Omrod/Alsfeld, Germany, 2008.

Berend, Ivan T. "Hungary: A Semi-succesful Peripheral Industrialization." In *The Industrial Revolution in National Context: Europe and the USA*, edited by Mikulas Teich and Roy Porter, 265–289. Cambridge: Cambridge University Press, 1996.

Berger, Jutta. "Chemische Mechanik und Kinetik: Die Bedeutung der mechanischen Wärmetheorie für die Theorie chemischer Reaktionen." *Annals of Science* 54, no. 6 (1997): 567–584.

Berger, Lynn. "Photography Distinguishes Itself: Law and the Emerging Profession of Photography in the 19th Century United States." PhD diss., Columbia University, 2016.

Berghoff, Hartmut, and Uwe Spiekermann. "Immigrant Entrepreneurship: The German-American Business Biography, 1720 to the Present: A GHI Research Project." *Bulletin of the German Historical Institute* 47 (Fall 2010): 69–82.

Bergmans, Oscar. *Notice historique sur l'École industrielle de Gand*. Ghent: Meyer-Van Loo, 1903.

Berman, Elizabeth P. *Creating the Market University: How Academic Science Became an Economic Engine*. Princeton, NJ: Princeton University Press, 2012.

Bertrams, Kenneth. "Converting Academic Expertise into Industrial Innovation: University-Based Research at Solvay and Gevaert, 1900–1970." *Enterprise and Society* 8 (2007): 807–841.

Bertrams, Kenneth. *Universités & entreprises: Milieux académiques et industriels en Belgique, 1880–1970*. Brussels: Le Cri, 2006.

Bertrams, Kenneth, Nicolas Coupain, and Ernst Homburg. *Solvay: History of a Multinational Family Firm*. Cambridge: Cambridge University Press, 2013.

Bessen, James, and Alessandro Nuvolari. "Knowledge Sharing among Inventors: Some Historical Perspectives." In *Revolutionizing Innovation: Users, Communities, and*

Open Innovation, edited by Dietmar Harhoff and Karim R. Lakhani, 135–156. Cambridge, MA: MIT Press.

Biagioli, Mario, and Peter Galison, eds. *Scientific Authorship: Credit and Intellectual Property in Science*. New York: Routledge, 2014.

Biagioli, Mario, Peter Jaszi, and Martha Woodmansee. "Introduction." In *Making and Unmaking Intellectual Property: Creative Production in Legal and Cultural Perspective*, edited by Mario Biagioli, Peter Jaszi, and Martha Woodmansee, 1–22. Chicago: University of Chicago Press, 2015.

Bijker, Wiebe E. *Of Bicycles, Bakelites, and Bulbs: Toward a Theory of Sociotechnical Change*. Cambridge, MA: MIT Press, 1995.

Blaszczyk, Regina Lee. *American Consumer Society, 1865–2005: From Hearth to HDTV*. Wheeling, IL: Harlan Davidson, 2009.

Bogert, Marston T. "Discussion." *Journal of Industrial and Engineering Chemistry* 8 (1916): 445–447.

Bonneuil, Christophe, and Dominique Pestre. "Introduction." In *Histoire des sciences et des savoirs*, t. 3: *Le siècle des technosciences*, edited by Christophe Bonneuil and Dominique Pestre, 9–24. Paris: Seuil, 2015.

Boudia, Soraya. *Marie Curie et son laboratoire: Sciences et industrie de la radioactivité en France*. Paris: Editions des archives contemporaines, 2001.

Bouvier, Yves Robert Fox, Pascal Griset, and Anna Guagnini, eds. *De l'atelier au laboratoire: Recherche et innovation dans l'industrie électrique, XIXe–XXe siècles*. Brussels: Peter Lang, 2011.

Bracha, Oren. *Owning Ideas: The Intellectual Origins of American Intellectual Property, 1790–1909*. New York: Cambridge University Press, 2016.

Brand, E. L. "Letter to the Editor: Shall We Make Our Own Plates?" *Philadelphia Photographer* 20 (1883): 189–190.

Brandt, Thomas. "Envisioning a National Infrastructure for Science: Academic Entrepreneurship in 1890s–1950s Norway." *Management and Organizational History* 12, no. 3 (2017): 216–236.

Brittain, James. "The Introduction of the Loading Coil: George A. Campbell and Michael I. Pupin." *Technology and Culture* 11, no. 1 (1970): 36–57.

Brock, William H. *Justus von Liebig: The Chemical Gatekeeper*. Cambridge: Cambridge University Press, 1997.

Brophy, Lee P., Wyndham D. Miles, and Rexmond C. Cochrane. *The Chemical Warfare Service: From Laboratory to Field*. Washington, DC: Office of the Chief of Military History, Department of the Army, 1959.

Brown, Allan "Bakelite—What It Is." *Plastics* 1, no. 1 (October 1925): 17.

Browne, Charles A., and Mary E. Weeks. *A History of the American Chemical Society: Seventy-Five Eventful Years*. Washington, DC: American Chemical Society, 1952.

Bruch, Rüdiger vom. "Der wissenschaftsgläubige Mensch." Reprinted in *Bürgerlichkeit, Staat und Kultur im Deutschen Kaiserreich*, selected writings edited by Hans-Christoph Liess, 11–24. Stuttgart: Franz Steiner Verlag, 2005.

Budinksi, Ken G., and Michael R. Budinski. *Materiaalkunde voor technici*. Den Haag: Academic Service, 2004.

Bush, Vannevar. "Gano Sillick Dunn, 1870–1953." *National Academy of Sciences Biographical Memoirs* 28 (1954): 31–44.

Buyse, Omer. *Méthodes américaines d'éducation générale et technique*. Paris: H. Dunod et E. Pinat, 1908.

Cahan, David. "Institutions and Communities." In *From Natural Philosophy to the Sciences: Writing the History of Nineteenth-Century Science*, edited by David Cahan, 291–328. Chicago: University of Chicago Press, 2003.

Cantwell, John, *Technological Innovation and Multinational Corporations*. Oxford: Blackwell, 1989.

Carlson, W. Bernard. "Academic Entrepreneurship and Engineering Education: Dugald C. Jackson and the MIT-GE Cooperative Engineering Course, 1907–1932." *Technology and Culture* 29, no. 3 (1988): 536–567.

Caron, François. "L'innovation collective." In *Ouvriers, villes et société: Autour d'Yves Lequin et de l'histoire sociale*, edited by Sophie Chauveau, Bruno Dumons, Olivier Faron, Gilles Pollet, and Michelle Zancarini-Fournel, 239–253. Paris: Nouveau monde, 2005.

Carswell, Thomas S. *Phenoplasts: Their Structure, Properties, and Chemical Technology*. New York: Interscience.

Cassier, Maurice. "Producing, Controlling, and Stabilizing Pasteur's Anthrax Vaccine: Creating a New Industry and a Health Market." *Science in Context* 21, no. 2 (2008): 253–278.

Cassier, Maurice, and Christiane Sinding. "'Patenting in the Public Interest': Administration of Insulin Patents by the University of Toronto." *History and Technology* 24, no. 2 (2008): 153–171.

Casson, Mark. *The Entrepreneur: An Economic Theory*. Cheltenham: Edward Elgar, 2003. First published 1982.

Casson, Mark, and Andrew Godley. "Entrepreneurship and Historical Explanation." In *Entrepreneurship in Theory and History*, edited by Youssef Cassis and Ioanna P. Minoglou, 25–60, Basingstoke: Palgrave MacMillan, 2005.

Cerveaux, Augustin. "Taming the Microworld: DuPont and the Interwar Rise of Fundamental Industrial Research." *Technology and Culture* 54, no. 2 (2013): 262–288.

Chandler, Alfred D., Jr. *Scale and Scope: The Dynamics of Industrial Capitalism*. Cambridge, MA: Belknap Press of Harvard University, 1990.

Chandler, Alfred D., Jr. *The Visible Hand: The Managerial Revolution in American Business*. Cambridge, MA: Belknap Press of Harvard University, 1977.

Chandler, Charles F. "Presentation Address." *Journal of Industrial and Engineering Chemistry* 6 (1914): 156–158.

Chang, Ha-Joon. "Intellectual Property Rights and Economic Development: Historical Lessons and Emerging Issues." *Journal of Human Development* 2 (2001): 287–309.

Channell, David F. *A History of Technoscience: Erasing the Boundaries between Science and Technology*. London: Routledge, 2017.

Chell, Elizabeth *The Entrepreneurial Personality: A Social Construction*. Hove, East Sussex: Routledge, 2008.

Choi, Hyungsub. "The Boundaries of Industrial Research: Making Transistors at RCA, 1948–1960." *Technology and Culture* 48, no. 4 (2007): 758–782.

Clark, William. *Academic Charisma and the Origins of the Research University*. Chicago: University of Chicago Press, 2008.

Clarke, Frank Wigglesworth. *A Report on the Teaching of Chemistry and Physics in the United States*. Washington, DC: Government Printing Office, 1881.

Clarke, Sabine. "Pure Science with a Practical Aim: The Meanings of Fundamental Research in Britain, circa 1916–1950." *Isis* 101, no. 2 (2010): 285–311.

Club der Amateur-Photographen in Wien. "Protokoll der aussenordentlichen Generalversammlung vom 18. Mai 1889, Abends 7 Uhr." *Photographische Rundschau* 3 (1889): 232–237.

Cody, Jeffrey W. *Exporting American Architecture, 1870–2000*. London: Routledge, 2003.

Cohoe, Wallace P. "Leo Hendrik Baekeland, 1863–1944: An Appreciation." *Chemical and Engineering News* 23, no. 3 (1945): 228–232.

Condit, Carl W. "The First Reinforced-Concrete Skyscraper: The Ingalls Building in Cincinnati and Its Place in Structural History." *Technology and Culture* 9, no. 1 (1968): 1–33.

Cornell, Thomas D. *Establishing Research Corporation: A Case Study of Patents, Philanthropy, and Organized Research in Early Twentieth-Century America*. Tucson, AZ: Research Corporation, 2004.

Cortada, James W. *Before the Computer: IBM, NCR, Burroughs, & Remington Rand and the Industry They Created, 1865–1956*. Princeton, NJ: Princeton University Press, 1993.

Cottrell, Frederick G. "The Research Corporation, an Experiment in the Public Administration of Patent Rights." *Journal of Industrial and Engineering Chemistry* 4 (1912): 864–867.

Crossick, Geoffrey. *An Artisan Elite in Victorian Society: Kentish London, 1840–1880.* London and Totowa, NJ: C. Helm and Rowman and Littlefield, 1978.

Crouch, Tom D. *The Bishop's Boys: A Life of Wilbur and Orville Wright.* New York: W. W. Norton, 1989.

Csiszar, Alex. *The Scientific Journal: Authorship and the Politics of Knowledge in the Nineteenth Century.* Chicago: University of Chicago Press, 2018.

Daemmrich, Arthur, and Leah Shaper. "The Gordon Research Conferences as Scientific Infrastructure." *Bulletin for the History of Chemistry* 33, no. 2 (2008): 94–102.

Dambre, Wouter. *August Wagener (1829–1896): Een leven voor het onderwijs.* Ghent: Archief RUG, 1987.

Dasgupta, Partha, and Paul A. David. "Toward a New Economics of Science." *Research Policy* 23 (1994): 487–521.

Daston, Lorraine, and H. Otto Sibum. "Introduction: Scientific Personae and Their Histories." *Science in Context* 16, nos. 1–2 (2003): 1–8.

David, Paul A. "The Historical Origins of 'Open Science': An Essay on Patronage, Reputation and Common Agency Contracting in the Scientific Revolution." *Capitalism and Society* 3, no. 2 (2008): Article 5.

Deacon, Desley, Penny Russell, and Angela Woollacott, eds. *Transnational Lives: Biographies of Global Modernity, 1700–Present.* Basingstoke: Palgrave Macmillan, 2010.

Deelstra, Hendrik. "Achtergronden van de wetten van 1890–1891 en de invloed op het natuurwetenschappelijk onderwijs in België." *Gewina* 16 (1993): 187–191.

Deelstra, Hendrik. "De fermentatiescheikunde in België als voorloper van de biochemie (1890–1900)." *Cerevisia* 29, no. 4 (2004): 211–215.

Deelstra, Hendrik. "De invloed van J. S. Stas op het scheikunde-onderwijs in de Belgische universiteiten." In *Jean-Servais Stas 1813–1891: Colloque international/International Colloquium*, edited by Robert Halleux and Anne-Cathérine Bernès, 83–92. Brussels: Palais des Académies, 1992.

Deelstra, Hendrik. "De scheikunde aan de universiteiten en hogescholen." In *Geschiedenis van de wetenschappen in België 1815–2000*, edited by Robert Halleux, Geert Vanpaemel, Jan Vandersmissen, and Andrée Despy-Meyer, 159–178. Brussels: La Renaissance du Livre, 2001.

Deelstra, Hendrik. *Het scheikunde-onderwijs aan de Rijksuniversiteit te Gent tijdens de 19de eeuw.* Brussels: Paleis der Academiën, 1986.

Delaere, André. *Notice sur l'École industrielle et commerciale de Gand*. Ghent: Annoot-Braeckman, 1902.

Delvaux, Klaas. "Emile Francqui: Een biografische bijdrage tot het onderzoek naar de impact van top-ondernemers op een kolonie." MA thesis, Ghent University, 2012.

DeMuth, Leopold. "Industrial Democracy at Work." *The Rotarian* 15, no. 4 (October 1919): 175–177.

Deneckere, Gita. *Uit de Ivoren Toren: 200 Jaar Universiteit Gent*. Ghent: Tijdsbeeld, 2017.

Depaepe, Marc. "De toelatingsvoorwaarden tot de universiteit in België: Een historisch overzicht." *Onze Alma Mater* 39, no. 2 (1985): 119–150.

Despy-Meyer, André. "Stas et l'enseignement supérieur." In *Jean-Servais Stas 1813–1891: Colloque international/International Colloquium*, edited by Robert Halleux and Anne-Cathérine Bernès, 75–82. Brussels: Palais des Académies, 1992.

De Vroede, Maurits. "Onderwijs 1840–1878" and "Onderwijs 1878–1914." In *Algemene Geschiedenis der Nederlanden*, dl. 13, edited by Dirk P. Blok, Walter Prevenier, and Daniel J. Roorda, 111–125 and 328–351. Haarlem: Fibula-Van Dishoeck, 1978.

[De Vylder, Gustave]. "Dr. D. V. Monckhoven." *Bulletin de l'Association belge de photographie* 10 (1883): 76–82.

De Weerdt, Denise. *De Gentse textielbewerkers en arbeidersbeweging tussen 1866 en 1881: Bijdrage tot de sociale geschiedenis van Gent*. Leuven: Nauwelaerts, 1959.

D'hoker, Mark. "Ontwikkeling van het nijverheids- en beroepsonderwijs voor jongens in België, ca. 1830–1914." PhD diss., Catholic University of Leuven, 1980.

Dhondt, Pieter. "Een tweevoudig compromis: Discussies over universitair onderwijs in het negentiende-eeuwse België, Leuven." PhD diss., Catholic University of Leuven, 2005.

Dhondt, Pieter. *Un double compromis: Enjeux et débats relatifs à l'enseignement universitaire en Belgique au XIXe siècle*. Ghent: Academia Press, 2011.

Dick, Brian, and Mark Jones. "The Commercialization of Molecular Biology: Walter Gilbert and the Biogen Startup." *History and Technology* 33, no. 1 (2017): 126–151.

Dienel, Hans-Liudger. *Ingenieure zwischen Hochschule und Industrie: Kältetechnik in Deutschland und Amerika, 1870–1930*. Göttingen: Vandenhoeck & Ruprecht, 1995.

Dinkelspiel, Frances. *Towers of Gold: How One Jewish Immigrant Named Isaias Hellman Created California*. New York: St. Martin's Press, 2008.

Diogo, Maria Paula, and Dirk Van Laak. *Europeans Globalizing: Mapping, Exploiting, Exchanging*. London: Palgrave Macmillan, 2016.

Diser, Lyvia. *Wetenschap Op de Proef: Laboratoria in Het Belgisch Overheidsbeleid, 1870–1940*. Leuven: Leuven University Press, 2016.

Dolby, R. G. A. "The Transmission of Two New Scientific Disciplines from Europe to North America in the Late Nineteenth Century." *Annals of Science* 34, no. 3 (1977): 287–310.

Donnelly, James. "Defining the Industrial Chemist in the United Kingdom, 1850–1921." *Journal of Social History* 29, no. 4 (1996): 779–796.

Donovan, Arthur. "Education, Industry, and the American University." In *Education, Technology and Industrial Performance in Europe, 1850–1939*, edited by Robert Fox and Anna Guagnini, 255–277. Cambridge and Paris: Cambridge University Press and Editions de la Maison des sciences de l'homme, 1993.

Drinkwater, W. E. A. "Velox." *Wilson's Photographic Magazine* 36 (1899): 80–84.

Duncan, Robert K. "On Certain Problems Connected with the Present-Day Relation between Chemistry and Manufacture in America," *Journal of Industrial and Engineering Chemistry* 3 (1911): 177–186.

Duranti, Marco. "Utopia, Nostalgia and World War at the 1939–40 New York World's Fair." *Journal of Contemporary History* 41, no. 4 (2006): 663–683.

Dutfield, Graham. *Intellectual Property Rights and the Life Science Industries: A Twentieth Century History*. Hackensack, NJ: World Scientific, 2009.

Ede, Andrew. *The Rise and Decline of Colloid Science in North America, 1900–1935: The Neglected Dimension*. Aldershot: Ashgate, 2007.

Eder, Josef M. *Des actions chimiques de la lumière colorée et de la photographie en couleurs naturelles*. Ghent: Annoot-Braeckman, 1881.

Eder, Josef M."Die Geschichte der Erfindung der Chlorsilbergelatine für Diapositive und positive Papierbilder mit chemischer Entwicklung." *Photographische Rundschau und Mitteilungen* 67 (1930): 855–856.

Eder, Josef M. *Geschichte der Photographie*, 4th ed. Halle: Knapp, 1932.

Eder, Josef M. "The History of the Invention of Chemically Developable Silver Chloride-Gelatin Emulsions for Transparencies and Positive Printing Papers." *Photographic Journal* 70 (1930): 391–392.

Eder, Josef M. *Photographie als Wissenschaft: Positionen um 1900*, edited by Maren Gröning and Ulrike Matzer. Munich: Fink, 2013.

Eder, Josef M., and Giuseppe Pizzighelli. "Beiträge zur Photochemie des Chlorsilbers." In *Sitzungsberichte der Akademie der Wissenschaften in Wien, Mathematisch-naturwissenschaftliche Klasse*, abt. 2, 83 (1881): 144–160.

Eder, Josef M., and Giuseppe Pizzighelli. "De la photographie par l'émulsion au chlorure d'argent et le développement chimique." Translated by Gustave De Vylder. *Bulletin de l'Association belge de photographie* 8 (1881): 225–235, 280–288, 321–328, 360–368, 396–402, 435–440, and 458–471.

Eder, Josef M., and Giuseppe Pizzighelli. *Die Photographie mit Chlorsilber-Gelatine mit chemischer Entwicklung*. Vienna: Verlag der Photographischen Correspondenz, 1881.

Edgerton, David. "Industrial Research in the British Photographic Industry, 1879–1939." In *The Challenge of New Technology: Innovation in British Business since 1850*, edited by Jonathan Liebenau, 106–133. Aldershot: Gower, 1988.

Edgerton, David. *The Shock of the Old: Technology and Global History since 1900*. Oxford: Oxford University Press, 2007.

[Editors]. "Editorial Notes." *Wilson's Photographic Magazine* 36 (1899): 237–240.

[Editors]. "Editorial Table: The Exhibit of the Nepara Chemical Company's Products at Detroit." *American Amateur Photographer* 7 (1895): 427.

[Editors]. "If We Were Giving Medals…We'd Pin One on …." *Modern Plastics* 13, no. 7 (March 1936): 27.

[Editors]. "New Era in Field of Phenol Resins." *Plastics* 2 (1926): 442.

[Editors]. "Our Table." *American Amateur Photographer* 11 (1899): 311–313.

[Editors]. "Personal Notes." *Journal of Industrial and Engineering Chemistry* 7 (1915): 638–639.

[Editors]. "Trade Names." *Plastics* 2 (1926): 92.

[Editors]. "Unfair Competition in Foreign Trade." *Chemical and Metallurgical Engineering* 33 (1926): 397.

[Editors]. "Willard Gibbs Medal Award." *Journal of Industrial and Engineering Chemistry* 5 (1913): 503–504.

Ellis, Carleton. *The Chemistry of Synthetic Resins*. New York: Reinhold, 1935.

Ellis, Carleton. *Synthetic Resins and Their Plastics*. New York: Chemical Catalog Company, 1923.

Ely, Richard T. *Property and Contract in Their Relations to the Distribution of Wealth*. New York: Macmillan, 1914.

Engel, Alexander. "Coloring the World: Marketing German Dyestuffs in the Late Nineteenth and Early Twentieth Centuries." In *Bright Modernity*, edited by Regina L. Blaszczyk and Uwe Spiekermann, 37–53. New York: Springer, 2017.

Erichsen, Hugo. "Society News: The Fifteenth Annual Meeting of the Photographers' Association of America." *American Amateur Photographer* 7 (1895): 420–426.

Essig, Mark. "Poison Murder and Expert Testimony: Doubting the Physician in Late Nineteenth-Century America." *Yale Journal of Law and the Humanities* 14 (2002): 177–210.

Etzkowitz, Henry. "Entrepreneurial Scientists and Entrepreneurial Universities in American Academic Science." *Minerva* 21, no. 2 (1983): 198–233.

Etzkowitz, Henry. "The Triple Helix and the Rise of the Entrepreneurial University." In *The Science-Industry Nexus: History, Policy, Implications*, edited by Karl Grandin, Nina Wormbs, and Sven Widmalm, 69–92. Sagamore Beach, MA: Science History Publications/USA, 2004.

Evans, Harold. *They Made America—From the Steam Engine to the Search Engine: Two Centuries of Innovators*. New York: Little, Brown, 2004.

Faber, Monika. "Josef Maria Eder und die wissenschaftliche Fotografie 1855–1918." In *Das Auge und der Apparat: Eine Geschichte der Fotografie aus den Sammlungen der Albertina*, edited by Monika Faber and Klaus A. Schröder, 142–169. Paris: Seuil/Albertina, 2003.

Factory Investigating Commission of the State of New York. *Second Report*, vol. 3: *Testimony and Proceedings*. Albany, NY: J. B. Lyons, 1913.

Faith, Thomas I. *Behind the Gas Mask: The U.S. Chemical Warfare Service in War and Peace*. Urbana: University of Illinois Press, 2014.

Fear, Jeffrey R. "Constructing Big Business: The Cultural Concept of the Firm." In *Big Business and the Wealth of Nations*, edited by Alfred Chandler, Franco Amatori, and Takashi Hikino, 546–574. Cambridge: Cambridge University Press, 1997.

Fisch, Jörg. "Zivilisation, Kultur." In *Geschichtliche Grundbegriffe: Historisches Lexicon zur politisch-sozialen Sprache in Deutschland*, vol. 7, edited by Otto Brunner, Werner Conze, and Reinhart Koselleck, 679–774. Stuttgart: Klett-Cotta, 1992.

Fisher, William W. "Geistiges Eigentum—ein ausuferndes Rechtsbereich: Die Geschichte des Ideenschutzes in den Vereinigten Staaten." In *Eigentum im internationalen Vergleich: 18.–20. Jahrhundert*, edited by Hannes Siegrist and David Sugarman, 264–291. Göttingen: Vandenhoeck und Ruprecht, 1999.

Fishman, Elliot A. "MIT Patent Policy, 1932–1946: Historical Precedents in University-Industry Technology Transfer." PhD diss., University of Pennsylvania, 1996.

Fisk, Catherine. *Working Knowledge: Employee Innovation and the Rise of Corporate Intellectual Property, 1800–1930*. Chapel Hill: University of North Carolina Press, 2009.

Forman, Paul. "The Primacy of Science in Modernity, of Technology in Postmodernity, and of Ideology in the History of Technology." *History and Technology* 23, no. 1/2 (2007): 1–152.

Fourcade, Marion. "Cents and Sensibility: Economic Valuation and the Nature of 'Nature.'" *American Journal of Sociology* 116, no. 6 (2011): 1721–1777.

Fox, Robert, and Anna Guagnini, eds. *Education, Technology and Industrial Performance in Europe, 1850—1939*. Cambridge and Paris: Cambridge University Press and Editions de la Maison des sciences de l'homme, 1993.

Fox, Robert, and Anna Guagnini. *Laboratories, Workshops, and Sites: Concepts and Practices of Research in Industrial Europe, 1800–1914*. Berkeley: Office for History of Science and Technology, University of California, 1999.

Foxlee, Edward W. "Velox for Professional Work." *Photographic Times* 31 (1899): 245–247.

Frerichs, F. W. "Remarks on Chemical Engineering Education." *Transactions of the American Institute of Chemical Engineers* 3 (1910): 151–152.

Frerichs, F. W. "Report of Committee on Chemical Engineering Education." *Transactions of the American Institute of Chemical Engineers* 3 (1910): 122–145.

Fressoz, Jean-Baptiste. "Beck Back in the 19th Century: Towards a Genealogy of Risk Society." *History and Technology* 23, no. 4 (2007): 333–350.

Friedel, Robert. "Perspiration in Perspective: Changing Perceptions of Genius and Expertise in American Invention." In *Inventive Minds: Creativity in Technology*, edited by Robert J. Weber and David N. Perkins, 11–26. New York: Oxford University Press, 1992.

Friedel, Robert. *Pioneer Plastic: The Making and Selling of Celluloid*. Madison: University of Wisconsin Press, 1983.

Friedman, Walter A. *Birth of a Salesman: The Transformation of Selling in America*. Cambridge, MA: Harvard University Press, 2004.

Furukawa, Yasu. *Inventing Polymer Science: Staudinger, Carothers, and the Emergence of Macromolecular Chemistry*. Philadelphia: University of Pennsylvania Press, 1998.

Gabriel, Joseph M. *Medical Monopoly: Intellectual Property Rights and the Origins of the Modern Pharmaceutical Industry*. Chicago: University of Chicago Press, 2014.

Galambos, Louis. "The Entrepreneurial Culture: Mythologies, Realities, and Networks in Nineteenth-Century America." Paper presented at The Institute for Applied Economics, Global Health, and the Study of Business Enterprise, Johns Hopkins University, 15 November 2018.

Galambos, Louis, and Jane Eliot Sewell. *Networks of Innovation: Vaccine Development at Merck, Sharp and Dohme, and Mulford, 1895–1995*. Cambridge: Cambridge University Press, 1997.

Galison, Peter, and Bruce Hevly, eds. *Big Science: The Growth of Large-Scale Research*. Stanford, CA: Stanford University Press, 1992.

Galvez-Behar, Gabriel. "Chapitre 9. Organisations et propriété scientifique dans l'entre-deux-guerres: Une comparaison transatlantique." Unpublished book manuscript chapter, 2017.

Galvez-Behar, Gabriel. "Controverses et paradoxes dans l'Europe des brevets au XIXe siècle." In *Innovations et transferts de technologie en Europe du Nord-Ouest aux XIXe et XXe siècles*, edited by Jean-François Eck and Pierre Tilly, 35–51. Brussels: Peter Lang, 2011.

Galvez-Behar, Gabriel. "Institutional Enterprise as a Compromise: The National Organization of Science in France." *Management and Organizational History* 12, no. 3 (2017): 237–260.

Galvez-Behar, Gabriel. "Louis Pasteur, entrepreneur: Pour une histoire économique des mondes savants." Unpublished working paper, 2016.

Garwood, Christine. "Green Crusaders or Captives of Industry? The British Alkali Inspectorate and the Ethics of Environmental Decision Making, 1864–95." *Annals of Science* 61, no. 1 (2004): 99–117.

Gasse v. Development & Funding Co. 135 New York Supplement 732 (N.Y. Sup. Ct. 1911).

Gaudillière, Jean-Paul, ed. "How Pharmaceuticals Became Patentable in the Twentieth Century." *History and Technology* 24, no. 2 (2008): 99–205.

Gerwen, Jacques van, and Ferry de Goey. *Ondernemers in Nederland: Varieties in ondernemen*. Amsterdam: Boom, 2008.

Giffard, Hermione. *Making Jet Engines in World War II: Britain, Germany, and the United States*. Chicago: University of Chicago Press, 2016.

Gillis, Jan. *Leo Hendrik Baekeland: Verzamelde oorspronkelijke documenten*. Brussels: Koninklijke Vlaamse Academie voor Wetenschappen, Letteren en Schone Kunsten van België, 1965.

Gispen, Kees. *New Profession, Old Order: Engineers and German Society, 1815–1914*. Cambridge: Cambridge University Press, 1989.

Gispen, Kees. *Poems in Steel: National Socialism and the Politics of Inventing from Weimar to Bonn*. New York: Berghahn, 2002.

Golan, Tal. *Laws of Men and Laws of Nature: The History of Scientific Expert Testimony in England and America*. Cambridge, MA: Harvard University Press, 2004.

Goldstein, Carolyn M. *Creating Consumers: Home Economists in Twentieth-Century America*. Chapel Hill: University of North Carolina Press, 2012.

Gooday, Graeme. "Liars, Experts and Authorities." *History of Science* 46 (2008): 431–456.

Gooday, Graeme. "'Vague and Artificial': The Historically Elusive Distinction between Pure and Applied Science." *Isis* 103, no. 3 (2012): 546–554.

Graham, Margaret B. W., and Bettye H. Pruitt. *R & D for Industry: A Century of Technical Innovation at Alcoa*. Cambridge: Cambridge University Press, 1990.

Grandin, Karl, Nina Wormbs, and Sven Widmalm, eds. *The Science-Industry Nexus: History, Policy, Implications*. Sagamore Beach, MA: Science History Publications/USA, 2004.

Greene, Jeremy A. "Attention to 'Details': Etiquette and the Pharmaceutical Salesman in Postwar America." *Social Studies of Science* 34, no. 2 (2004): 271–292.

Gröning, Maren. "Die Photographische Gesellschaft als Motor für die Gründung der k. k. Lehr- und Versuchsanstalt für Photographie und Reproductionsverfahren." In *Die Explosion der Bilderwelt: Die Photographische Gesellschaft in Wien, 1861–1945*, edited by Michael Ponstingl, 167–175. Vienna: Christian Brandstätter Verlag and Albertina Museum, 2011.

Grosvenor, William M. "The Relation of the Manufacturer to Our Patent System." *Industrial and Engineering Chemistry* 5, no. 8 (1913): 685–692.

Guagnini, Anna. "A Bold Leap into Electric Light: The Creation of the Società Italiana Edison 1880–1886." *History of Technology* 32 (2014): 155–189.

Guagnini, Anna. "Ivory Towers? The Commercial Activity of British Professors of Physics and Engineering, 1880–1914." *History and Technology* 33, no. 1 (2017): 70–108.

Guagnini, Anna. "Patent Agents, Legal Advisers and Guglielmo Marconi's Breakthrough in Wireless Telegraphy." *History of Technology* 24 (2002): 171–202.

Guise-Richardson, Cai. "Redefining Vulcanization: Charles Goodyear, Patents, and Industrial Control, 1834–1865." *Technology and Culture* 51, no. 2 (2010): 357–387.

Gustin, Bernard Henry. "The Emergence of the German Chemical Profession, 1790–1867." PhD diss., University of Chicago, 1975.

Haber, Ludwig. *The Chemical Industry 1900–1930: International Growth and Technological Change*. Oxford: Oxford University Press, 1971.

Hall, Bronwyn H., Christian Helmers, Mark Rogers, and Vania Sena. "The Choice between Formal and Informal Intellectual Property: A Review." *Journal of Economic Literature* 52, no. 2 (2014): 375-423.

Hamlin, Christopher. "Scientific Method and Expert Witnessing: Victorian Perspectives on a Modern Problem." *Social Studies of Science* 16, no. 3 (1986): 485–513.

Hannaway, Owen. "The German Model of Chemical Education in America: Ira Remsen at Johns Hopkins (1876–1913)." *Ambix* 23, no. 3 (1976): 145–164.

Hård, Mikael. "German Regulation: The Integration of Modern Technology into National Culture." In *The Intellectual Appropriation of Technology: Discourses on Modernity, 1900–1939*, edited by Mikael Hård and Andrew Jamison, 33–68. Cambridge, MA: MIT Press, 1998.

Harvey, Mortimer. "Further Studies on Phenolic Hexamethylenetetramine Compounds." PhD diss., Columbia University, 1920.

Harvey, Mortimer, and Leo H. Baekeland. "Further Studies on Phenolic Hexamethylenetetramine Compounds." *Journal of Industrial and Engineering Chemistry* 13 (1921): 135–141.

Harwood, Jonathan. "Institutional Innovation in Fin de Siècle Germany." *British Journal for the History of Science* 2, no. 27 (1994): 197–211.

Hausman, Willam J., Peter Hertner, and Mira Wilkins. *Global Electrification: Multinational Enterprise and International Finance in the History of Light and Power, 1878–2007.* New York: Cambridge University Press, 2008.

Haynes, William. *American Chemical Industry: The Chemical Companies.* Vol. 6. New York: Van Nostrand, 1949.

Haynes, William. "Leo Hendrik Baekeland." In *Great Chemists,* edited by Edouard Farber, 1181–1190. New York: Interscience, 1961.

Haynes, William. *This Chemical Age: The Miracle of Man-Made Materials.* New York: Knopf, 1942.

Hemming, Emile. *Molded Electrical Insulation and Plastics.* New York: Ward Clausen, 1914.

Hemming, Emile. *Plastics and Molded Electrical Insulation.* New York: Chemical Catalog Company, 1923.

Hens, Jules, and Laurent Roosens. *De Belgische fotochemische industrie: Met behartiging van de chemische nijverheid (1879–1939).* Brussels: Lieven Gevaert-Leerstoel, 2008.

Herf, Jeffrey. *Reactionary Modernism: Technology, Culture, and Politics in Weimar and the Third Reich.* Cambridge: Cambridge University Press, 1984.

Hershberg, James G. *James B. Conant: Harvard to Hiroshima and the Making of the Nuclear Age.* New York: Knopf, 1993.

Herstein, Bernard. "Patents and Chemical Industry in the United States." *Industrial and Engineering Chemistry* 4 (1912): 328–333.

Hesse, Bernhard C. "Shall We Have Annual Patent Renewal Fees?" *Journal of Industrial and Engineering Chemistry* 11 (1919): 697–698.

Hesse, Bernhard C. "The United States Patent System, Its Use and Abuse: Some Suggestions as to Desirable Improvement." *Journal of Industrial and Engineering Chemistry* 1 (1909): 203–204.

Hidvégi, Mária. *Anschluss an den Weltmarkt: Ungarns elektrotechnische Leitunternehmen 1867–1949.* Göttingen: Vandenhoeck und Ruprecht, 2016.

Hilkey, Judy. *Character Is Capital: Success Manuals and Manhood in Gilded Age America.* Chapel Hill: University of North Carolina Press, 1997.

Hintz, Eric S. "The 'Monopoly' Hearings, Their Critics, and the Limits of Patent Reform in the New Deal." In *Capital Gains: Business and Politics in Twentieth-Century America*, edited by Kim Phillips-Fein and Richard R. John, 61–79. Philadelphia: University of Pennsylvania Press, 2016.

Hintz, Eric S. "Portable Power: Inventor Samuel Ruben and the Birth of Duracell." *Technology and Culture* 50, no. 1 (2009): 24–57.

Hollestelle, Marijn. "De academische polymeerwetenschap in Nederland, 1945–2011." In *Universiteit, publiek en politiek: Het aanzien van de Nederlandse universiteiten, 1800–2010*, edited by Leendert J. Dorsman and Peter J. Knegtmans, 83–110. Hilversum: Verloren, 2012.

Homburg, Ernst. "Chemistry and Industry: A Tale of Two Moving Targets." *Isis* 109, no. 3 (2018): 565–576.

Homburg, Ernst. "The Emergence of Research Laboratories in the Dyestuffs Industry, 1870–1900." *British Journal for the History of Science* 25, no. 1 (1992): 91–111.

Homburg, Ernst. *Van beroep "Chemiker": De opkomst van de industriële chemicus cn het polytechnische onderwijs in Duitsland.* Delft: DUP, 1993.

Hooker, Elon H. "An Appreciation of Dr. Baekeland." *Journal of Industrial and Engineering Chemistry* 8 (1916): 183–184.

Hooker, Elon H. "Discussion." *Journal of Industrial and Engineering Chemistry* 8 (1916): 447–449.

Hounshell, David A. "Edison and the Pure Science Ideal in 19th-Century America." *Science* 207, no. 4431 (8 February 1980): 612–617.

Hounshell, David A. "The Evolution of Industrial Research in the United States." In *Engines of Innovation: U.S. Industrial Research at the End of an Era*, edited by Richard S. Rosenbloom and William J. Spencer, 13–85. Boston: Harvard Business School Press, 1996.

Hounshell, David A., and John K. Smith Jr. *Science and Corporate Strategy: Du Pont R&D, 1902–1980.* Cambridge: Cambridge University Press, 1988.

Howe, James L. "The Aim and Future of Natural Science." *Science* 16, no. 404 (31 October 1890): 239–244.

H. S. "In Memoriam: Ir. Georges LOPPENS." *Annales de l'Association des ingénieurs sortis des écoles spéciales de Gand* 14 (1939): 125–127.

Hughes, Thomas P. *American Genesis: A Century of Invention and Technological Enthusiasm, 1870–1970.* New York: Viking, 1989.

Hughes, Thomas P. *Elmer Sperry: Inventor and Engineer.* Baltimore: Johns Hopkins University Press, 1971.

Hughes, Thomas P. *Networks of Power: Electrification in Western Society, 1880–1930.* Baltimore: Johns Hopkins University Press, 1983.

Hultzsch, Kurt. *Chemie der Phenolharze.* Berlin: Springer, 1950.

Inkster, Ian. "Indigenous Resistance and the Technological Imperative: From Chemistry in Birmingham to Camphor Wars in Formosa, 1860s–1914." In *Technology and Globalisation: Networks of Experts in World History*, edited by David Pretel and Lino Camprubi, 41–74. London: Macmillan, 2018.

International Congress of Pure and Applied Chemistry. *Eighth International Congress of Applied Chemistry*, vol. 27: *Discussions.* Concord, NH: Rumford Press, 1913.

Investigation of Concentration of Economic Power: Hearings before the Temporary National Economic Committee, Part 3, Patents: Proposals for Changes in Law and Procedure. Washington, DC: USGPO, 1939.

Israel, Paul. *Edison: A Life of Invention.* New York: Wiley, 1998.

Israel, Paul. "The Flash of Genius: Defining Invention in the Era of Corporate Research" (conference paper, 2006). cdn.law.ucla.edu/SiteCollectionDocuments/workshops%20and%20colloquia/israel-the%20flash.pdf.

Israel, Paul. *From Machine Shop to Industrial Laboratory: Telegraphy and the Changing Context of American Invention, 1830–1920.* Baltimore: Johns Hopkins University Press, 1992.

Israel, Paul, and Robert Rosenberg. "Patent Office Records as a Historical Source: The Case of Thomas Edison." *Technology and Culture* 32, no. 4 (1991): 1094–1101.

Jacobsen, Paul. "Signs of Progress." *Trefoil* 3, no. 4 (April 1938): 9.

Jarausch, Konrad H. "Higher Education and Social Change: Some Comparative Perspectives." In *The Transformation of Higher Learning 1860–1930: Expansion, Diversification, Social Opening, and Professionalization in England, Germany, Russia, and the United States*, edited by Konrad H. Jarausch, 9–56. Chicago: University of Chicago Press, 1983.

Jasanoff, Sheila. *Science at the Bar: Law, Science, and Technology in America.* Cambridge, MA: Harvard University Press, 1995.

Jenkins, Reese V. *Images and Enterprise: Technology and the American Photographic Industry, 1839 to 1925.* Baltimore: Johns Hopkins University Press, 1975.

Jensen, William B., and Julia Kuhlmann. "Leopold Pfaundler and the Origins of the Kinetic Theory of Chemical Reactions." *Bulletin for the History of Chemistry* 37 (2012): 29–41.

John, Richard R. "Robber Barons Redux: Antimonopoly Reconsidered." *Enterprise and Society* 13 (2012): 1–38.

Johnson, Ann. "What If We Wrote the History of Science from the Perspective of Applied Science? Review Essay of Ndiayde's *Nylon and Bombs*, Westwick's *Into the Black* and Lécuyer's *Making Silicon Valley*." *Historical Studies in the Natural Sciences* 38 (2008): 610–620.

Johnson, Jeffrey A. "The Academic-Industrial Symbiosis in German Chemical Research, 1905–1939." In *The German Chemical Industry in the Twentieth Century*, edited by John E. Lesch, 15–56. Dordrecht: Kluwer, 2000.

Joint Hearings before the Committees on Patents of the Senate and House of Representatives, Sixty-Sixth Congress, First Session, on S.3223 and H.R. 9932. November 5, 1919. Washington, DC: USGPO, 1919.

Jones, Daniel P. "American Chemists and the Geneva Protocol." *Isis* 71, no. 3 (1980): 426–440.

Jones, Daniel P. "The Role of Chemists in Research on War Gases in the United States during World War I." PhD diss., University of Wisconsin, 1969.

Jordan, John M. *Machine-Age Ideology: Social Engineering and American Liberalism, 1911–1939*. Chapel Hill: University of North Carolina Press, 1994.

Joseph, Steven, Paul Faelens, and Laurent Roosens. *D. van Monckhoven 1834–1882*. Antwerp: Provinciaal Museum voor Fotografie, 1982.

Kaataja, Sampsa. "University Researchers Contributing to Technology Markets 1900–85: A Long-Term Analysis of Academic Patenting in Finland." *Minerva* 49, no. 4 (2011): 447–460.

Kargon, Robert H., Karen Fiss, Morris Low, and Arthur P. Molella. *World's Fairs on the Eve of War: Science, Technology & Modernity, 1937–1942*. Pittsburgh: University of Pittsburgh Press, 2015.

Kargon, Robert H., and Elizabeth Hodes. "Karl Compton, Isaiah Bowman, and the Politics of Science in the Great Depression." *Isis* 76, no. 3 (1985): 301–318.

Kargon, Robert H., and Scott G. Knowles. "Knowledge for Use: Science, Higher Learning, and America's New Industrial Heartland, 1880–1915." *Annals of Science* 59, no. 1 (2002): 1–20.

Katzir, Shaul. "Technological Entrepreneurship from Patenting to Commercializing: A Survey of Late Nineteenth and Early Twentieth Century Physics Lecturers." *History and Technology* 33, no. 1 (2017): 109–125.

Kaufmann, Carl B. "Grand Duke, Wizard and Bohemian: A Biographical Profile of Leo Hendrik Baekeland (1863–1944)." MA thesis, University of Delaware, 1968.

Kevles, Daniel J. "George Ellery Hale, the First World War, and the Advancement of Science in America." *Isis* 59, no. 4 (1968): 427–437.

Khan, B. Zorina. *The Democratization of Invention: Patents and Copyrights in American Economic Development, 1790–1920*. Cambridge: Cambridge University Press, 2005.

Kingsley, Hope. "Gelatine Silver Print." In *Encyclopedia of Nineteenth-Century Photography*, edited by John Hannavy, 573-575. New York: Routledge, 2008.

Kissling, Richard. "Wissenschaft und Industrie." *Chemiker-Zeitung* 34 (1910): 762.

Kleeberg, Werner. "Ueber die Einwirkung des Formaldehyds auf Phenole." *European Journal of Organic Chemistry* 263, no. 3 (1891): 283–286.

Klein, Ursula. "Artisanal-Scientific Experts in Eighteenth-Century France and Germany." *Annals of Science* 69, no. 3 (2012): 303–306.

Klein, Ursula. *Nützliches Wissen: Die Erfindung Der Technikwissenschaften*. Göttingen: Wallstein, 2016.

Klein, Ursula. "Technoscience avant la lettre." *Perspectives on Science* 13, no. 2 (2005): 226–266.

Kline, Ronald. "Construing 'Technology' as 'Applied Science': Public Rhetoric of Scientists and Engineers in the United States, 1880–1945." *Isis* 86, no. 2 (1995): 194–221.

Kline, Ronald. *Steinmetz: Engineer and Socialist*. Baltimore: Johns Hopkins University Press, 1992.

Kline, Ronald, and Thomas Lassman. "Competing Research Traditions in American Industry: Uncertain Alliances between Engineering and Science at Westinghouse Electric, 1886–1935." *Enterprise and Society* 6 (2005): 601–645.

Klinge, Matti. "Die Universitätslehrer." In *Geschichte der Universität in Europa. Band III: vom 19. Jahrhundert zum Zweiten Weltkrieg 1800–1945*, edited by Walter Rüegg, 113–143. Munich: Beck, 2004.

Knight, David, and Helge Kragh, eds. *The Making of the Chemist: The Social History of Chemistry in Europe, 1789–1914*. Cambridge: Cambridge University Press, 1998.

Knoedler, Janet, and Anne Mayhew. "Thorstein Veblen and the Engineers: A Reinterpretation." *History of Political Economy* 31 (1999): 255–272.

Knorre, Georg Karl von. "Über die Bedeutung der Elekrochemie für die Technik." In *Sitzungsberichte des Vereins zur Beförderung des Gewerbfleißes*, edited by W. Wedding, 129–149. Berlin: Leonhad Simion, 1900.

König, Wolfgang. *Der Gelehrte und der Manager: Franz Reuleaux (1829–1905) und Alois Riedler (1850–1936) in Technik, Wissenschaft und Gesellschaft*. Stuttgart: Steiner, 2014.

König, Wolfgang. "Engineering Professors as Entrepreneurs: The Case of Franz Reuleaux (1829–1905) and Alois Riedler (1850–1936)." *History and Technology* 33, no. 1 (2017): 53–69.

Kranakis, Eda. "Hybrid Careers and the Interaction of Science and Technology." In *Technological Development and Science in the Industrial Age*, edited by Peter Kroes and Martijn Bakker, 177–204. Dordrecht: Kluwer, 1992.

Kranakis, Eda. "Patents and Power: European Patent-System Integration in the Context of Globalization." *Technology and Culture* 48, no. 4 (2007): 689–728.

Kuznick, Peter J. "Losing the World of Tomorrow: The Battle over the Presentation of Science at the 1939 New York World's Fair." *American Quarterly* 46, no. 3 (1994): 341–373.

Ladas, Stephen P. *Patents, Trademarks, and Related Rights: National and International Protection*. Cambridge, MA: Harvard University Press, 1975.

Laird, Pamela W. *Pull: Networking and Success since Benjamin Franklin*. Cambridge, MA: Harvard University Press.

Lamoreaux, Naomi R. "Entrepreneurship in the United States, 1865–1920." In *The Invention of Enterprise: Entrepreneurship from Ancient Mesopotamia to Modern Times*, edited by David S. Landes, Joel Mokyr, and William J. Baumol, 367–400. Princeton, NJ: Princeton University Press, 2010.

Lamoreaux, Naomi R., and Kenneth L. Sokoloff. "Afterword." In *Financing Innovation in the United States: 1870 to the Present*, edited by Naomi R. Lamoreaux and Kenneth L. Sokoloff, 469–474. Cambridge, MA: MIT Press, 2007.

Lamoreaux, Naomi R., and Kenneth L. Sokoloff. "The Decline of the Independent Inventor: A Schumpeterian Story?" NBER Working Paper 11654. Cambridge, MA: National Bureau of Economic Research, 2005.

Lamoreaux, Naomi R., and Kenneth L. Sokoloff. "Inventors, Firms, and the Market for Technology in the Late Nineteenth and Early Twentieth Centuries." In *Learning by Doing in Markets, Firms, and Countries*, edited by Naomi R. Lamoreaux, Daniel M. G. Raff, and Peter Temin, 19–60. Chicago: University of Chicago Press, 1999.

Landau, Ralph, and Nathan Rosenberg. "Successful Commercialization in the Chemical Process Industries." In *Technology and the Wealth of Nations*, edited by Nathan Rosenberg, Ralph Landau, and David C. Mowery, 73–119. Stanford, CA: Stanford University Press, 1992.

Lavoisier, Antoine. *Oeuvres de Lavoisier*, vol. 4. Paris: Imprimerie Impériale, 1868.

Layton, Edwin T. *The Revolt of the Engineers: Social Responsibility and the American Engineering Profession*. Baltimore: Johns Hopkins University Press, 1986. First published in 1971.

Lebach, Hans. "Bakelite and Its Applications." *Journal of the Society of Chemical Industry* 32, no. 11 (1913): 559–564.

LeCain, Timothy. "The Limits of "Eco-efficiency": Arsenic Pollution and the Cottrell Electrical Precipitator in the US Copper Smelting Industry." *Environmental History* 5, no. 3 (2000): 336–351.

Lécuyer, Christophe. "The Making of a Science Based Technological University: Karl Compton, James Killian, and the Reform of MIT, 1930–1957." *Historical Studies in the Physical and Biological Sciences* 23, no. 1 (1992): 153–180.

Lenoir, Timothy. "Revolution from Above: The Role of the State in Creating the German Research System, 1810–1910." *American Economic Review* 88, no. 2 (1998): 22–27.

Leslie, Stuart W. *Boss Kettering*. New York: Columbia University Press, 1983.

Leslie, Stuart W. *The Cold War and American Science: The Military-Industrial-Academic Complex at MIT and Stanford*. New York: Columbia University Press, 1993.

Lidbury, F. A., Lawrence Addicks, Leo H. Baekeland, C. G. Atwater, Calvert Townley, George R. Smith, D. B. Rushmore, Oscar T. Crosby, and John H. Finney. "Discussion on 'Water Power and Defense' (Whitney), 'The Water Power Situation, Including Its Financial Aspect' (Dunn), Washington, DC, April 26, 1916." *Transactions of the American Institute of Electrical Engineers* 35 (1916): 457–477, 1145–1165.

Liebenau, Jonathan. *Medical Science and Medical Industry: The Formation of the American Pharmaceutical Industry*. Baltimore: Johns Hopkins University Press, 1987.

Liebenau, Jonathan. "Patents and the Chemical Industry: Tools of Business Strategy." In *The Challenge of New Technology: Innovation in British Business since 1850*, edited by Jonathan Liebenau, 135–150. Farnham: Gower/Ashgate, 1988.

Loppens, Georges, and Honoré De Swarte. Process of retting or steeping flax or hemp. US Patent 576,867, filed 26 November 1895, and issued 9 February 1897.

Lovelace, B. F., and Charles C. Thomas, eds. *Annual Chemical Directory of the United States*. Baltimore: Williams and Wilkins, 1918 and 1919.

Lowen, Rebecca S. *Creating the Cold War University: The Transformation of Stanford*. Berkeley: University of California Press, 1997.

Lucier, Paul. "The Professional and the Scientist in Nineteenth-Century America." *Isis* 100, no. 4 (2009): 699–732.

Lucier, Paul. *Scientists & Swindlers: Consulting on Coal and Oil in America, 1820–1890*. Baltimore: Johns Hopkins University Press, 2008.

Lundgren, Anders. "The Transfer of Chemical Knowledge: The Case of Chemical Technology and Its Textbooks." *Science and Education* 15 (2006): 761–778.

Lüppo-Cramer, Hinricus. "Biographie von Josef Maria Eder." In Josef M. Eder, *Geschichte der Photographie*, 1043–1053. Halle: Knapp, 1932.

Lynch, David, *The Concentration of Economic Power*. New York: Columbia University Press, 1946.

MacLeod, Christine. *Heroes of Invention: Technology, Liberalism and British Identity, 1750–1914*. New York: Cambridge University Press, 2007.

MacLeod, Christine, and Gregory Radick. "Claiming Ownership in the Technosciences: Patents, Priority and Productivity." *Studies in History and Philosophy of Science* 44 (2013): 188–201.

MacLeod, Julius. "Théodore Swarts." In *Université de Gand. Liber memorialis: Notices biographiques*, vol. 2, 181–186. Ghent: Vanderpoorten, 1913.

Maestrejuan, Andrea. "Inventors, Firms, and the Market for Technology during the Kaiserreich, 1877–1914." PhD diss., University of California, Los Angeles, 2009.

Makowka, O. [Wissenschaft und Technik]. *Chemiker-Zeitung* 34 (1910): 849–850.

Manny, Frank A. "American Schools as Seen by a Belgian Educator." *Elementary School Teacher* 9 (1909): 322–326.

Mantels, Ruben. *Gent: Een geschiedenis van universiteit en stad, 1817–1940*. Brussels: Mercatorfonds, 2013.

Marcovich, Anne, and Terry Shinn. "Regimes of Science Production and Diffusion: Towards a Transverse Organization of Knowledge." *Scientiae Studia* 10 (2012): 33–64.

Marder, William, Estelle Marder, and Robert G. Duncan. *Anthony, the Man, the Company, the Cameras: An American Photographic Pioneer: 140 Year History of a Company from Anthony to Ansco, to GAF*. [Plantation,FL]: Pine Ridge, 1982.

Marsch, Ulrich. "Transferring Strategy and Structure: The German Chemical Industry as an Exemplar for Great Britain." In *The German Chemical Industry in the Twentieth Century*, edited by John E. Lesch, 217–241. Dordrecht: Kluwer, 2000.

Marsch, Ulrich. *Zwischen Wissenschaft und Wirtschaft: Industrieforschung in Deutschland und Grossbritannien 1880–1936*. Paderborn: Verlag von Ferdinand Schöningh, 2000.

Martin, Ben R. "Are Universities and University Research under Threat? Towards an Evolutionary Model of University Speciation." *Cambridge Journal of Economics* 36, no. 3 (2012): 543–565.

Martin, Robert W. *The Chemistry of Phenolic Resins*. New York: Wiley, 1956.

Marvin, Abba. "Electrochemical Patents, and Some Needs of the United States Patent System." *Transactions of the American Electrochemical Society* 13 (1908): 431–445.

Marx, Leo. "Technology: The Emergence of a Hazardous Concept." *Technology and Culture* 51, no. 3 (2010): 561–577.

Masquelez, Auguste. "L'institut industriel, agronomique et commercial du Nord de la France." In *Association française pour l'avancement des sciences, Compte rendu de la 3me session, 1874*, 62–67. Paris: Association française pour l'avancement des sciences, 1875.

Matthis, Alfred R. *Léo-H. Baekeland 1 863–1944: Professeur, docteur en sciences, chimiste, inventeur et grand industriel*. Brussels: Office de publicité, 1948.

Maxim, Hudson. "The Warfare of the Future." *Science* 28, no. 728 (11 December 1908): 820–831.

McCormick, Richard L. "The Discovery That Business Corrupts Politics: A Reappraisal of the Origins of Progressivism." *American Historical Review* 86 (1981): 247–274.

McFarland, J. Horace, "A Convenient and Rapid Printing Process." *Photographic Times* 29 (1897): 221–223.

Mees, Kenneth. *From Dry Plates to Ektachrome Film: A Story of Photographic Research*. New York: Ziff-Davis, 1961.

Mees, Kenneth. *The Fundamentals of Photography*. Rochester, NY: Eastman Kodak Company, 1928. First published in 1920.

Mees, Kenneth. "Leo Hendrik Baekeland and Photographic Printing Papers." *Chemistry and Industry* 74 (1955): 1134–1138.

Mees, Kenneth. *The Organization of Industrial Scientific Research*. New York: McGraw-Hill, 1920.

Megson, N. J. L. "Aldehyde Phenolic Condensations from a Chemical Standpoint." *Transactions of the Faraday Society* 32 (1936): 336–345.

Meikle, Jeffrey L. *American Plastic: A Cultural History*. New Brunswick, NJ: Rutgers University Press, 1995.

Meikle, Jeffrey L. "Materia Nova: Plastics and Design in the U.S., 1925–1935." In *The Development of Plastics*, edited by Susan Mossman and Peter Morris, 38–53. Cambridge: Royal Society of Chemistry, 1994.

Meiksins, Peter. "The 'Revolt of the Engineers' Reconsidered." *Technology and Culture* 29, no. 2 (1988): 219–246.

Mercelis, Joris. "Commercializing Academic Knowledge and Reputation in the Late Nineteenth and Early Twentieth Centuries: Photography and Beyond." *History and Technology* 33, no. 1 (2017): 23–52.

Mercelis, Joris. "Corporate Secrecy and Intellectual Property in the Chemical Industry through a Transatlantic Lens, c.1860–1930." *Entreprises et histoire* 1 (2016): 32–46.

Mercelis, Joris. "Jacob Hasslacher." In *Immigrant Entrepreneurship: German-American Business Biographies, 1720 to the Present*, vol. 3: *From the End of the Gilded Age to the*

Progressive Era, 1893–1918, edited by Giles Hoyt. Washington, DC: German Historical Institute. Last modified 25 September 2014. http://www.immigrantentrepreneurship.org/entry.php?rec=48.

Mercelis, Joris. "Public Knowledge and Private Enterprise: Nineteenth-Century Photography and the Practicability of Open Innovation." Paper presented at the annual meeting of the Society for the History of Technology, Singapore, 22–26 June 2016.

Mercelis, Joris. "The Scientist and the Advertisement: *Reklamegutachten* in Imperial Germany." *History of Science* (forthcoming, 2020).

Mercelis, Joris, Gabriel Galvez-Behar, and Anna Guagnini. "Commercializing Science: Nineteenth- and Twentieth-Century Academic Scientists as Consultants, Patentees, and Entrepreneurs." *History and Technology* 33, no. 1 (2017): 4–22.

Mersch, Christian. *Die Welt der Patente: Soziologische Perspektiven auf eine zentrale Institution der globalen Wissensgesellschaft*. Bielefeld: Transcript Verlag, 2013.

Merton, Robert K. "A Note on Science and Democracy," reprinted as "The Normative Structure of Science." In Robert K. Merton, *The Sociology of Science: Theoretical and Empirical Investigations*, 267–278. Chicago: University of Chicago Press, 1973.

Merton, Robert K. "Priorities in Scientific Discovery: A Chapter in the Sociology of Science." *American Sociological Review* 22, no. 6 (1957): 635–659.

Messamore, Barbara J. "Review of 'David Lambert and Alan Lester (eds), Colonial Lives across the British Empire: Imperial Careering in the Long Nineteenth Century, Cambridge: Cambridge University Press, 2006.'" *Journal of Historical Biography* 2 (Autumn 2007): 115–119.

Metlay, Grischa. "Reconsidering Renormalization: Stability and Change in 20th-Century Views on University Patents." *Social Studies of Science* 36, no. 4 (2006): 565–597.

Meyer-Thurow, Georg. "The Industrialization of Invention: A Case Study from the German Chemical Industry." *Isis* 73, no. 3 (1982): 363–381.

Miller, David P. "Intellectual Property and Narratives of Discovery/Invention: The League of Nations' Draft Convention on 'Scientific Property'and Its Fate." *History of Science* 46 (2008): 299–342.

Miller, David P. *James Watt, Chemist: Understanding the Origins of the Steam Age*. Pittsburgh: University of Pittsburgh Press, 2015.

Miller, David P. "The Political Economy of Discovery Stories: The Case of Dr Irving Langmuir and General Electric." *Annals of Science* 68, no. 1 (2011): 27–60.

Mirowski, Philip. *Science-Mart: Privatizing American Science*. Cambridge, MA: Harvard University Press, 2011.

Misa, Thomas J. "The Changing Market for Chemical Knowledge: Applied Chemistry and Chemical Engineering in the Delaware Valley, 1851–1929." *History and Technology* 2, no. 3 (1985): 245–268.

Mitchell, Lawrence E. *The Speculation Economy: How Finance Triumphed over Industry*. San Francisco: Berrett-Koehler, 2008.

Mody, Cyrus C. M. *Instrumental Community: Probe Microscopy and the Path to Nanotechnology*. Cambridge, MA: MIT Press, 2011.

Mody, Cyrus C. M. *The Long Arm of Moore's Law: Microelectronics and American Science*. Cambridge, MA: MIT Press, 2016.

Mokyr, Joel. *The Gifts of Athena: Historical Origins of the Knowledge Economy*. Princeton, NJ: Princeton University Press, 2002.

Monaghan, Frank. *Official Guide Book of the New York World's Fair, 1939*. New York: Exposition Publications, 1939.

Moore, Edward B. "The United States Patent System, Its Use and Abuse." *Journal of Industrial and Engineering Chemistry* 1 (1909): 198–199.

Moreau, Alfred. *Traité de la concurrence illicite*. Brussels: Bruylant-Christophe, 1904.

Morgan, J. Livingston R. "The Dissociation of Lead Nitrate." *Journal of Physical Chemistry* 9 (1904): 416–420.

Morrell, Jack. "WH Perkin, Jr., at Manchester and Oxford: From Irwell to Isis." *Osiris* 8 (1993): 104–126.

Morris, Susan W. "Commerce and Academe: American Universities as Hosts of Entrepreneurial Science, 1880–1920." In *Knowledge Management and Intellectual Property: Concepts, Actors and Practices from the Past to the Present*, edited by Stathis Arapostathis and Graham Dutfield, 183–199. Northampton, MA: Edward Elgar, 2013.

Morris, Susan W. "Resource Networks: Industrial Research in Small Enterprises, 1860–1930." PhD diss., Johns Hopkins University, 2003.

Mory, Austin V. H. "Nature Was Notably Aided." *Journal of Chemical Education* 6 (1929): 607–621.

Mory, Austin V. H. "Synthetic Resins." *Industrial and Engineering Chemistry* 19 (1927): 1106–1109.

Mowery, David C. "*Plus ça change*: Industrial R&D in the Third Industrial Revolution." *Industrial and Corporate Change* 18, no. 1 (2009): 1–50.

Mowery, David C., Richard R. Nelson, Bhaven N. Sampat, Arvids A. Ziedonis. *Ivory Tower and Industrial Innovation: University-Industry Technology Transfer before and after the Bayh-Dole Act*. Stanford, CA: Stanford Business Books, 2004.

Murmann, Johann P. *Knowledge and Competitive Advantage: The Coevolution of Firms, Technology, and National Institutions*. New York: Cambridge University Press, 2003.

Nader, David, and Spasoje Marcinko. "The Rise and Fall of 'Chlorine Chambers' against Cold and Flu." In *An Element of Controversy: The Life of Chlorine in Science, Medicine, Technology and War*, edited by Hasok Chang and Catherine Jackson, 296–323. S.l.: British Society for the History of Science, 2007.

Nagy, Brian, and Franz Lohrke. "Only the Good Die Young? A Review of Liability of Newness and Related New Venture Mortality Research." In *Historical Foundations of Entrepreneurship Research*, edited by Hans Landström and Franz Lohrke, 185–204. Cheltenham: Edward Elgar, 2010.

Nasby, Judith. *Rolph Scarlett: Painter, Designer, and Jeweller*. Montréal: McGill-Queen's University Press, 2004.

National Academy of Sciences. *Report for the Fiscal Year 1935–1936*. Washington, DC: USGPO, 1937.

Newman, Richard S. *Love Canal: A Toxic History from Colonial Times to the Present*. New York: Oxford University Press, 2016.

New York State Court of Appeals. Simmons-Baker Company v. Development & Funding Company. Appeal Book, vol. 56, 1917.

Noble, David F. *America by Design: Science, Technology, and the Rise of Corporate Liberalism*. New York: Knopf, 1988.

Noyes, William A. "Julius Stieglitz, 1867–1937." *National Academy of Sciences Biographical Memoirs* 21 (1941): 275–314.

Nye, Mary J. "Mine, Thine, and Ours: Collaboration and Co-authorship in the Material Culture of the Mid-twentieth Century Chemical Laboratory." *Ambix* 61, no. 3 (2014): 211–235.

Ó Gráda, Cormac. "Did Science Cause the Industrial Revolution?" *Journal of Economic Literature* 54, no. 1 (2016): 224–239.

Oldenziel, Ruth. *Making Technology Masculine: Men, Women and Modern Machines in America, 1870–1945*. Amsterdam: Amsterdam University Press, 1999.

Oldfield Revision and Codification of the Patent Statutes. Hearing on H.R. 23417, before the Committee on Patents, House of Representatives. Washington, DC: USGPO, 1912.

Olsen, J. C. "A Course in Chemical Engineering Education." *Transactions of the American Institute of Chemical Engineers* 4 (1911): 250–277.

Onghena, Sofie. "Altruïstisch ambtenaar of heroïsch genie? Het gepropageerde beeld van provinciale en academische directeurs van bacteriologische laboratoria in België (ca. 1900–1940)." *Studium* 2, no. 4 (2009): 191–210.

Osterhammel, Jürgen. *Die Verwandlung der Welt*. Munich: Beck, 2009.

Owens, Larry. "MIT and the Federal 'Angel': Academic R&D and Federal-Private Cooperation before World War II." *Isis* 81, no. 2 (1990): 188–213.

Owens, Larry. "Patents, the 'Frontiers' of American Invention, and the Monopoly Committee of 1939: Anatomy of a Discourse." *Technology and Culture* 32, no. 4 (1991): 1076–1093.

Paquot, Catherine. *Henri Sainte Claire Deville: Chimie, recherche et industrie*. Paris: Vuibert, 2005.

Parlette, Ralph A. *The University of Hard Knocks: The School That Completes Our Education*. Chicago: Parlette-Padget, 1916. First published in 1914.

Patterson, Gary. *A Prehistory of Polymer Science*. Berlin: Springer, 2012.

Pelikan, Kurt A. "The Plastics Industry Abroad." *Plastic and Molded Products* 6 (1930): 703, 705, 726–729.

Penrose, Edith T. *The Economics of the International Patent System*. Baltimore: Johns Hopkins University Press, 1951.

Pestre, Dominique. "Regimes of Knowledge Production in Society: Towards a More Political and Social Reading." *Minerva* 41, no. 3 (2003): 245–261.

Pestre, Dominique. "Scientists in Time of War: World War II, the Cold War, and Science in the United States and France." *French Politics, Culture and Society* 24, no. 1 (2006): 27–39.

Photographische Gesellschaft in Wien, "Prämien-Verleihung der Photographischen Gesellschaft," *Photographische Korrespondenz* 35 (1898): 85–86

Picard, Edmond, and Xavier Olin. *Traité des brevets d'invention et de la contrefaçon industrielle, précédé d'une théorie sur les inventions industrielles*. Paris: Durand, 1869.

Ponstingl, Michael, ed. *Die Explosion der Bilderwelt: Die Photographische Gesellschaft in Wien, 1861–1945*. Vienna: Christian Brandstätter Verlag and Albertina Museum, 2011.

Prindle, Edwin J., Elihu Thomson, Lawrence V. Redman, R. L. Stinchfield, Edwin A. Hill, Willis R. Whitney, Arthur D. Little, J. M. Francis, T. Hart Anderson, T. B. Wagner, John U. Lloyd, Leo H. Baekeland. "Symposium on Annual Patent Renewal Fees." *Journal of Industrial and Engineering Chemistry* 11 (1919): 930–938.

Pritchard, Henry B. *The Photographic Studios of Europe*. London: Pipe and Carter, 1882.

Pupin, Michael. *From Immigrant to Inventor*. New York: Scribner's Sons, 1923.

Raegener, Louis. "Some Defects in the Practice of Our Patent System and Suggested Remedies." *Journal of Industrial and Engineering Chemistry* 1 (1909): 201–203.

Ramsay, William, and William MacNab, eds. *Seventh International Congress of Applied Chemistry, Section 11: Law, Political Economy, and Legislation Affecting Industry*. London: Partridge and Cooper, 1910.

Rasch, Manfred. "Auf dem Weg zum Diensterfinder: Zur kommerziellen Nutzung von Forschungsergebnissen aus Kaiser-Wilhelm-Instituten." In *"Dem Anwenden muss das Erkennen vorausgehen": Auf dem Weg zu einer Geschichte der Kaiser-Wilhelm-/Max-Planck-Gesellschaft*, edited by Dieter Hofmann, Birgit Kolboske, and Jürgen Renn, 219–242. 2nd extended ed. Berlin: Neopubli, 2015.

Raschig, Fritz. "Die wissenschaftliche und technische Bedeutung der Teerphenole." *Zeitschrift für angewandte Chemie* 25 (1912): 1939–1948.

Rassow, Berthold. "Was haben wir Chemiker in Amerika gelernt?" *Zeitschrift für angewandte Chemie* 26 (1913): 705–708.

Redaktion der *Chemiker-Zeitung*. [Wissenschaft und Industrie]. *Chemiker-Zeitung* 34 (1910): 762.

Redman, Lawrence V. "Research as a Fixed Charge." *Industrial and Engineering Chemistry* 24 (1932): 112–115.

Redman, Lawrence V., and Austin V. H. Mory. *The Romance of Research*. Baltimore: Williams and Wilkins in cooperation with the Century of Progress Exposition, 1933.

Redman, Lawrence V., Archie J. Weith, and Frank P. Brock. Phenolic condensation product article and method of manufacture. U.S. Patent 1,424,738, filed 8 September 1921, and issued 1 August 1922.

Redman, Lawrence V., Archie J. Weith, and Frank B. Brock. "Synthetic Phenol Resins." *Journal of Industrial and Engineering Chemistry* 8 (1916): 1077.

Redman, Lawrence V., Archie J. Weith, and Frank B. Brock. "Synthetic Resins." *Journal of Industrial and Engineering Chemistry* 6 (1914): 3–16.

Reed, Germaine M. *Crusading for Chemistry: The Professional Career of Charles Holmes Herty*. Durham, NC, and Athens, GA: Forest History Society and University of Georgia Press, 1995.

Reed, Peter. "The Alkali Inspectorate 1874–1906: Pressure for Wider and Tighter Pollution Regulation." *Ambix* 59, no. 2 (2012): 131–151.

Reich, Leonard S. *The Making of American Industrial Research: Science and Business at GE and Bell, 1876–1926*. Cambridge: Cambridge University Press, 1985.

Reich, Leonard S. "Research, Patents, and the Struggle to Control Radio: A Study of Big Business and the Uses of Industrial Research." *Business History Review* 51 (1977): 208–235.

Reimer, Thomas. "Bayer & Company in the United States: German Dyes, Drugs, and Cartels in the Progressive Era." PhD diss., Syracuse University, 1996.

Reinhardt, Carsten. *Forschung in der chemischen Industrie: Die Entwicklung synthetischer Farbstoffe bei BASF und Hoechst, 1863 bis 1914*. Freiberg: TU Bergakademie Freiberg, 1997.

Reports of Cases Decided in the Court of Appeals of the State of New York, vol. 220. Albany, NY: J. B. Lyon.

Reynolds, Terry S. "Defining Professional Boundaries: Chemical Engineering in the Early 20th-Century." *Technology and Culture* 27, no. 4 (1986): 694–716.

Rhees, David. "The Chemists' Crusade: The Rise of an Industrial Science in Modern America, 1907–1922." PhD diss., University of Pennsylvania, 1987.

Rhodes, E. O. "The American Way in Industrial Research: Mellon Institute and Other Institutions." *Industrial and Engineering Chemistry* 31 (1939): 549–556.

Richards, Joseph W. "Niagara as an Electrochemical Centre." Early Publications of the Lehigh Faculty. Paper 140. Bethlehem, PA: Lehigh University, 1902. http://preserve.lehigh.edu/early-faculty-publications/140.

Robbins, Jane E. "Solving the Patent Problem: Cognition, Communication, and the National Academy of Sciences in the Evolution of University Patent Policy, 1917–1966." PhD diss., University of Pennsylvania, 2004.

Roberts, Lissa, Simon Schaffer, and Peter Dear, eds. *The Mindful Hand: Inquiry and Invention from the Late Renaissance to Early Industrialisation.* Amsterdam: Koninklijke Nederlandse Akademie van Wetenschappen, 2007.

Rocke, Alan J. "Theory versus Practice in German Chemistry: Erlenmeyer beyond the Flask." *Isis* 109, no. 2: 254–275.

Rodgers, Daniel T. "In Search of Progressivism." *Reviews in American History* 10 (1982): 113–132.

Rogers, Allen, Milton C. Whitaker, Leo H. Baekeland, and Robert K. Duncan. "Discussion of Professor Duncan's Paper." *Journal of Industrial and Engineering Chemistry* 3 (1911): 183–186.

Rohkrämer, Thomas. *Eine andere Moderne? Zivilisationskritik, Natur und Technik in Deutschland 1880–1933*. Paderborn: Schöningh, 1999.

Roqué, Xavier. "Marie Curie and the Radium Industry: A Preliminary Sketch." *History and Technology* 13, no. 4 (1997): 267–291.

Rosenberg, Nathan. "Chemical Engineering as a General Purpose Technology." In *General Purpose Technologies and Economic Growth,* edited by Elhanan Helpman, 167–192. Cambridge, MA: MIT Press, 1998.

Rosenberg, Nathan, and W. Edward Steinmueller. "Engineering Knowledge." *Industrial and Corporate Change* 5, no. 22 (2013): 1129–1156.

Rowland, Henry Augustus. "A Plea for Pure Science." *Science* 2, no. 29 (24 August 1883): 242–250.

Rugen, W. H. "The 'X' in Research." *Trefoil* 3, no. 4 (April 1938): 25.

Russell, Edmund. *War and Nature: Fighting Humans and Insects with Chemicals from World War I to Silent Spring*. Cambridge: Cambridge University Press, 2001.

Rydell, Robert W. "The Fan Dance of Science: American World's Fairs in the Great Depression." *Isis* 76, no. 4 (1985): 525–542.

Sá, Creso M., and Andrew Kretz. "Technology Commercialization as University Mission: Early Historical Developments at the University of Toronto." *Technology and Culture* 57, no. 1 (2016): 119–143.

Sacré, Edmond. *A la mémoire de Gustave De Vylder*. Ghent: Annoot-Braeckman, 1896.

Sáiz, Patricio. "Why Did Corporations Patent in Spain? Some Historical Inquiries." Economic History Working Paper 02/2010, Universidad Autónoma de Madrid, 2010. http://www.uam.es/departamentos/economicas/analecon/especifica/mimeo2/wp_2010_02.pdf.

Sandage, Scott A. *Born Losers: A History of Failure in America*. Cambridge, MA: Harvard University Press, 2005.

Sarachek, Bernard. "American Entrepreneurs and the Horatio Alger Myth." *Journal of Economic History* 38 (1978): 439–456.

Saunier, Pierre-Yves. *Transnational History: Theory and History*. Houndmills: Palgrave Macmillan, 2013.

Schatzberg, Eric. "From Art to Applied Science." *Isis* 103, no. 3 (2012): 555–563.

Schatzberg, Eric. *Technology: Critical History of a Concept*. Chicago: University of Chicago Press, 2018.

Scherer, Frederic M. "The Political Economy of Patent Policy Reform in the United States." *Journal of Telecommunications and High Technology Law* 7 (2009): 167–216.

Schumpeter, Joseph A. *The Theory of Economic Development*. Cambridge, MA: Harvard University Press, 1934.

Servos, John W. "Changing Partners: The Mellon Institute, Private Industry, and the Federal Patron." *Technology and Culture* 35, no. 2 (1994): 221–257.

Servos, John W. *Physical Chemistry from Ostwald to Pauling: The Making of a Science in America*. Princeton, NJ: Princeton University Press, 1996.

Shapin, Steven. *Never Pure: Historical Studies of Science as if It Was Produced by People with Bodies, Situated in Time, Space, Culture, and Society, and Struggling for Credibility and Authority*. Baltimore: Johns Hopkins University Press, 2010.

Shapin, Steven. *The Scientific Life: A Moral History of a Late Modern Vocation*. Chicago: University of Chicago Press, 2008.

Shreve, R. Norris, and Brage Golding. "Oil-Soluble Phenolic Resins." *Industrial and Engineering Chemistry* 43 (1951): 134–141.

Simon, Ariel. "Reinventing Discovery: Patent Law's Characterizations of and Interventions upon Science." *University of Pennsylvania Law Review* 157, no. 6 (2009): 2175–2232.

Slaton, Amy. *Reinforced Concrete and the Modernization of American Building, 1900–1930*. Baltimore: Johns Hopkins University Press, 2001.

Slotten, Hugh R. "Humane Chemistry or Scientific Barbarism? American Responses to World War I." *Journal of American History* 77 (1990): 476–498.

Smith, Crosbie, and Norton M. Wise. *Energy and Empire: A Biographical Study of Lord Kelvin*. Cambridge: Cambridge University Press, 1989.

Smith, George D. *From Monopoly to Competition: The Transformations of Alcoa, 1888–1986*. Cambridge: Cambridge University Press, 1988.

Smith, John K., Jr. "The Scientific Tradition in American Industrial Research." *Technology and Culture* 31, no. 1 (1990): 121–131.

Snelders, Henricus A. M. "Dissociation, Darwinism and Entropy: A Case Study from the History of Physical Chemistry." *Janus Leiden* 64 (1977): 51–75.

Snelders, Henricus A. M. "Jean-Servais Stas (1813–1891) en de scheikunde van zijn tijd." In *Jean-Servais Stas 1813–1891: Colloque international/International Colloquium*, edited by Robert Halleux and Anne-Cathérine Bernès, 5–19. Brussels: Palais des Académies, 1992.

Spero, Ellan F. "An Entrepreneurial Opportunity in Process: Creating an Industrial Fellowship in Early Twentieth Century America." *Management and Organizational History* 12, no. 3 (2017): 199–215.

Stas, Jean S. "Recherches de statique chimique au sujet du chlorure et du bromure d'argent: Troisième partie." *Anneles de chimie et de physique*, 5e série, 3 (1874): 289–327.

Steen, Kathryn. *The American Synthetic Organic Chemicals Industry: War and Politics, 1910–1930*. Chapel Hill: University of North Carolina Press, 2014.

Stephan, Paula E. *How Economics Shapes Science*. Cambridge, MA: Harvard University Press, 2012.

Stieglitz, Julius. "Chairman's Address," *Journal of Industrial and Engineering Chemistry* 5 (1913): 504–505.

Stieglitz, Julius. "Synthetic Pharmaceuticals and the Patent Law." *Transactions of the American Institute of Chemical Engineers* 11 (1918): 369–382.

Swanson, Kara W. "The Emergence of the Professional Patent Practitioner." *Technology and Culture* 50, no. 3 (2009): 519–548.

Swanson, Kara W. "'Great Men,' Law, and the Social Construction of Technology." *Law and Social Inquiry* 43, no. 3 (2018): 1093–1112.

Swayne, Noah H. *Twenty-Five Year Record, Class of Ninety-Three, Yale College*. New Haven, CT: Tuttle, Morehouse and Taylor, 1918.

Swinfin, Stephen T. "Chemists Employed in the Manchester Area, 1902–1936." *Annals of Science* 69, no. 2 (2012): 239–256.

Szöllösi-Janze, Margit. "Science and Social Space: Transformations in the Institutions of Wissenschaft from the Wilhelmine Empire to the Weimar Republic." *Minerva* 43, no. 4 (2005): 339–360.

Tate, A. C. "The Protective Power of a Good Trade Mark." *Plastics* 1 (1925): 53.

Teixeira, Pedro. "Two Continents Divided by the Same Trends? Reflections about Marketization, Competition, and Inequality in European Higher Education." In *The University under Pressure*, edited by Elizabeth P. Berman and Catherine Paradeise, 489–508. Bradford: Emerald Group, 2016.

Thackray, Arnold. "University-Industry Connections and Chemical Research: An Historical Perspective." In *University-Industry Research Relationships*, edited by National Science Board, 193–233. Washington, DC: National Science Board, 1982.

Thackray, Arnold, Jeffrey L. Sturchio, P. Thomas Carroll, and Robert Bud. *Chemistry in America, 1876–1976: Historical Indicators*. Dordrecht: Reidel, 1986.

Thomas, Arthur W., Stephen P. Burke, Colin G. Fink, Wm. D. Turner, and Arthur W. Hixson. "Leo Hendrik Baekeland." *Science* 100, no. 2585 (14 July 1944): 22–24.

Thomas, Robert E. *Salt & Water, Power & People: A Short History of Hooker Electrochemical Company*. Niagara Falls, NY: [Hooker Electrochemical Company], 1955.

Thurlow, Nathaniel. Process of producing camphor. US Patent 698,761, filed 6 August 1901, and issued 29 April 1902.

Tocqueville, Alexis de. *Democracy in America*. New York: Bantam Dell, 2002. First published in 1835.

Tomory, Leslie. *Progressive Enlightenment: The Origins of the Gaslight Industry, 1780–1820*. Cambridge, MA: MIT Press, 2012.

Travis, Anthony S. *The Rainbow Makers: The Origins of the Synthetic Dyestuffs Industry in Western Europe*. Bethlehem, PA: Lehigh University Press, 1993.

Trescott, Martha Moore. *The Rise of the American Electrochemicals Industry, 1880–1910*. Westport, CT: Greenwood Press, 1981.

Trischler, Helmuth, and Martin Kohlrausch. *Building Europe on Expertise: Innovators, Organizers, Networkers*. New York: Palgrave Macmillan, 2014.

Turkington, Victor H. Coating articles. US Patent 2,125,827, filed 24 December 1936, and issued 2 August 1938.

Turkington, Victor H. Coating composition. US Patent 2,076,507, filed 28 February 1929, and issued 6 April 1937.

Turkington, Victor H. Floor covering. US Patent 1,988,616, filed 9 October 1930, and issued 22 January 1935.

Turkington, Victor H. Shock resistant material. US Patent 1,958,452, filed 10 November 1928, and issued 15 May 1934.

Turkington, Victor H. Varnish compositions. US Patent 2,075,839, filed 22 June 1934, and issued 6 April 1937.

Turkington, Victor H., and Ivey Allen Jr. "Oil-Soluble Phenolic Resins: Influence of Substituents on Properties." *Industrial and Engineering Chemistry* 33 (1941): 966–971.

Turkington, Victor H., and William H. Butler. Drying oil composition. US Patent 1,988,615, filed 9 May 1930, and issued 22 January 1935.

Turkington, Victor H., and William H. Butler. Phenolic resinous products and compositions containing them. US Patent 2,070,148, filed 30 July 1932, and issued 9 February 1937.

Turkington, Victor H., Robert J. Moore, William H. Butler, and Ralph C. Shuey. "Phenolic Resin Varnishes: Influence of Linseed Oil on Film Properties." *Industrial and Engineering Chemistry* 27 (1935): 1321–1325.

Turkington, Victor H., Ralph C. Shuey, and William H. Butler. "Phenol Resinoids in Oil Varnishes." *Industrial and Engineering Chemistry* 22 (1930): 1177–1180.

Turkington, Victor H., Ralph C. Shuey, and William H. Butler. "Viscosity Increase and Gelation in Phenolic Resin Varnish Cooking." *Industrial and Engineering Chemistry* 23 (1931): 791–797.

Tweedale, Geoffrey. "Geology and Industrial Consultancy: Sir William Boyd Dawkins (1837–1929) and the Kent Coalfield." *British Journal for the History of Science* 24, no. 4 (1991): 435–451.

Ulrich, R. D. *Macromolecular Science: Retrospect and Prospect*. New York: Plenum Press, 1978.

United States Tariff Commission. *Census of Dyes and Coal-Tar Chemicals, 1919*. Washington, DC: USGPO, 1921.

United States Tariff Commission. *Census of Dyes and of Other Synthetic Organic Chemicals, 1927–29*. Washington, DC: USGPO, 1928–1931.

United States Tariff Commission. *Report to the President of the United States…in the Matter of Synthetic Phenolic Resin*. Washington, DC: USGPO, 1931.

University of Michigan. "Fellowships in Chemical Engineering, 1917–1918." *Journal of Industrial and Engineering Chemistry* 9 (1917): 207.

Usselman, Steven W. *Regulating Railroad Innovation: Business, Technology, and Politics in America, 1840–1920*. Cambridge: Cambridge University Press, 2002.

Usselman, Steven W. "Research and Development (R&D)." In *The Oxford Encyclopedia of the History of American Science, Medicine, and Technology*, edited by Hugh Slotten. Oxford University Press, 2015. http://www.oxfordreference.com/view/10.1093/acref/9780199766666.001.0001/acref-9780199766666.

Van den Belt, Henk. "Robert Merton, Intellectual Property, and Open Science: A Sociological History for Our Times." In *The Commodification of Academic Research: Science and the Modern University*, edited by Hans Radder, 187–230. Pittsburgh: University of Pittsburgh Press, 2010.

Van Keuren, David K. "Science, Progressivism, and Military Preparedness: The Case of the Naval Research Laboratory, 1915–1923." *Technology and Culture* 33, no. 4 (1992): 710–736.

Van Monckhoven, Désiré. *De photographie met broomzilvergelatine*. Translated from the French by Edward Verschaffelt. Ghent: J. Vuylsteke, 1880.

Vanpaemel, Geert. "Jean-Servais Stas en de experimentele wetenschappen in België." In *Jean-Servais Stas 1813–1891: Colloque international/International Colloquium*, edited by Robert Halleux and Anne-Cathérine Bernès, 45–52. Brussels: Palais des Académies, 1992.

Vanpaemel, Geert. *Wetenschap als roeping: Een geschiedenis van de Leuvense Faculteit voor Wetenschappen*. Leuven: Universitaire Pers Leuven, 2017.

Vanpaemel, Geert, and Brigitte Van Tiggelen. "The Profession of Chemistry in Nineteenth-Century Belgium." In *The Making of the Chemist: The Social History of Chemistry in Europe, 1789–1914*, edited by David Knight and Helge Kragh, 191–206. Cambridge: Cambridge University Press, 1998.

Van Tiggelen, Brigitte. "De la chaire aux laboratoires: Louis Henry et la professionnalisation de la recherche en sciences naturelles en Belgique." *Gewina* 16 (1993): 192–203.

Van Tiggelen, Brigitte. "De scheikundige theorieën." In *Geschiedenis van de wetenschappen in België 1815–2000*, edited by Robert Halleux, Geert Vanpaemel, Jan Vandersmissen, and Andrée Despy-Meyer, 179–194. Brussels: La Renaissance du Livre, 2001.

Van Tiggelen, Brigitte, and Hendrik Deelstra. "Belgium: From Industry to Academia: The Belgian Chemical Society, 1887–1914." In *Creating Networks in Chemistry: The Founding and Early History of Chemical Societies in Europe*, edited by Anita K. Nielsen and Sona Strbanova, 23–42. Cambridge: Royal Society of Chemistry, 2008.

Verbruggen, Christophe. *De stank bederft onze eetwaren: De reacties op industriële milieuhinder in het 19de-eeuwse Gent*. Ghent: Academia Press, 2002.

Vermeir, Koen, and Dániel Margócsy. "States of Secrecy: An introduction." *British Journal for the History of Science* 45, no. 2 (2012): 153–164.

Veysey, Laurence R. *The Emergence of the American University*. Chicago: University of Chicago Press, 1970.

Vietinghoff-Scheel, Eduard J. von. [Wissenschaft und Technik]. *Chemiker-Zeitung* 34 (1910): 850.

Vogel, Hermann W. "Ueber die verschiedenen Modifikationen des Bromsilbers und Chlorsilbers." *Berichte der deutschen chemischen Gesellschaft* 16 (1883): 1170–1179.

Vogel, Hermann W., Ellerslie Wallace, and Edward L. Wilson. *The Progress of Photography since the Year 1879*. Philadelphia: Edward L. Wilson, 1883.

Voskuhl, Adelheid. "Engineering Philosophy: Theories of Technology, German Idealism, and Social Order in High-Industrial Germany." *Technology and Culture* 57, no. 4 (2016): 721–752.

Wadhwani, R. Daniel, Gabriel Galvez-Behar, Joris Mercelis, and Anna Guagnini. "Academic Entrepreneurship and Institutional Change in Historical Perspective." *Management and Organizational History* 12, no. 3 (2017): 175–198.

Wadhwani, R. Daniel, and Geoffrey Jones. "Schumpeter's Plea: Historical Reasoning in Entrepreneurship Theory and Research." In *Organizations in Time: History, Theory, Methods*, edited by Marcelo Bucheli and R. Daniel Wadhwani, 192–216. Oxford: Oxford University Press, 2014.

Wadhwani, R. Daniel, and Christina Lubinski. "Reinventing Entrepreneurial History." *Business History Review* 91, no. 4 (2017): 767–799.

Wakeman, Reginald L. *The Chemistry of Commercial Plastics*. New York: Reinhold, 1947.

Walker, William H. "The University and Industry." *Journal of Industrial and Engineering Chemistry* 8 (1916): 63–65.

Wall, Edward J. *Photographic Emulsions: Their Preparation and Coating on Glass, Celluloid and Paper, Experimentally and on the Large Scale*. Boston: American Photographic Publishing, 1929.

Walmsley, William H. "An Admirable Printing Process for Photo-Micrographs." *Photographic Times* 29 (1897): 284–286.

Ware, Mike. "Positives: Minor Processes." In *Encyclopedia of Nineteenth-Century Photography*, edited by John Hannavy, 1154–1161. New York: Routledge, 2008.

Warnerke, Leon. "Photographic Technical Instruction on the Continent of Europe." *Journal and Transactions of the Royal Photographic Society of Great Britain* 15 (1891): 25–37.

Wautier, Kristel, and Danny Segers. "Nothing Succeeds like Success": Het levensverhaal van Leo Baekeland, de uitvinder van het bakeliet." *Heemkundige kring De Oost-Oudburg v.z.w., jaarboek* 44 (2007): 125–178.

Weber, Max. *Die protestantische Ethik und der Geist des Kapitalismus: Vollständige Ausgabe*, edited and introduced by Dirk Kaesler. Munich: Beck, 2006.

Weiner, Charles. "Patenting and Academic Research: Historical Case Studies." *Science, Technology, and Human Values* 12, no. 1 (1987): 50–62.

Weith, Archie. "Polymerization and Synthetic Resins." In Robert E. Burk, Howard E. Thompson, Archie J. Weith, and Ira Williams, *Polymerization and Its Applications in the Fields of Rubber, Synthetic Resins, and Petroleum*, 184–221. New York: Reinhold, 1937.

Welter, Friederike. "All You Need Is Trust? A Critical Review of the Trust and Entrepreneurship Literature." *International Small Business Journal* 30, no. 3 (2012): 193–212.

Wengenroth, Ulrich. "Germany: Competition Abroad—Cooperation at Home, 1870–1990." In *Big Business and the Wealth of Nations*, edited by Alfred Chandler, Franco Amatori, and Takashi Hikino, 139–175. Cambridge: Cambridge University Press, 1997.

Wentzel, Fritz. *Memoirs of a Photochemist*. Philadelphia: American Museum of Photography, 1960.

Wenzel, Henry, Jr. "How to Work Velox Paper." *American Amateur Photographer* 9 (1897): 493–498.

Wenzel, Henry, Jr. "Velox Again." *American Amateur Photographer* 10 (1898): 69–70.

Wetzel, Walter. *Naturwissenschaften und chemische Industrie in Deutschland: Voraussetzungen und Mechanismen ihres Aufstiegs im 19. Jahrhundert*. Stuttgart: Steiner, 1991.

Whitaker, Milton. "The New Chemical Engineering Course and Laboratories at Columbia University." *Transactions of the American Institute of Chemical Engineers* 5 (1912): 150–169.

White, Richard. "Information, Markets, and Corruption: Transcontinental Railroads in the Gilded Age." *Journal of American History* 90 (2003): 19–43.

Whittemore, Jr., Gilbert F. "World War I, Poison Gas Research, and the Ideals of American Chemists." *Social Studies of Science* 5, no. 2 (1975): 135–163.

Whitten, David O. "The Depression of 1893." *EH.net Encyclopedia*, edited by Robert Whaples. http://eh.net/encyclopedia/article/whitten.panic.1893.

Wilkins, Mira. *The Emergence of Multinational Enterprise: American Business Abroad from the Colonial Era to 1914*. Cambridge, MA: Harvard University Press, 1970.

Wilmot, Sarah. "Pollution and Public Concern: The Response of the Chemical Industry in Britain to Emerging Environmental Issues, 1860–1901." In *The Chemical Industry in Europe*, edited by Ernst Homburg, Anthony S. Travis, and Harm G. Schröter, 133–159. Dordrecht: Kluwer, 1998.

Wise, George. *Willis R. Whitney, General Electric, and the Origins of U.S. Industrial Research*. New York: Columbia University Press, 1985.

Wisniak, Jaime. "Henri Etienne Saint-Claire Deville: A Physician Turned Metallurgist." *Journal of Materials Engineering and Performance* 13 (2004): 117–128.

Witt, Otto N., and Georg Pulvermacher. *Bericht: V. internationaler Kongress für angewandte Chemie, Berlin, 2.-8. Juni 1908*, vol. 4. Berlin: Deutscher Verlag, 1904.

Wolf, Mechthild. *Im Zeichen von Sonne und Mond: Von der Frankfurter Münzschneiderei zum Weltunternehmen Degussa AG*. Frankfurt am Main: Degussa AG, 1993.

Wolfrom, Melville. "Harold Hibbert." *National Academy of Sciences Biographical Memoirs* 32 (1958): 125–181.

Woodbury, David O. *A Measure for Greatness: A Short Biography of Edward Weston*. New York: McGraw-Hill, 1949.

Woodhull, John F. "What Specialization Has Done for Physics Teaching." *Science* 31, no. 802 (13 May 1910): 729–731.

Wyllie, Irvin G. "The Businessman Looks at the Higher Learning." *Journal of Higher Education* 23, no. 6 (1952): 295–344.

Young, Ralph C. "The Sainte-Claire Deville Hot-Cold Tube and Some of Its Applications." *Journal of Chemical Education* 20 (1943): 378–380.

Index